The Tomb of the Unknown Soldier, Modern Mourning, and the Reinvention of the Mystical Body

LAURA WITTMAN

The Tomb of the Unknown Soldier, Modern Mourning, and the Reinvention of the Mystical Body

UNIVERSITY OF TORONTO PRESS
Toronto Buffalo London

Toronto Buffalo London
www.utppublishing.com
Printed and bound by CPI Group (UK) Ltd, Croydon, CR0 4YY

ISBN 978-1-4426-4339-0

Printed on acid-free, 100% post-consumer recycled paper with vegetable-based inks.

Toronto Italian Studies

Library and Archives Canada Cataloguing in Publication

Wittman, Laura, 1968–
The tomb of the unknown soldier, modern mourning, and the reinvention of the mystical body / Laura Wittman.

(Toronto Italian studies)
Includes bibliographical references and index.
ISBN 978-1-4426-4339-0

1. Unknown military personnel – Social aspects – Italy. 2. Unknown military personnel – Political aspects – Italy. 3. Soldiers' monuments – Social aspects – Italy. 4. War memorials – Social aspects – Italy. 5. Mourning customs – Italy. 6. Human body – Symbolic aspects – Italy. I. Title. II. Series: Toronto Italian studies

NA9330.I82W58 2011 355.1'60945 C2011-900826-2

This book was published with the aid of a grant from the Division of Literatures, Cultures, and Languages, Stanford University.

University of Toronto Press acknowledges the financial assistance to its publishing program of the Canada Council for the Arts and the Ontario Arts Council.

Canada Council for the Arts Conseil des Arts du Canada
ONTARIO ARTS COUNCIL CONSEIL DES ARTS DE L'ONTARIO

University of Toronto Press acknowledges the financial support for its publishing activities of the Government of Canada through the Book Publishing Industry Development Program (BPIDP).

For my parents,
Luciana Flavia Conti Harwood
and
Douglas James Harwood

Contents

PART II EMBODIMENT AND SPECTACLE

Acknowledgments

This project would not have been possible without the generous support of a number of funders and institutions. I was able to do research in Italy and France thanks to a National Endowment for the Humanities Summer Research Stipend, a Summer Research Stipend from the American Academy in Rome, a Faculty Career Development Award and a Humanities and Arts Fellowship from the University of California, Santa Barbara, and the generous research funds provided by Stanford University. I would also like to thank the Division of Literatures, Cultures, and Languages, and the Department of French and Italian at Stanford University for their continued support. The Rome Prize Fellowship provided me with a full year in Rome and an extraordinary intellectual community, in which the earliest versions of this project began to germinate. A Summer Research Grant from the Beinecke Rare Book and Manuscript Library at Yale University was also invaluable to the beginning of this project.

My research benefited from the expertise and help of the staff at numerous archives and libraries. In Rome, I wish to thank the Archivio Centrale dello Stato, the Biblioteca Giovanni Spadolini, and the Biblioteca di Storia Moderna e Contemporanea. I am also especially grateful for the generous welcome I received at the Archivi e Biblioteche, Il Vittoriale degli Italiani, Gardone, which houses d'Annunzio's papers and library. In Paris and Vincennes, I am very grateful for the help of the Bibliothèque Nationale de France, including the Archives et Manuscrits, as well as the Service Historique de la Défense. For many years of invaluable research help, I wish to thank the staff of Yale University Libraries, as well as the University of California Libraries at Santa Barbara and Berkeley. The Stanford University Libraries staff, especially in the interlibrary loan office, have also been extremely helpful. Finally, I owe special thanks to Sarah Sussman at the Stanford Libraries for finding and purchasing books that furthered this project.

I also wish to thank the institutions and people that provided me with illustrations and generously gave permission to use them: Lobster films, the Centro Documentazione Mondadori, the Archivi e Biblioteche, Il Vittoriale degli Italiani, the Biblioteca di Storia Moderna e Contemporanea, the Archivio Centrale dello Stato, Stanford University Library Special Collections, and the University of California Northern Regional Library Facility. A special thanks to Luca Spaventa for the photo of Redipuglia. Thank you also to Diego Bastianutti for his translations of Ungaretti, and to Alexander Nemser and Nariman Skakov for their translation of Tarkovsky. For allowing me to cite Merini's poetry, I am grateful to Emanuela Carniti, Simonetta Carniti, Susanna Flavia Carniti, and Giuliano Grittini.

I am grateful to all my students for their stimulating conversations. I wish to thank especially my research assistants, Vetri Nathan and Lorenzo Giachetti (who also helped with English translations throughout the book), for their patience and assistance. Thank you also to Serena Ferrando for her expertise on Alda Merini.

My colleagues at Stanford have been extraordinarily insightful mentors, commenting on this project in its many versions, and offering rich resources for intellectual growth. Thank you especially to Robert Harrison, Hans Ulrich Gumbrecht, Joshua Landy, and Carolyn Springer for their critiques of this project and for their ongoing support. Colleagues elsewhere have also been generous readers and critics of various versions of this book. Thank you in particular to Jon Snyder and to Didier Maleuvre at the University of California, Santa Barbara, for their careful readings of early drafts. For their ongoing mentorship throughout the years, I wish to thank my dissertation adviser, Paolo Valesio, as well as Ernesto Livorni.

For taking the time to read the whole manuscript carefully and make invaluable suggestions for revisions, my heartfelt thanks go to Barbara Spackman and Claudio Fogu. Thank you also to Catherine Brice for sharing her impressive knowledge of modern Italian history and its archives, to Lucia Re for her illuminating comments on talks based on this book, to Thomas Harrison for ongoing discussions of my research, and to Dan Edelstein for helping me formulate the book proposal. An especially warm thanks goes to Mia Fuller, who has time and again shared her expertise on modern Italy and taken the time to read and critique my work, and to offer her wisdom. Very warm thanks also to Richard

Wittman for his ideas and help, especially with photographs. For their emotional support, as well as their many intellectual insights, I wish to thank Dana Renga, Allison Cooper, Elizabeth Leake, Linda Pellecchia, and David Stone. Warmest thanks for friendship, advice, and discussions of PTSD to Joe Helms and Pamela Sinclair. Mary Margaret Jones and Doug Argue have been a family away from home, and I thank them for years of nurturing and stimulating conversations. My husband, Joshua Marker, has offered me the encouragement needed to bring this project to completion; I thank him for his patience and wisdom.

The Tomb of the Unknown Soldier, Modern Mourning, and the Reinvention of the Mystical Body

Introduction

On 4 November 1921, in 'the greatest patriotic celebration of united Italy,' a 'sea of people,' impressively silent, gathered in Piazza Venezia in Rome to bury the unidentifiable remains of one of the 650,000 Italian soldiers who died in the First World War. The ceremonial burial of one anonymous body, gathered from the battlefield and given a proper tomb, was a cathartic response to mass death and the irretrievable loss of so many bodies, and the event had immense cultural and popular appeal. The Unknown Soldier was at once a representation of the body of the nation and of the human body, both felt to be ruptured, perhaps permanently, by the war and by modernity. The Tomb of the Unknown Soldier was a new memorial invented by the Italians, the French, and the British, in the last years of the war and in its immediate aftermath, though it is clear that the concept of an Unknown Soldier was widespread in all combatant nations, including the United States and Germany. By the summer of 1920, the Italian, French, and British military were digging up corpses from the major battlefields of the war, in search of one who might be transported home to become *the* Unknown Soldier. Public debates about when and where he should be buried, and reactions to this new memorial were intense. Thus, the French rushed to inaugurate their Tomb of the Unknown Soldier on 11 November 1920, but did not bury him permanently under the Arc de Triomphe until January 1921, for fear of dissent. The British buried their Unknown Soldier in Westminster on 11 November 1920, but also decided to make Sir Edwin Luytens' celebrated cenotaph – an empty tomb – permanent, in order to defuse arguments about the symbolic status of the anonymous body. In Italy, for reasons that included the nation's recent unification, its late industrialization, the erosion of popular Catholicism, and the incipient rise of Fascism, the new Memorial was controversial enough to delay inauguration until 1921. The Italian case is thus most representative of the new relationship between the body, mourning, and the nation that this book explores, for it is here that we find the greatest divergence between official scripts and a host of other interpretations of the Memorial, ranging from spontaneous popular expressions of grief to complex literary and philosophical critiques. It is only by focusing on these popular and unofficial reactions, in Italy and elsewhere, that we can understand why the idea of an Unknown Soldier was so successful that by the end of the twentieth century most modern nations had inaugurated one.[1]

Recent events in Italy confirm the appeal of the Unknown Soldier Memorial. On 17 November 2003, more than 300,000 people once again gathered in Piazza Venezia, this time to pay their respects to the nineteen Italians killed in Nasiriyah, Iraq, on 12 November: 'for the second time,' a commentator wrote in *La Repubblica*, 'a sea of people, filled with deep emotion,' gathered to 'touch with their own hands what happened.' Even though the majority of Italians opposed the war in Iraq, for all of 17 November, and all night until 6:30 a.m. on 18 November, the public waited for as much as four hours in the rain, in a line that stretched all the way to the Colosseum, to see the coffins lined up inside the Vittoriano, near the same crypt that houses the Unknown Soldier's Tomb (above and behind the externally visible stone coffin). Press coverage not only emphasized the identification of the nineteen dead with the Unknown Soldier but also, echoing the press in 1921, drew from it a reflection on 'the instrumentalization of mourning.' In 1921, as in 2003, the declaration of a national day of mourning and the choice of the Vittoriano as its stage had obvious nationalist and predominantly bellicist implications: at stake was Italy's willingness to fight for its piece of the peace, be it the territories in Dalmatia that it wished to annex after the First World War, or more recently its status as the economic ally of the United States in the reconstruction of Iraq. In both instances, however, the anonymity and the vulnerable, exposed body of the Unknown Soldier, metonymically extended to all combatants including the nineteen killed in Iraq, was depicted in newspapers and in interviews with the public as a 'reproach' against 'bellicist nationalism,' and more broadly against any institutional appropriation of mourning.[2]

What is the value of this reproach? At its core lies a debate about the meaning of death and the individual's attempt, in an era of secularism, democracy, and mass culture, to wrest that meaning from institutions and vice versa. Undoubtedly in 1921 the tension between official history

Figure 1. The Italian Unknown Soldier Memorial in 2003. © Laura Wittman

and local or personal interpretations of the war experience was exacerbated by the unprecedented scale of human loss in the First World War, which many felt was absurd or traumatic to the point that it could not be 'understood,' in the sense later used by Claude Lanzmann to describe the Shoah's resistance to narrative, psychological, or redemptive closure. Yet it is evident that both the institutional normalization of death and its questioning result from broader cultural changes in the nineteenth and twentieth centuries: the rise of nationalism, scientific and technological innovation, secularization and religious pluralism, the development of psychoanalysis and anthropology, and multiculturalism all contribute to a crisis in mourning. In short, by the end of the First World War, death was something that Western culture could no longer interpret adequately, and this is what the Unknown Soldier Memorial sought to expose and explore. People were fascinated by the disquieting corpse at its centre, dead yet still alive, individual yet extended to all, a real body yet also the symbol of a new sort of transcendence, still in the making. The fact that Italy's Unknown Soldier Memorial has been the site of three suicides (two real and one fictional) is emblematic, for these were attempts to endow individual physical death with a public meaning, and hence, to reconnect the existential and the political.[3]

How much such a meaning was and still is up for debate is shown by a strange fact: between 1921 and the present, the popularity of the Unknown Soldier has been extraordinarily consistent, but his anonymous body has received radically divergent interpretations. In the late 1930s Mussolini could celebrate through the Unknown Soldier Italy's dead in its colonial war in Ethiopia, whereas in 1945 the same Unknown Soldier came to represent the partisans' 'non-military' sacrifice to free Italy from Fascist oppression. In 1940, in occupied France, students converged on the Unknown Soldier Memorial on 11 November in a gesture of deliberate dissent against the Germans, even as the same Memorial was the site for paramilitary right-wing extremist rallies in Paris in 1935. As mentioned, Italians used their Unknown Soldier Memorial in 2003 to mourn dead soldiers and civilians from the war against Iraq. In apparent contradiction, earlier in 2003, when the largest rallies in the world *protesting* the war in Iraq took place in Italy, they too converged on the Vittoriano and the tomb at its centre, not only as an image of the nationalism they opposed, but simultaneously as an image of the dehumanization and suffering caused by war, which they sought to emphasize. In 1998, Great Britain added to its Unknown Soldier a Memorial to Innocent Victims Worldwide, making explicit the extension of his anonymous body to non-combatants and to other nations, races, and genders. The Memorial to the Unknown Deserter, inaugurated in Potsdam in 1990, extends the concept even further, to commemorate those who oppose or perhaps merely flee war. What is again surprising, as in Italy, is the reuse of the same Memorial instead of the creation of a new one. This is also the case in the 1955 Finnish film, *The Unknown Soldier*, based on the novel by Väinö Linna, which tells an ambiguous national history of simultaneous resistance and aggression. More provocatively, Michael Verhoeven chose to entitle *The Unknown Soldier* his documentary about the controversial late 1990s' 'Wehrmacht Exhibition,' which explored both the participation and dissent of 'common soldiers' in the 'final solution.' In the United States, the Unknown Soldier Memorial at Arlington Cemetery, inaugurated in 1921, did not become controversial until after the Vietnam War, when it sparked debate about who – families or the government – owned the dead bodies of soldiers. Such apparent contradictions within each country are echoed worldwide: Saddam Hussein erected a massive monument to the Unknown Iraqi Soldier in 1982 in the context of the war against Iran; Canada inaugurated its Unknown Soldier in 2000, and New Zealand repatriated an unknown soldier in 2004. We cannot then but ask: How can the same Memorial receive such opposite political interpretations? How can it appear, in turn, bellicist and pacifist? And why insist on recycling this particular type of memorial instead of inventing a new one for each new cause? The flexibility and endurance of the Unknown Soldier suggest that its core is a quintessentially modern confrontation with death that is still with us and cannot be subsumed to any specific political or religious agenda. Evidently, the Memorial is able to incorporate within the bellicist, heroic monumentalization of death, which reached its apex in the nineteenth century, a different kind of mourning, which is not eliminated by nationalism but remains implicit within it.[4]

Historians have written about the shape and ideology of post–First World War inaugurations of the Italian, French, and British Unknown Soldier Memorials, concentrating for the most part on how different political agents sought to channel public mourning. Overwhelmingly, this new memorial is considered to be a political success story, whereby the state was able to overcome party lines to give discontented, traumatized populations an outlet for grief that otherwise would have caused social instability and revealed long-standing cultural divides within each nation, and among them. Even as this is, to be sure, a substantial part of the story, I intend to question this view in two related ways, one chronological and the other conceptual. First, concentrating almost exclusively on the years 1918–21 does not allow for a broader cultural, symbolic, and philosophical history that can explain the enduring and apparently contradictory appeal of the Memorial that I just outlined. Second, focusing on how political debates frame the question of mourning says little about the anthropological origin of the idea, and does not explore the psychological and spiritual power of this anonymous corpse. Grief, both public and private, cannot be reduced to a passive blank slate to be shaped by ideology; it also has a more inward history, both individual and collective, that is traced in all the unofficial, alternative, and imaginative interpretations of the Unknown Soldier that I will analyse. Concentrating on these more symbolic and poetic manifestations, I show that grief repeatedly fissures the myths of unity both of the nation and of histories written from the top down. What emerges is a more figural, multifaceted, and dynamic sense of the Memorial's history.[5]

Consequently, this study draws on a vaster and more eclectic set of materials than previous ones, and focuses on the recurrence of archetypal motifs across time and national boundaries. In particular, focus on the 1920s has obscured the extent to which the Unknown Soldier idea drew from public and private discussions about mourning and commemoration that began almost as soon as the war did, by 1915–16: significant aspects of this idea were in the popular imagination, especially that of combatants, and appear in their letters and diaries, well before it was debated in government chambers. Moreover, this reflected a crisis in the West's understanding of death that was at least a century old: discussions not only included increasingly overt condemnations of past memorials – the Roman and Parisian Pantheons were both targeted – as insufficient or outdated, but they also drew on images of grief from the recent and ancient past. Finally, and perhaps most important, the genesis and success of the Unknown Soldier was not limited to war memorials and related ceremonies: from the beginning, in Europe and America, the Unknown Soldier was also the protagonist of imagined biographies, satirical cartoons, popular fiction, avant-garde poems, philosophical plays, and sociological studies, which would continue to be published throughout the twentieth century. I thus use 'the Unknown Soldier,' 'the Tomb of the Unknown Soldier,' and 'the Unknown Soldier Memorial' to refer to an archetypal image that emerges from this wide range of sources. Also, as we will see, the Unknown Soldier as a body was often thought of as separable from the Memorial. At the same time, I will analyse specific examples, focusing on the Italian case, but putting it in a comparative context: these will be referred to as 'the Italian Unknown Soldier Memorial,' 'the Italian Unknown Soldier,' and so forth. It is important to realize that not only in written commentary but in the actual construction and inauguration, each nation understood its particular Unknown Soldier to be related – not unproblematically – to other Unknown Soldiers. Literary texts, but also public reactions reported in the press show that the official national rhetoric depicting the Unknown Soldier as Italian, or French, or British was subverted in a variety of ways that pointed to a deeper common humanity. Overall, the wide range of materials I consider reveal that within apparent, and only temporary political success, the Unknown Soldier Memorial has an appeal that is psychologically visceral, ethically ambiguous, and culturally destabilizing. It may be termed cathartic, as historians have written, but without this larger context of figural history, we cannot understand how and why.[6]

Yet this remains a provocative question, which brings to the fore the Memorial's strangeness: a monument to anonymity was a shocking acknowledgment of modern warfare's unprecedented reduction of the individual to an expendable cog in the machine; and the attribution of anonymity to a single body was not only an entirely new type of memorial, but one that emphasized physical suffering, transforming it into spectacular abjection, forcing us to witness the price of war. This is not to say that the Memorial was pacifist, nor that it rejected a heroic glorification of suffering. What was remarkable, rather, emerges from contemporary discussions, as well as from fictional and historical writings about its purpose: it sought to commemorate the psychological and social conflict of veterans, and of a culture that rejected mass death as incomprehensible and dehumanizing yet needed desperately to feel that so many had not died in vain, that suffering and death could still have a common, communicable meaning. As such, the Unknown Soldier is a contradictory, anti-monumental monument, whose appeal resurfaces when we grasp that death must have dignity, and be properly mourned, even when it is in the name of causes that are hotly debated and perhaps unjust. It also shows us that the dead body's materiality, however violent or absurd the person's death may have been, reorganizes cosmologies, giving existential depth to political life.[7]

This new kind of mourning, as I will show respectively in Parts I and II of this book, finds expression in the Unknown Soldier's two most salient features: anonymity and embodiment. As we begin to interpret these, a contrast is useful, for memorials to the unknown dead have had a very long history: the empty tomb dates back as far as Thucydides, whereas the ossuary holding multiple unidentified bodies developed during the American Civil War; both ideas were used during the First World War. Significantly, both types of memorial compensate for physical loss (the absence or disintegration of the bodies) by providing a list of names, which may be located at the Memorial itself or, increasingly after the First World War, in the home town of the deceased. In other words, in both cases, anonymity is seen as something that affects the body alone, in a way that can be redeemed by the higher, and more social, identity provided by the monument itself and by the name; this social and linguistic identity is what cenotaphs and ossuaries celebrate as lasting, whereas bodies decay and pass away. In a stark reversal, in the case of the Unknown Soldier, physical presence (a single body, retrieved from the battlefield in all of its parts, and recomposed) compensates for the loss of a name. Civic, cultural, and linguistic identity is now what is missing, whereas the monument celebrates the deeper or more primitive selfhood of the body. This explains why inscriptions on the Unknown Soldier Memorial were reduced to a minimum; and even those, it was suggested, should be broken, as though to indicate that words had been erased, and were no longer significant. Speeches, too, were reduced to a minimum, and the British are credited with inaugurating collective silence as their replacement. It also explains why the Memorial's proponents and creators were so concerned with protecting the body that it contained from the elements, to the point that in the Italian case, it was moved after a few years further into the building surrounding it. All these factors were salient again in 2003 mourning for Italy's dead in Nasiriyah, especially silence and the integrity of bodies.[8]

This reversal of past memorials is telling, for it shows that unlike the empty tomb, the Unknown Soldier Memorial is not so much about anonymous bodies, as we might initially think, as it is about anonymous people. In other words, anonymity is no longer an accident, but an existential state, a positive quality rather than a deprivation – a situation created by war, and modernity, perhaps, but one that cannot be reversed, and must be confronted and accepted, to the point that it is part of what the Memorial celebrates, even if it also decries it. As we will see in Part I, shared anonymity here seeks to replace the familial relationships that traditionally defined the nation. Further, as the state, the public, and individual writers all try to understand this new existential anonymity, comparing it and still confusing it with older definitions, they come to focus on the notion of sacrifice. Remembering, among many other models, medieval nobles who chose unmarked graves to indicate their humility, the sacrifice of their social status to God, they depict the Unknown Soldier as having sacrificed his name: but to whom and for what purpose? These become hotly contested issues, along with the nature of sacrifice itself, including, of course, its relationship to religion. A central question here is how and at what cost the Unknown Soldier's anonymity becomes 'distributive' and applicable to all – not just the dead but the living as well.[9]

Similarly, again in contrast with our assumptions, and with the traditional ossuary, the Unknown Soldier Memorial does not enact the body's passing away so that the spirit, and the enduring dignity of the name, may transcend it. Rather, dignity now resides in the body itself, which is no longer seen as passive but active; the body is the agent that transcends social constructions and names (if the term 'transcendence' can still be used for this downward push to the chthonic, the inchoate origin); in doing so it creates a different kind of dignity which is linked to embodied suffering, yet also to physical joy. In other words, insofar as the Memorial is still heroic, what it makes larger than life is not the spirit but embodiment itself, as the experience of suffering, violence, joy, and at bottom, mortality. As we will see in Part II, the body within the Unknown Soldier Memorial is fascinating for its makers and its audience alike: it is gory, abject, barely human or animalistic, horrific in the suffering it reveals, and still terrifying in the latent violence it might unleash; yet all of this enacts and makes spectacular the experience of human mortality. Just as anonymity redefines sacrifice, the active embodied suffering of the Unknown Soldier sparks a debate about spectacle: its ritual, cathartic power is evident, but who is capable of using it wisely and how? Central to this debate is understanding how and why the Unknown Soldier's body could enact a new kind of mourning, and do so repeatedly, becoming a sort of meta-memorial – a critical comment on our fraught relationship to death and dying.[10]

The new community created by the idea of anonymous sacrifice, and the spectacle of mortality generated by linking anonymity to a single, concrete body together configure the Unknown Soldier as a modern mystical body. In Christian tradition, the *corpus mysticum* originally referred to the body of Christ, hence, to the mystery of spirit become flesh and destined to be transformed again into spirit after death. Between the Middle Ages and the end of the eighteenth century, however, this concept shifted drastically in two related ways. First, *corpus mysticum* referred more and more to the Church as a community united in Christ and, with time, indicated also that community's land, eventually leading to the depiction of the nation as a body. Second, this shift to community and then nation was also a turn away from the concreteness of the flesh: originally part of the spirit and of salvation, as seen in the concept of the resurrection of the body, the flesh became the impure part of the *corpus mysticum*, associated with decay and death that the mind must transcend. I will return to this history in my discussions of both sacrifice and spectacle, but the important point now is that by the time of the First World War, the mystical body had become abstract: it thus lost its ties to individual experiences of death and spiritual struggle, which were shared via symbol and ritual, and was instead linked to transcendence understood conceptually as a greater, permanent, homogeneous 'whole.' The Unknown Soldier Memorial counters this abstraction, seeking to make body, ritual, and experience relevant to the community and to the political sphere. In doing so it returns to some of the older meanings of *corpus mysticum* and seeks to reinvent them in a modern context. In particular, the Unknown Soldier Memorial attempts to express the sacredness of the flesh, and its inseparability from the spirit, in terms that are secular or multicultural and pan-religious, even if they are inspired by Christianity. Further, focusing on mortality as shared experiences rather than as a concept, the Memorial makes room for individual variations, diversity, and communication. In Chapter 6, I conclude this book with four recent works – two novels, a poem, and a film – that reflect on the Unknown Soldier as a modern mystical body. Like the Memorial itself, much of what they have to say is negative, as they decry our culture's disregard for and objectification of the body and link it to our failure to mourn. Nonetheless, in different ways, they each draw from the Memorial a sense of the body's grace that is inseparable from knowledge and acceptance of its mortality, and they seek to affirm politics that will respect that grace.[11]

Before I turn to the details of the Memorial's invention, and the issues of anonymity and sacrifice, on the one hand, and embodiment and spectacle, on the other, I would like to offer three brief observations about nationalism and mourning. First, I am proposing that within the heroic model of mourning associated with bellicist nationalism, a different mourning, linked to new understandings of identity and embodiment, emerges. This may sound as if veterans, or their families, or intellectuals and writers, or political activists, opposed their vision to the nationalistic one imposed by the state. Such a polarity, however, is too simple to describe the ambiguity of the Memorial itself, and of the many cultural players who created it and reacted to it. As Jay Winter has argued in his seminal *Sites of Memory, Sites of Mourning: The Great War in European Cul-*

tural History, while First World War memorials in general are certainly not pro-war, neither can they be said to be entirely pacifist: as Winter's book shows, many people felt that old models of heroism were now empty rhetoric, but at the same time could not give up the need to see the death of their loved ones as meaningful, often in the context of their culture and nation; and, even as their leaders had much at stake in claiming that the war had been justified, they, too, could not but acknowledge its excess as unprecedented, and perhaps incomprehensible. Similarly, it is not the case that certain nations had a more pacifist or more existential interpretation of the Memorial than others, but rather that each instantiation of the Unknown Soldier idea struggled in its own way to create a new sort of mourning. Bellicist or heroic mourning was not the purview of the state alone, of a specific political party, or of a given nation; and innovation came from many different quarters, official and unofficial. In the Unknown Soldier Memorial, old and new understandings of death coexist and confront each other not, or not only, as the reflection of different social or political factions, but as the expression of a widely shared inner psychological tension. At stake was a question that is still with us: is nationalism necessarily bellicist, or can national culture and its institutions, including public mourning, be reinvented along more pacifist lines? Another way of putting this would be: what traces of pacifism, or also of dissent from bellicist mourning, can we discern as we question the hegemonic and isolated national histories that have until now dominated interpretations of the Unknown Soldier Memorial?[12]

My second remark has to do with the rise of Fascism in Italy. Numerous historians have associated the appeal of Fascism with the weakness of the Italian state after the First World War and, in particular, with its inability to create a better future out of wartime loss or to make a more modern and more just society that veterans could identify with – in other words, with its failure to mourn and commemorate. As Emilio Gentile has shown, the unique 'sacralization of politics' inaugurated by Fascism finds one of its main expressions in immense collective mourning rituals. Indeed, the Fascist cult of the dead is central to the creation of a religion of the nation precisely because, as Gentile writes, it reflects 'ineradicable' 'cultural and historical passions,' that is, it draws on tradition and reinvents it. The Unknown Soldier Memorial is an inspiration for this reinvention. As we will see in more detail later, what made the Unknown Soldier Memorial inspiring for Fascist politics was not its more traditional heroic side, but rather its insistence that the meaning of sacrifice, and the power of suffering were absent, perhaps undefinable. Fascism could then fill this gap that had been opened, and it filled it not with the nation of old, but with the body of Mussolini, which became a metonymy for the new Italy. Recent work has shown how carefully Mussolini managed the identification of his body with that of the Unknown Soldier in the months following his March on Rome in October 1922. But, as I mentioned above, in 1945, after Fascism fell, and after the liberation, the anti-Fascist partisans were even faster at reclaiming the Memorial for their dead. Moreover, Fascist uses of the Memorial did not prevent the French, the British, or the Americans from continuing to honour it. We must therefore be wary of conflating the Fascist cult of the dead with the Unknown Soldier Memorial, and take care to separate proto-Fascist elements present in its invention in the years 1916–21 from its later uses. At stake is another very topical question: is the alliance of religion and totalitarianism, such as was seen under Fascism, inevitable? Or is it possible to develop understandings of death that respond to people's need for the sacred and for its communal expression outside of the 'religion of the nation' paradigm?[13]

My last remark brings us even closer to the present, and to the United States. The polarization between bellicism and pacifism is most evident in our culture today, in particular in the wake of the second Iraq war. Our failure to grasp the sacred dimension of mourning is less visible, though for that reason perhaps even more pernicious, as Robert Harrison shows so well in *Dominion of the Dead.* It is, in any case, clear that for many who opposed the war, and still oppose United States policies in Iraq, the question of how to respect the military involved in it and especially veterans and the deceased remains a difficult and pressing one; how to honour countless civilian dead, including Iraqi ones, is an even more controversial issue. War memorials are not very popular these days, perhaps because we sense they would provoke tremendous dissent. Yet the Unknown Soldier Memorial reminds us that mourning is necessary even when, perhaps most of all when, it is politically fraught; furthermore, its emphasis on anonymity and embodiment highlights our own failure to address these very same issues, as mangled bodies and military as well as civilian deaths are minimized in our media. We see this in two very recent and contradictory decisions made by the Obama administration: after an eighteen-year ban, photographs will be allowed of military coffins if families permit it (though, of course, not of body bags or bodies); however, the practice of refusing the families of soldiers who commit suicide a Presidential letter of condolence will continue. The Unknown Soldier Memorial forces us to reconsider how the political and the existential aspects of mourning are intertwined; only in confronting this can a genuine discussion about the traditional concept of a 'just war' take place: is it still in any way viable, and if not, what can replace it in our era of globalization, nuclear weapons, and terrorism? As a monument that attempted to bring home some of the realities of modern, technological, mass warfare, which the First World War inaugurated, the Unknown Soldier Memorial still has much to say to us, if we are willing to hear it. In this, I believe the Unknown Soldier Memorial can be compared to the Vietnam Memorial in Washington, DC, for both are able to acknowledge war's meaningless violence, its excess, and all the political, social, and human questions that it raises, without succumbing to cynicism, with genuine hope that we will learn from the dead here commemorated. For a contemporary American to understand how moving the Unknown Soldier Memorial was in the wake of the First World War, it might be useful to witness how the families and friends of those who fell in Vietnam come to the wall of names, so vast that it borders on anonymity, to search for and touch the one name significant to them. Similarly, in 1921 and again in 2003, countless Italians waited for hours and days to touch the Unknown Soldier's coffin, and through that touch to feel that the body within was that of their relative or friend, however anonymous it might seem outwardly. In all these cases I am struck by the importance of touch, for what is asserted by it is not identification with a bellicist cause, or even a pacifist one taken in the abstract: this touch indicates, rather, a solidarity between the living and the dead which goes deeper than any ideology. These memorials invite us not only to respect this solidarity, but also to investigate it and to learn from it.[14]

PART I

Anonymity and Sacrifice

INTRODUCTION TO PART I: THE RETURN OF THE DEAD

From antiquity until the First World War, it was held that those who died, and especially those who died violently at war, would return to haunt the living unless they received proper burial. This entailed gathering and identifying their remains, as well as a ceremony, and a grave marker that loved ones, comrades, and society could honour. In practice, only nobles or officers received ceremonial burials, whereas common soldiers were left to mass graves on the battlefield or given quick burials in local communities. With the First World War, however, 'the growing number of the dead was accompanied by a greater awareness of each one individually. As though, just at the inception of death on a mass scale, the individual became all the more precious.' Each combatant, noble or not, officer or not, expected now to be remembered. Even though policies varied as to whether this would be done in situ, or via repatriation, all combatant nations in the Great War promised each of their soldiers a proper burial. How impossible this promise was to keep became clear almost as soon as the war started, as trench warfare, combined with massive artillery and machine guns, created a 'no man's land' between the front lines where casualties were unprecedentedly numerous, and very difficult to retrieve. Before the war even ended, governments were faced with a twofold problem. On the one hand, the dead were so many that individual burial and ceremony, let alone repatriation, was out of the question; worse yet, without this, one could never be sure that the missing were actually dead. What was at stake, as we are about to see in Part I, was that proper burial required identification and repatriation, that is, re-establishing the connection between a dead body (initially perceived by mourners as a frightening 'thing' which both is and is not 'my loved one') and some sort of enduring identity ('the loyal soldier,' 'the courageous husband'). On the other hand, especially as the war wore on, in many cases the bodies of the dead were so damaged and piled together that they lost their individual integrity, and could not be separated from each other, or from the mud of the trenches. As we will see in Part II, proper burial was even more impossible in this case because it required gathering a corpse, whereas nothing was left but disparate body parts. Yet mourners needed to feel that if some essential humanity was lost in the trenches, at least after death it could be reasserted: 'they died like animals' could be transformed into 'we remember them as men.' The Unknown Soldier Memorial responded to both of these problems, but I will now focus on the first, and specifically on how and why his anonymity became a new sort of enduring identity.[1]

To answer this question, let me divide it into two more parts, which I will address in turn. First, as regards identification, how is it that mourners were willing to say 'my loved one' is 'anonymous'? What qualities did 'anonymous' still retain, to transform him into 'everyman' instead of a dead thing devoid of spirit? What sort of recognition – 'this is he, this is the person I mourn' – could take place here? Second, concerning repatriation, what role would 'anonymous' have in postwar society? What sort of home could 'everyman' return to and inhabit? What understanding of death, or 'message from beyond the grave' would he bring? Before I consider how the Unknown Soldier seeks to answer these questions, it is important to realize that they were imagined to be addressed by the dead to the living, not vice versa: it was the dead who demanded recognition and a home, and who 'returned' and haunted the living, in the popular and literary imagination. To understand the source of these questions, and what haunting meant, the story of the 'living unknown soldier' is especially telling.[2]

In February 1918, a French soldier who had lost his memory was found in the Lyon train station; he would be known, since a name had to be adopted, though it was not his own, as Anthelme Mangin. By 1922, the press referred to him as a 'poilu inconnu vivant' (a living unknown soldier), and associated him with the Unknown Soldier under the Arc de Triomphe in Paris. Almost three hundred families responded with recognition to his photo in the newspapers, and dozens were considered good enough cases to be given a personal interview with him, leading to a handful of positive identifications. He never recognized any of them. Though his case was far from unique, it became famous because two families competing to claim him as their own eventually went to trial; a verdict was not reached until 1939, and it was under appeal when Mangin died in early 1942, still in a mental hospital, as he had been for over twenty years. In Italy, the case of 'the amnesiac of Collegno' opposed two wives who both recognized their lost husband in a man arrested for petty theft in 1926; though he sided with Mrs Canella, who came from a wealthy family and identified him as a professor, by 1931, fingerprints would confirm him to be the husband of Mrs Bruneri, and a swindler. Nonetheless, the Canella family would continue for years to claim him as their own.[3]

As Jean-Yves Le Naour, who has recently reconstructed the entire story of Mangin, points out, both of these causes célèbres reflect anxieties that were pervasive in Europe as the First World War ended. After the war, these two amnesiacs inspired mediations on identity and its loss in famous Modernist writers such as Pirandello, Anouilh, and Giraudoux. But their stories also existed in fiction before they became a reality, most famously in Balzac's *Le Colonel Chabert.* The even earlier historical story of Martin Guerre was the subject of numerous works of fiction, including two by Dumas père, as well as the 1982 film, *The Return of Martin Guerre.* The common theme in these works is the return of the dead, and specifically of a lost soldier, who comes home to find that he has been forgotten or replaced, faces suspicion and rejection, and discovers to his horror that his home, both familial and national, has radically changed, and not for the better. At times, realism is abandoned in such stories, as the soldier comes back to life from beyond the grave; more often, the returning soldier is ambiguously a dream, a ghost, or an actual survivor, mistakenly thought to be have been dead. This trope appears in four of the five authors, all First World War veterans, that are central to Part I: Gabriele d'Annunzio, Jean Giono, Blaise Cendrars, and Abel Gance – in Giuseppe Ungaretti, our fifth author, the dead haunt the living as well, but are not explicitly soldiers. Le Naour rightly sees here the expression of survivors' guilt, both personal and political, as those who return decry the failures of individual memory and of a collective understanding of the war.[4] He echoes Winter, who identified the 'return of the dead' theme, noting that 'their aim [was] to see if their sacrifices had been in vain.'[5] Both Winter and Le Naour also detect in these stories an astounding 'power of denial, the retention of hope in a loved one's survival,' and the latter points out that the Vatican, which had taken charge of the exchange of soldiers and information between combatant nations, received 16,000 requests for data on missing soldiers between 1915 and 1918.[6] After the war ended, countless soldiers continued to be missing. The word itself, most markedly in French (*disparu*) but also in English, shifted in meaning and became synonymous with dead, even as people wished to deny it. In Italian, *caduto* (fallen) exhibited similar ambivalence, as it avoided making death explicit and implied that the soldier might yet stand up again. It is clear that all over Europe, mourning became uncertain, incomplete, and potentially endless, as people could neither accept nor refuse the death of their loved ones.[7] Even before the end of the war, private exhumations had to be forbidden, as governments strove (and failed) to give all the corpses still on the battlefield an identity and a grave.[8] In the end, however, with a few rare exceptions, the missing who did not turn up were listed, without any differentiation, on war memorials along with dead whose remains had been retrieved.[9] At times families fought the inclusion of their missing relative's name on the list, but it was deemed that names were the property of the state, and they were overruled.[10] Nonetheless, stories about miraculous, horrific, or comical returns persisted, all the way to Joël Calmettes television movie about Mangin, *Le soldat inconnu vivant* (2004), and Bertrand Tavernier's *La vie et rien d'autre* (1989). Set right after the First World War, the latter shows two women who search desperately for their men in mental hospitals, and also sort through piles of personal effects retrieved from disinterred bodies. In a parallel story line, the military search for a body that will become the Unknown Soldier's.

Without a doubt, the Tomb of the Unknown Soldier responded to this twofold anxiety described by Winter and Le Naour alike: families could imagine his burial to be that of their lost relative, putting their anxiety about his possible return to rest, and his memorial would ensure that he finally received the honours he deserved and that society be transformed by his memory. I will shortly get to the details of how this took place, for it was not uncontroversial, and quite a feat of the imagination, to make one burial stand for so many, and to put in its mouth, as it were, an acceptable agenda for social change. Before that, and because it is essential to understanding how the Unknown Soldier's anonymity worked, I would like to emphasize a different aspect of the return of the dead, which is most salient in the case of the 'living unknown soldier.' It has to do with the identity ascribed to the dead and even more the missing, well before they could be buried and speak the true lessons of the war. To understand what the Unknown Soldier's anonymity says and how, we must first ask: who speaks here, and from whence?

From this perspective what is most remarkable about the 'living unknown soldier' is his amnesia. As noted by Le Naour, the First World War saw a growing vocabulary to describe psychological and 'nervous' 'lesions,' which included the enduring British coinage *shell shock*, the French *obusite* and *commotion*, and the Italian *psicosi traumatica*. These, however, indicated growing confusion and controversy about the psychological state of combatants and veterans, and about its relation to physical violence. To begin with, especially before the war ended, few wanted to admit such states were caused by wartime violence; rather often they were seen as the result of a 'feminine' 'nervous' or 'hysterical' constitution, most likely to be far more prevalent among the enemy's soldiers than among one's own. Furthermore, traumatized soldiers, and particularly those suffering from amnesia, were more often than not suspected of simulating, to avoid a return to the front, even as some doctors began to recognize that even those who had no verifiable physical lesions were genuinely ill. In fact, as I will elaborate later, Freud's shift around 1896 away from the 'seduction theory' of trauma, and thus away from attributing it to concrete physical violence, made the very definition of the term difficult, especially as regards war trauma, where concrete causes were obvious but evidently not the only ones.[11] Problems of definition notwithstanding, as Eric Leed has shown, drawing from the diaries and correspondence of combatants, symptoms of 'discontinuity' and 'estrangement,' of 'loss of self' tied to 'the death of comrades' were dominant, and reflected psychological traumas that might become acute, as in the case of total amnesia, but in reality affected a large majority of soldiers, and extended to civilians as well.[12] As he sums it up, 'the symbols of invisibility, death, burial, and pollution are particularly apt descriptions of individuals who are for a moment passing between categories [...] in war, these are not symbols but experiences. [...] war experience is nothing if not a transgression of categories. [...] a "structural death."'[13] What is crucial here is the experience of existing 'between categories,' between a 'structural death' and an actual physical death, but also between pre-war and postwar identity, and ultimately between sanity and insanity. In essence, the war experience questions all of these dual oppositions, and it is clear that wartime trauma is met with suspicion precisely for this reason. Its capacity to interrupt categorization and question both rationality and society is encapsulated in the remarks of an *ardito* in 1918 (the *arditi* were the feared Italian shock troops, wooed by both the left and the right after the war): 'With the end of the war, we are precisely those who have no direction any more, those surrounded by the abyss, those without bread. Every one of us [...] is obliged to exclude the possibility of picking up our lives at the point at which they were interrupted in 1915.'[14] A century later, we are familiar with Michel Foucault's argument that nineteenth-century culture locked up as insane those who threatened its cohesion around the values of capitalism, nationalism, and materialism. Undoubtedly, First World War shell shock was marginalized for similar reasons. But, as Foucault argues, the insane were not so much politically subversive as they were subversive of politics, and civilization, per se: they represented what is beyond society, outside of the bounds of culture, at the very margins of what it is to be human.[15] Thus, the Unknown Soldier's 'amnesia' or estrangement has a double valence. On the one hand, following more traditional models, he is a stranger like Montesquieu's Persian, who can critique one culture, or one specific historical moment, from the perspective of another (e.g., he is a small-town boy who is patriot but not a nationalist – he does not understand or 'remember' international politics and big business).[16] Another related, but not identical traditional model attributes to him the critical distance or estrangement of an ancestor, who rises above the squabbles of the present (e.g., the soldier's suffering at war makes him much older than his actual years, he is wiser than his leaders who had an armchair view of the war).[17] On the other hand, the Unknown Soldier's association with Mangin suggests, and Leed's work on estrangement confirms, that a new model was gradually developing. Death did not give him the authority of a stranger or an ancestor, but made him a far more radical outsider: an 'alien' in the strong sense of the word, which combines the culturally 'inhuman' and mental alienation or 'illness' so great as to abolish the normal ego. Thus, anticipating reactions to the Second World War and genocide, as an alien the Unknown Soldier forces us to see the utter absurdity of war that undoes all the claims of civilization, making not only art and literature impossible, but also justice, personal morality, and psychological cohesion. The Unknown Soldier's anonymity, in its more radical modern form, speaks from a 'no place' that is beyond or between what culture can understand; and who speaks here is equally a 'non ego' who questions not just social identities, but psychological and spiritual ones as well.[18]

Once we grasp that such extreme alienation is what demands recognition and a home, we see why survivors felt so powerfully haunted, and also how this feeling went deeper than guilt or denial over the war itself. For indeed, alienation, as a crisis of ego and culture was, to be sure, catalysed by the Great War, but it already dominated nineteenth-century decadent and Symbolist Europe. I will address this history later, but for the moment let me return to Mangin, the 'living unknown soldier,' and draw out two more details of his story that connect him to pre-war fears about the loss of identity. In the late 1930s, he was subjected to an induced high fever, which was thought to work like an extreme, forced hypnosis, removing psychological repressions; it was hoped that at least some of the deeper elements of his identity – a childhood story, a home town – might emerge, but none did. Though Mangin evoked pity in his doctors and the public, and at times tenderness in those who insisted on recognizing him, he was seen as something less than human in his amnesiac state, as evidenced, for example, by the fact that he was routinely required to strip naked so that prospective families might 'recognize' the 'marks' of their loved one on his body.[19]

The fascination with fever, and altered states of mind in general, comes from late nineteenth-century spiritualism and early psychology. In Joris-Karl Huysmans' canonical *Against Nature* (1884), the hero, des Esseintes, combats a mysterious illness which brings with it not only the disintegration of his identity but also confused spiritual aspirations which traditional religion cannot answer. He seeks via complex liqueurs, heady perfumes, induced synaesthesia, and fever itself to find his core self, and thus a cure, but by the end of the novel fails entirely. One of the basic messages of this complex novel can be boiled down to des Esseintes' enraged observation that 'la psychologie du mysticisme était nulle' – the psychology of mysticism was useless.[20] At the very time when psychoanalysis was emerging from psychopathology, Huysmans was only one among many to question one of its basic tenets, namely, that there even was *a psyche*, a legitimate subject or identity that alienation departed *from*. Mangin, and the Unknown Soldier, posed the same question, as they confounded rather than upheld the medical establishment. We should remember that, in those same years, André Breton was working with shell-shocked patients, and would come to see their words as the expression of a collective and personal unconscious whose liberation would break all previous cultural forms and be the prelude to a new sort of human. Breton's allegiance to Dada, until 1924, can be seen as a preliminary destructive phase (which most of the avant-gardes also had) of dismantling the illusions of any stability – cultural, but especially psychological, as the very idea of mental 'health' came under fire. Freud himself would in these years come to discuss the death instinct as counter to the ego, and different from the pleasure principle he had emphasized earlier; however, by the early 1930s he would also oppose its encouragement or exploration, on pacifist grounds. Overall, though, there was ample cultural ground to perceive the Unknown Soldier's anonymity as a sort of existential alienation that was not an illness, but was on the contrary 'more real' than conventional identity. Crucially then, its recognition could be not only destructive, though it might seem so initially; it might also be the prelude to a new humanity. In Italy in particular, the idea of an *uomo nuovo*, or 'new man' that might emerge from the trauma of war not only took up the call for spiritual progress that was pervasive in turn-of-the-century Europe, but returned to the idea that Italy would not be truly unified until political unity could become incarnate in a 'new Italian.'[21]

Mangin's not-quite-human body evinces a similar ambivalence of the destructive and the constructive. It calls to mind another side of early psychology, not spiritualism but criminology. Most famously in the works of Cesare Lombroso, nineteenth-century psychology drew from the desire to identify the criminal temperament via readable physical characteristics; phrenology was only one aspect of this: birthmarks, scars, and tattoos were also all considered significant. As a number of scholars have shown, the body was here objectified as readable in response to the great anxiety over identity that I have briefly outlined above. In the end, however, this resulted in bringing alienation to the body itself, that is, to put it most succinctly, in exacerbating the split between mind and body, between animate and inanimate, between ethics and biology, drawing attention most of all to the failure of Western science, and culture, in defining life itself. Was there something that was not simply biological 'automatism' or even evolution, understood not as survival of the human, but of far more basic biological processes (which were at the time understood mechanistically, not holistically)?[22] As Thomas Harrison has shown in *1910*, that year was an apex for the feelings of alienation I mentioned above, as evidenced, in particular, by a startling number of artists' and intellectuals' suicides; at the same time, Harrison argues, this existential mood was central to the development of Expressionism in literature and the visual arts. As seen, for example, in Munch's famous *The Scream*, artists here emphasized the conflict between the body as disanimated automaton, on the one hand, and on the other, as radically self-expressive or self-generative (hence, expressionist).[23] Well before the First World War, they asked, would the body be reduced to a machine, or would it generate its own meanings, which might or indeed should threaten the supremacy of the rational mind, but could be creative in a wholly new way? I will discuss the expressionist quality of the Unknown Soldier's body and draw out its philosophical implications in Part II, but for the moment what I wish to emphasize is how a self-generative identity, as opposed to an externally readable one, posed serious ethical and criminal problems in the wake of the First World War.

The desire to find readable marks of Mangin's true identity was paralleled in the case of the Unknown Soldier, as we shall see shortly. Just as the law in the Mangin case considered these physical marks 'hard evidence' in contrast with 'anecdotal' personal recognition, it is rather remarkable that the public and the state alike wished to see on the Unknown Soldier's bones some sort of readable, intelligible trace of suffering, as though this were, again, the only 'hard evidence' in contrast with the purportedly 'anecdotal' knowledge that soldiers in the First World

War had suffered unspeakably. What was at stake was, to a great extent, the ethical and social value of radical alienation, for to say that it could be creative as well as destructive was not yet to answer the question of whether it was good. In particular, discussions of the Unknown Soldier Memorial show great awareness that distinctions between sane/insane, self/other, and rational/irrational break down in the face of wartime estrangement; at the same time, they show an equally strong desire to maintain another set of distinctions, between lawful and unlawful violence, between duty or sacrifice and murder, between cowardice and courage. Anonymity came from 'no place' and was 'non ego.' Nonetheless, ideally at least, it still made intelligible a minimal, basic, difference between good and evil. The extent to which such an ideal could be maintained once anonymity was in play will be revealed to us as we examine the details of the Unknown Soldier's identification and repatriation.

Figure 2. 'Versailles is not eternal peace for the dead.' © Biblioteca di Storia Moderna e Contemporanea

1 A Unanimous Idea

Unanimity and Confused Bones

> Versailles is not eternal peace for the dead. They feel their very bones confused again by all of surviving humanity. [...] tombs open and the dead emerge from them, rising up in pride or malediction [...] poets and political weasels, leave the dead in peace, the just peace they dream of and invoke. and in their name build the temple of justice whose foundations they have set with blood.[1]

This caption accompanied a 1919 Italian cartoon from *Il secolo illustrato* depicting three skeletons in uniform standing in a cemetery, rising from the grave to admonish the living. It captures many of the issues regarding anonymity, identification, and the Unknown Soldier as 'everyman,' that haunted the wartime and postwar imagination, in Italy and beyond. As we have seen, the dead returned to demand proper burial: even as this traditionally required naming, and a marker or ceremony that would ascribe meaning to death, the chaos of the First World War made this nearly impossible. The dead's alienation required a recognition that was other than naming and demanded a new ethics that could take into account the meaninglessness of war. In this light, the first salient element of the cartoon above is its attack against 'poets and political weasels': the dead soldiers demand to be '[left] in peace,' implying that their burial has failed because it has been used instrumentally by postwar factions. This was part of a widespread feeling that 'plans for massive national monuments' had 'failed' to be true to the war experience, and to rise above petty politics.[2] This was not an attack on the Unknown Soldier Memorial, which at the time had not yet been officially proposed, but rather an expression of impatience with leaders who promised appropriate public mourning, but failed to deliver. Echoed in articles and cartoons by the popular phrase 'we do not want monuments,' it indicated a rejection of the heroic model, which was considered both undemocratic and full of empty rhetoric, providing false glory in lieu of true recognition. In it, some 'names' endured more than others, appropriating the suffering and courage of the nameless; even lists of names, no matter how much they tried, could not be complete, nor could they erase social hierarchies or political agendas, and thus they too were powerless to express an alienation that extended to all.

The second noticeable element of the cartoon above is that these soldiers are not rising up from a war memorial, but from a cemetery, showing the failure not just of monuments, but of traditional burial as well. As Philippe Ariès has demonstrated, in the nineteenth century mourning was increasingly privatized as scepticism over institutional authority, whether of the Church or the state, grew. Winter, in turn, documents numerous examples of such private mourning during the First World War. Yet his work and that of many others reveals that private mourning was not enough, and that, specifically, this was the first war in which all classes of combatants expected official remembrance.[3] Moreover, what they wanted, predominantly, was a public recognition of their private suffering, of the daily, mundane, at times even trite battle between fear and loyalty. The reality of trench warfare, for them, was that there could no longer be any heroes in the sense of leadership, extraordinary strength and courage, or decisive action. What had to be remembered, officially but without the pomp of past monuments, were the myriad tiny battles of each soldier's journey – alienation extended to all, but without a capital A. Commemoration was thus in a double bind, made visually clear in the cartoon: soldiers left battlefield cemeteries, seeking public recognition, but as monuments seemed too hierarchical and politicized, they found themselves once again stuck in the no man's land evoked by Leed. In the popular imagination, they were 'remobilized,' caught between categories, dead and buried, yet still in need of a new burial, no longer at war but, as they are in this cartoon, still in uniform and not yet at peace.

From this place of unheroic alienation what did they ask for? The third remarkable element of the cartoon above is its reference to the Treaty of Versailles and the 'temple of justice.' For a number of different reasons, neither the Italian, French, nor British populations were happy with the Treaty of Versailles, which was seen as the result of political compromises that pleased no one. There was a clear element of bellicist nationalism in these complaints, as all three countries fought with each other over reparations, and many felt Germany had not been sufficiently crippled. Italians also felt cheated of the territories in Dalmatia that they had wanted, as was emphasized by an earlier cartoon in *Il secolo illustrato.* Beyond nationalism, however, in all three countries the complaint was that reparations did not reach those who deserved them, but stayed in the hands of big businessmen and politicians, who tried to distract people with empty ceremonies. And there were, of course, those who pointed out that such punishment of Germany was more vengeful than peaceful, and went against veterans' desire to put an end to all wars. Even as the mixture of nationalism and pacifism was different in each country, and evolved in the postwar years, the cartoon claims that true commemoration has to rise above 'Versailles' and politics, and above nations: its appeal to 'humanity' calls for a unanimous understanding of why soldiers did not die in vain, even though it does not make explicit what a 'just peace' would be or how unanimity is to be achieved.[4]

Why a just peace was so desirable, even as it was politically and psychologically fraught, emerges in the last and perhaps most important element of the cartoon. Both visually and textually, it insists that the true meaning of First World War deaths, and proper commemoration, is to be found in confused bones. This was a powerful, recurrent image at the time because, as we are about to see, in it coalesced some very contradictory feelings. On the one hand, as Leed has shown, soldiers identified with each other on the battlefield to an extraordinary degree, to the point that veterans increasingly felt inseparable from their dead comrades. This 'chorality' was intensely physical, as soldiers held the dying and the dead, 'confusing' their limbs together in the mud of the trenches. At the same time, chorality became the ideal unanimity betrayed by squabbles back home. In the cartoon we are discussing, 'bones confused again by all of surviving humanity' express an ambivalent fascination with, and fear of, repeating in death the chorality of the trenches. For this image suggests that the living desecrate the dead by mixing up their bones when they disinter and rebury them. At the same time, however, since battlefield burials (not to mention bodies simply left there to decay) often resulted in such mixing, the living are in fact repeating the fusion, both physical and psychological, of soldiers, living and dead, in the trenches. The point, then, is that the living must learn to repeat well, that is, to respect a chorality that is horrific yet fundamental, rather than trying to impose monuments upon it. Only in this way can chorality become a solid basis both for justice at home and for pacifism, as in their common experience of suffering and death, soldiers from all nations are the same.

On the other hand, the cartoon acknowledges that these bones 'rise up in pride or malediction.' The very inability to distinguish between pride and malediction, and the ominous aspect of the second, betray the fear that chorality might bring with it endless mourning, the reliving of a bond that will never progress beyond trauma into healing. As Leed writes,

> the psychic dangers of identification with the men of one's unit are obvious, and yet this was the reaction that was often necessary given the emotional severance from the home. This identification ensured that the death of every comrade would be a loss of self. No longer could the death of a friend be justified with the comfort this was loss that preserved the life of the whole. Each departure could be compensated for only by an intensification of the bonds to those who remained. [...] At the end of the process one's own demise could be welcomed as a resolution to an insupportable, continuous, mourning.[5]

Central to the image of confused bones, then, is an anxious question: is it possible to remember them and honour them without giving in to a pathological refusal to put an end to mourning? Can 'pride and malediction' eventually be separated? Though the phases of grief have been described in various ways, both before the First World War and after, it is generally agreed that initial feelings of shock, denial, anger, and guilt must eventually lead to acceptance and some sort of understanding; this also involves the mourner's gradual separation from the dead, and the re-establishment of distinct identities. But as we have seen, wartime alienation called for a new kind of mourning, and rejected standard pathology: who was to judge, then, what an appropriate end for mourning might be, both in terms of its goal and of its timing? For many, both acceptance of death and separation from the dead had to be reinvented. This was the deeper purpose of extending wartime chorality, and the more existential aspect of anonymity, to all survivors: not just pacifism or justice, but the common creation of a new process for grief. To be sure, the extension of confused bones to peacetime and all of humanity might just draw everyone back into the war experience, into 'insupportable [...] mourning,' in a kind of 'not forgetting' that was in danger of becoming a refusal to understand, interpret, and historicize. But this was the risk that had to be run in order to mourn confused bones, and chorality itself, from a new perspective that would not betray the lessons of the war.

The Unknown Soldier Memorial addresses this issue directly, as I now hope to show, in that it represents unanimity, and its link to a public alienation beneath heroic names and beyond political factions, via the reconstruction of the Unknown Soldier's skeleton. Telling images from the contemporary press claimed that only a proper reassembly of soldiers' broken bones would allow them to 'speak,' and give a true voice to their experience; similarly, the Unknown Soldier's suffering, they wrote, would become 'legible' only once his bones were gathered. These images recur in fiction, poetry, and wartime diaries. What was at stake in all these evocations of the Unknown Soldier's skeleton was the desire to make unanimous grief or confused bones become intelligible. We will see in Chapter 3 what these bones actually said, what sort of knowledge or home they founded, once reassembled; in Chapters 1 and 2, though, I want to focus on the process of gathering and reassembling, for what we see here is the struggle between two different forms of legibility, which are also different kinds of identification, and different understandings of unanimity in grief. Remarkably, Mircea Eliade's discussion of the practice of 'contemplating one's own skeleton' as part of the shaman's initiation evinces a similar dualism: on the one hand, 'in the spiritual horizon of hunters and herdsmen bone represents the very source of life, both human and animal. To reduce oneself to the skeleton condition is equivalent to re-entering the womb of this primordial life, that is, to a complete renewal, a mystical rebirth. On the other hand, [...] reduction to the skeleton has, rather, an ascetic and metaphysical value – anticipating the work of time, reducing life by thought to what it really is, an ephemeral illusion in perpetual transformation. Such contemplations, it should be noted, have remained alive even within Christian mysticism.' In the context of the First World War, the latter reading of bones, as ascetic purification and transcendence, was tied to traditional monuments; in contrast, the former reading emphasized a kind of primordial and visceral vitality in the bones of dead soldiers, and demanded a new kind of memorial that would transfer this vitality to the living.[6]

The struggle between a transcendent reading of bones and a more visceral one is integral to the Memorial's cultural origins (Chapter 1), and runs through the different phases of its conception (Chapter 2), colouring also its initial official interpretations (Chapter 3). (These three chapters form Part I. After examining the Memorial's execution and reception, with the focus especially on divergences from planned scripts,

I will turn, in Part II, to my proposal for understanding the Memorial's enduring appeal.) Chapter 1 discusses the putatively unanimous invention of the Unknown Soldier as a new form of memorial appropriate to the First World War, linked to a shared primal experience of mourning in the trenches. Chapter 2 divides the Memorial's conception into four moments. First came the choice of a body, or more accurately, of a set of bones, and the emphasis was on randomness, or better, on the idea that the Unknown Soldier was not chosen *by* someone but manifested himself. Second came the moment of identification proper, as these particular bones were now 'recognized' by the living. Third, the physical shape of the tomb itself was designed to reflect confused bones. And finally, fourth, an inscription was chosen – as minimalist as possible – to indicate that bones spoke for themselves. Finally, Chapter 3 considers official interpretations of what the Unknown Soldier's bones said, which brought sacrifice into play. This had two moments, first ritual absolution, which sought to guarantee the Unknown Soldier's status as sacrificial victim, and second, initiation, which endowed him with new authority at once religious and political.

The Origins of the Unknown Soldier

When it comes to determining who first imagined a 'Tomb of the Unknown Soldier,' what is most striking is that there seem to be many answers rather than one. Jean-François Jagielski, in his *Le soldat inconnu,* sums it up best when he notes that throughout the twentieth century and beginning as soon as the various memorials were created, the Italians, the French, and the British all claimed paternity for the idea. For him, these somewhat nationalistic squabbles should be left behind in order to focus on the more explicit aspect of the story, which is the transformation into law and monument an idea that evidently had many cultural 'origins': Jagielski's book on France's Unknown Soldier focuses on debates among politicians, both official and unofficial, and how they shaped both the Monument and its inauguration. Here we see how unanimity concerning its concrete shape, and its value as well as its origin, was politically constructed. Similarly, Alessandro Miniero's recent book, *Da Versailles al Milite ignoto,* which has the advantage of comparing the Italian story with the French and British ones, focuses on how political authorities gradually came to the Unknown Soldier idea. From this perspective, the Unknown Soldier Memorials in all three countries stem from the authorities' recognition that the other celebrations investigated by Miniero (Victory and Armistice parades, starting in 1919) were insufficient, especially as regards creating a feeling of unity. While these political histories are an important part of the story, which I will bring into my discussion, neither book has the space to investigate another remarkable, if less visible element of state archives regarding the Memorial: the feeling among politicians that this idea was one they could not afford to be against, for its popular appeal was overwhelming. Thus, repeatedly after the Memorial's inauguration politicians report in this vein: 'must inform you spontaneity [sic] public demonstrations that exceeded our own expectations. All parties participated [sic] ceremony.' More significant, before the inauguration, the Italian Ministry of the Interior requested and received reports from every province and major city regarding the popularity of the Memorial and the likelihood that its inauguration would cause unrest, as the Unknown Soldier's humility might be seized upon as a critique of those in power. My central purpose is to interpret the symbolic power of the Unknown Soldier and, thus, to understand this popular appeal by focusing not only on politics but also on the cultural crisis, as well as the primordial connection to the dead, whence the idea emerged.[7]

This is a different type of origin, and it is much harder to pinpoint. Striking here is how the more fluid world of the symbolic imagination is retrospectively and at times wilfully narrowed down to a single starting point. Thus, it seems evident that the idea of burying 'a soldier' in a special symbolic fashion that would honour those who were not receiving proper burial, who remained trapped in no man's land, was popular with combatants in Italy, France, and England by 1916, if not earlier, and in any case well before the idea entered political debate. But the search for a specific 'first proposal' remains compelling. According to Jagielski, the Prussian War veteran Francis Simon was the first to mention burying 'a little soldier, glorious and anonymous' (*petit soldat, glorieux et anonyme*) in the Pantheon as a 'symbol' of the French nation, in a November 1916 speech. Simon mentions his 'illustrious bones' (*illustres ossements*), inaugurating what, as we will see, is a consistent link between bones and unanimity. However, the first legal and political proposal for an Unknown Soldier in France took place in November 1918 and did not mention Simon's speech, presenting the idea as something that combatants in general desired. Similarly, in the case of England, Father David Railton is credited (most recently, by Miniero, but also by Neil Hanson) with imagining as early as July 1916, when he was a chaplain during the Battle of the Somme, that 'an unknown soldier could be sent home for burial' from France. But this is based on his own later testimony, which enters the written record only in 1920, when the Dean of Westminster, to whom Father Railton wrote of it, brought it to the King's attention, and to that of Lloyd George. More important, perhaps, Railton did not claim to have invented the idea, as British politicians later wished to suggest, but instead gave as its origin an actual tomb that he saw near Armentières, where a 'nameless British soldier' had been buried recently (sometime in 1916, we may infer). Railton emphasizes the 'simple white cross' of this tomb, evoking a minimalism which, as we will see, many associated with a mourning that would be true to the bare bones of combatants rather than covering them up with decorations.[8]

The Italian case is murkier, and more interesting, mainly because individual allegiances, national identity, and Church authority were far more than elsewhere precarious, and in conflict with each other, exacerbating political and social conflict. Thus, the official and more political origin of the Unknown Soldier in Italy never quite meets with its imaginary one, for reasons that will become clearer as we proceed. In Italy the first official proposal for an Unknown Soldier Memorial is attributed to Giulio Douhet, an aviator and veteran who campaigned for it in the summer of 1920 in his veterans' newspaper, *Il Dovere*. He opened his discussion by submitting that '*Il Dovere* has always asserted that the true winner of the war was the Soldier, the humble Soldier' and by contrasting his loyalty and courage to the indifference of 'incapable political and military leaders.' Adding that this unrecognized heroism, 'cannot be materialized if not in the corpse of an unknown Soldier,' he called for an anonymous soldier to be buried in the Pantheon, next to the Italian kings. Remarkably, Douhet insisted repeatedly, 'the apotheosis of the soldier does not in any way mean the apotheosis of war,' concluding that 'everyone wishes that different peoples become brothers.' Finally, like Simon and Railton, Douhet presented the idea as a demand from the soldiers themselves, rather than as his own. It is thus quite notable how, already in 1921, in discussions of the Memorial in the Italian parliament, Douhet's name, but even more his pacifism, are elided, even as much of his vocabulary regarding heroic humility is borrowed piecemeal. When we move to Italian histories of the Memorial, Douhet is indicated as its inventor, with a remarkable erasure of contemporary, if not preceding, proposals in France and England, and generally, his pacifism is downplayed, as is his rather acerbic critique of the Italian government. To a greater degree than in France or England, then, an enormous effort is made to present the Unknown Soldier as a government idea, perhaps inspired by a veteran such as Douhet, but most important, entirely free of any conflict between soldiers and their leaders or between pacifism and bellicism.[9]

There is, however, a second, more indirect or genealogical origin to the Unknown Soldier idea in Italy, which has not been mentioned by historians. The poet and veteran Gabriele d'Annunzio asserted the following at the end of the *Notturno*, his poetic-diaristic evocation of the war experience:

> While in sadness I transcribed the example of the nameless peasant who entered the ford dropping to his knees in the midst of the current, and, sacrificing himself, sculpted his sacrifice into the water, inside the basilica of Aquileia, a mourning mother was choosing between the eleven nameless coffins the one to be buried in the monument. [...]
>
> And why wasn't the Great Sufferer's coffin enveloped by the flag of the Timavo, the flag I keep, the one once named 'the shroud of sacrifice,' 'the foot soldier's banner,' the one draped across the coffins of my dead of Fiume, lined up against the earth?
> At first, the image of only one dead hero was incised upon it; now it bears the image of all the dead, as all those who died for their *Patria* and in the *Patria* resemble each other just as Giovanni Randaccio in his stone ark resembles the unknown soldier gathered inside four boards.[10]

These passages are from 'Annotazione,' published as a postface only in the 1921 version of the *Notturno*. However, the image of the peasant he refers to comes from earlier portions, written in 1916 and partially published at that time, when d'Annunzio was convalescing from a seaplane accident. Similarly, d'Annunzio's mourning for his comrade Randaccio dates back to 1917, and the 'dead of Fiume' refers to his paramilitary takeover of that city in 1919–20, which was a protest against Italy not receiving territories in Dalmatia, but also against the current government's acceptance of that situation. I will return to all three of these references in more detail later, and to the *Notturno*, as it was an extraordinarily popular evocation of the Great War in its immediate aftermath and remains one of d'Annunzio's most appreciated works. For the moment, I wish to take seriously d'Annunzio's claim that these three images were an inspiration for the Unknown Soldier idea, which preceded both Douhet's proposal and government discussions.[11]

As regards the first, d'Annunzio invented in the *Notturno* an image that would become recurrent in Italian culture, that of the *miles patiens*, as

Paolo Valesio has termed him, that is, the peasant-soldier in his uniquely humble suffering.[12] This Italian peasant who accepts death by drowning in order not to reveal to the enemy that the river can be forded is close to the Unknown Soldier not only because his sacrifice is anonymous, but also because of his social class, which separates him from officers or career military men and connects him to the earth. Also, as Miniero points out, the image of the peasant-soldier was how the Italian press responded to a popular desire to imagine the Unknown Soldier as 'a known person,' anonymous yet representative of 'the character of the Italian people.' As regards the dead of Fiume, d'Annunzio is describing here his use of a tattered Italian flag in mourning them, the same flag he draped over the coffin of his comrade Randaccio during the war. I will get back to this flag when I discuss Fiume in detail in Chapter 5. What is more immediately remarkable is that the Unknown Soldier can for d'Annunzio – as for Douhet – represent those who died in the First World War but, more uncomfortably for the state, also those who died in paramilitary fighting. Moreover, since the flag in question bore the traces of Randaccio's suffering and death, like the Shroud of Turin, the Unknown Soldier represents all the dead in a remarkably embodied fashion, whose links to the incarnation I will examine. Finally, as regards Randaccio himself, d'Annunzio's 'Annotazione' points us to another of his wartime texts, *La riscossa.* Published in May 1918 by the Print Secretary's Office, it was widely distributed among Italian soldiers; it was also simultaneously published in English translation as *The Rally*, and distributed to British soldiers as well. Its last chapter is a funeral oration for Randaccio that d'Annunzio gave in 1917, on the occasion of Randaccio's reburial in Aquileia (he had been initially buried on the battlefield at Monfalcone). D'Annunzio's speeches at both burials were widely reported in the press, appearing, for instance, in the *Illustrazione Italiana* of 17 June 1917 and the *Domenica del Corriere* of 24 June 1917.[13] In his oration, D'Annunzio asserted:

> But this day cannot be mine, comrades. This can only be the votive day of the martyr of Aquileia, the day sacred to the hero of the Timavo. [...] Giovanni Randaccio [...] was a son of the earth, a creature of the soil and of the rock, of the mud and of the dust. [...] he was the true maker of victory. In sum, he was the foot soldier.
>
> He was like you [...] [His] heroism was a flash of lightning. [...] Yours is like your bones, it is the internal core, it is always there; [...]
>
> '[Y]ou are all heroes,' Giovanni Randaccio cried out to you [...]
>
> You, people of the fields, people of different trades, people of the work-

> shop and of the office, peasants, workers, bourgeois from every walk of life and in every profession, grown wild like those who wait in caves, [...] stooping in the filthy burrows that smell of sewer and grave, [...] you, people foul and heavy beneath the earth, you in that moment were but a swift flame, a splendid soul, as on Resurrection Day.[14]

We find here three elements central to the Unknown Soldier. First, the naming of '*il fante*' (the foot soldier) as representative of all true heroes, in which '*operaio della vittoria*' (maker of victory) is a prelude to Douhet's '*il vero vincitore*' (the true winner). Second, the soldier is associated not just with the peasant ('*gente dei campi*,' or farmers, noticeably the first category mentioned), but with a list of vocations (more than professions) that tries to be encyclopedic while excluding precisely those who would normally be heroes, the nobility. Third, 'the soldier's humility and courage,' are represented by his bones, and by a reading of them that eschews transcendence and purity in order to emphasize a kind of resurrection that is inseparable from the earth, the mud, the mass grave, and even the sewer. The importance of this humble and primal chorality is attested to by the fact that d'Annunzio had already, in a rather new show of humility brought on by the war, evoked it in an entry from his war diaries dated 1915 and published in 1916. Here he identified with a soldier sitting next to him in the trenches: 'We are both here on the bench, one next to the other. It seems our destinies are about to be linked, intertwined. He is young, and I am no longer young. Both of us could be dead before noon on Tuesday: a fistful of scorched flesh, a few blackened bones, some shrivelled cartilage, a crushed skull with a few gold teeth sparkling in the sludge.'[15] We return here to the mud of the trenches, in which the bones of comrades become indistinguishable from each other. Most important, this chorality is once again not 'from above,' tied to abstraction and purity, but 'from below,' a descent into the chthonic. This descent, represented by 'the foot soldier's banner' or Randaccio's flag, unites this soldier, the dead of Fiume, Randaccio, and the nameless peasant, and is d'Annunzio's most important contribution to the genealogy of the Unknown Soldier Memorial.[16]

This contribution was indirectly acknowledged as the mayor of Udine invited d'Annunzio to participate in exhumations in preparation for the choice of the Unknown Soldier in the area of Castagnevizza, where he had seen his comrade Randaccio fall (as written about in *La Tribuna* on 27 October 1921).[17] Later, d'Annunzio was also invited by the authorities in Rome to attend the inaugural ceremony as a special guest. But

Figure 3. D'Annunzio in 1916. © Archivi del Vittoriale

Figure 4. Giovanni Randaccio in 1916. © Archivi del Vittoriale

Figure 5. D'Annunzio before the Banner of Randaccio at Fiume (*L'illustrazione italiana* [1921]). © University of California Northern Regional Library Facility

the Ministry of the Interior archives reveal that d'Annunzio's link to the Unknown Soldier, even more than Douhet's, was an uncomfortable one for the authorities: they invited him largely because they feared he might otherwise turn up and undermine the scripted ceremony by making too visible the divisions between those who had suffered and their leaders, or by emphasizing the mud and sewer of the trenches instead of a cleaner resurrection.[18] Moreover, d'Annunzio's takeover of Fiume had exacerbated political dissent in Italy, as it claimed that Italy would remain 'mutilated' not only without its territories in Dalmatia, but also if wartime chorality did not lead to major social change. Though d'Annunzio was forced to abandon Fiume by December 1920, the Italian state evidently feared that the inauguration of the Unknown Soldier Memorial would provoke some action from Fiume veterans or from d'Annunzio himself. A report from Naples to the Ministry of the Interior claims that the arditi were planning to March on Rome with d'Annunzio at their head to honour the Unknown Soldier. In the end, d'Annunzio refused both invitations, and did not turn up with the arditi, saying that he preferred to pay homage to the Unknown Soldier via his book, the *Notturno*, originally scheduled to be published on 4 November 1921, the day of the Unknown Soldier Memorial's inauguration (it ended up being a few weeks late, but significant excerpts appeared in the newspapers on that day, side by side with coverage of the inauguration of the Unknown Soldier Memorial). It is clear that d'Annunzio felt that in this and other writings and speeches, he would be more free to express his vision of the Memorial than he would have been in active participation, which would also have been tacit approval of the state's vision. And indeed, d'Annunzio's absence was noted by the press, and generally seen as a rebuke to politicians who did not understand wartime chorality. D'Annunzio was, and still is, too charismatic to be ignored and yet too divisive to be included: even in 2003, when Italy mourned its dead from Nasiriyah, Iraq, at the Unknown Soldier Memorial, 'd'Annunzian rhetoric' was adduced to be a symptom of Italians' failure, yet again, to achieve a modern political discourse. But back in June 1921, most tellingly, when Minister of War Rodinò discussed the Unknown Soldier with the Parliamentary Commission created to plan the Memorial, he echoed word for word d'Annunzio's 1917 oration for Randaccio, giving the same unusual (and a little archaicizing) list of vocations the soldier would represent as 'homage to all our dead, to people of the fields, to people of different trades, to people of the workshop and of the office, to workers, to bourgeois, to people from every walk of life and in every profession.'[19]

One last point regarding the Italian origin of the Unknown Soldier idea is the connection between Douhet and d'Annunzio. The language of Douhet's proposal echoes imagery from the *Notturno*, and his admiration for the poet is clear in *Il Dovere*'s regular reports about his activities at Fiume and his development of new mourning rituals for his soldiers. Further, like d'Annunzio, Douhet was an early proponent of air power, and even liked to compete with the poet by coming up with ever more dangerous and symbolic airplane stunts, modelled after d'Annunzio's famous flight over Vienna to drop patriotic leaflets. Douhet was court-martialled for criticizing General Cadorna's strategy, only to be vindicated later as the overwhelming importance of air power became clear. Cadorna was widely held responsible for Italy's defeat at Caporetto. This was no doubt the most traumatic event for Italians in the First World War, all the more so because it was initially blamed on the lack of courage and desertion of soldiers, only to be later attributed, more or less implicitly, to the tactical errors of their leaders. To associate the Unknown Soldier with this sketchy history – be it Caporetto or Fiume – was obviously something the Italian state wanted to avoid: hence parliamentary discussions failed to mention Douhet or d'Annunzio. Remarkably, instead of contesting this, Douhet and d'Annunzio, like Simon and Railton, attribute the idea to the war experience itself, and in particular to a battlefield chorality that is primordial, visceral, and antagonistic to more official and abstract notions of unanimity.[20]

Further complicating the issue of origins is the creation of cenotaphs, among them a temporary one in Paris, and one that became permanent in London. The use of an empty tomb to commemorate multiple dead whose bodies were lost (a very old idea, most famous in the West for its use in Greek antiquity) was planned by the French to counterbalance the patriotic victory celebrations of 14 July 1919 with a more sombre mourning on 13 July and in the following week; their cenotaph was placed under the Arc de Triomphe, thus anticipating the location of the Unknown Soldier. The British immediately took up the idea, and Luytens, who had been collaborating since 1917 with the War Graves Commission, designed the Whitehall cenotaph which, within four days of its inauguration, was so popular people wanted it to be permanent. It became such, and thus at the inauguration of the Unknown Soldier Memorial more than a year later, crowds went both to Westminster and to Whitehall, connecting the two monuments. As mentioned, the cenotaph is not quite yet the Unknown Soldier, for it lacks a spectacular and uncannily distributive body; it does, however, evoke anonymity, albeit one that is still more historically contingent and less existential or visceral. As a list of names that have been separated from bodies that are missing, the empty tomb shows that one of the horrors of war is to not receive proper burial. But the Unknown Soldier does eventually receive proper burial (symbolically at least) yet still remains anonymous, for this is no longer a (possibly reversible) result of the war, but an expression of the modern human condition. What remains important about the cenotaph, as it carries over, but not entirely, to the Unknown Soldier Memorial, is its democratic component: equality in death was, at the time, still new and controversial.[21]

As I already suggested, a similar emphasis is found in d'Annunzio, Douhet, Simon, and Railton, who grasp the subversive, provocative aspect of associating heroism (in d'Annunzio's case, that of Renaissance warriors and other historic figures, in Douhet's, that of the Italian royalty, in Simon's, that of French heroes buried in the Pantheon) with a *fante* (foot soldier) or *contadino* (peasant), an unknown soldier, an unknown Tommy, *un petit soldat* (a little soldier), *un fantassin* (a foot soldier). Their language eschews the lofty and stresses the everyday, in a way that popular response to the Unknown Soldier Memorial would push even further: in France he became a *poilu* (literally, a 'hairy' soldier, the implication being 'uneducated peasant'); in England, there were complaints of his being termed an 'unknown warrior' on his tombstone instead of an 'unknown comrade,' such that recently this had to be counterbalanced by 'unknown victim'; in Italy, *contadino* (peasant) and *soldato* (soldier, as opposed to officer) complemented the more Roman *milite*. Overall, this goes against memorials of the past, which always seek an exemplar, a case of exceptional heroism; arguably, even past cenotaphs see heroism as lost to history and human memory, yet still possible beneath anonymity. With the Unknown Soldier, it is to a great degree anonymity itself – the everyday – that becomes heroic. This, I believe, reflects the reality of trench warfare, where, more than in any other war, commanders and their men faced the same fear, were mired in the same mud, and had equally to abandon or to bury the bodies of comrades as they fled. Such improvised mourning, as documented by both Winter and Leed, was traumatic yet pervasive, and it is the true imaginative origin of the Unknown Soldier Memorial. This explains why Simon, Railton, Douhet, and d'Annunzio do not present the Unknown Soldier Memorial as their own invention, but as a creation of the soldiers themselves and almost an imposition, a demand on the part of the everyday itself, that cannot be avoided.[22]

In these two types of origin, one more political, the other more popular, different aspects of unanimity are emphasized. One is debated, constructed, made clear, brought into the light of day; the other is inchoate, instinctive, and largely implicit, or latent. Also, the first pertains to 'civilian' or 'civic' discourse, whereas the second participates in the ritualistic, mythical discourse of the trenches. The first has a 'plebiscitary character,' for which it is necessary to create 'much propaganda'; the second activates a visceral reaction to a body 'thrown naked and unarmed against the reinforced concrete and the hard steel' and to 'humble bones.' I have termed the first a 'constructed unanimity' and the second a 'choral unanimity.' From the point of view of the state, but also of many other authorities such as medical ones, the latter should give way to the former: popular but disorganized desires for a new type of memorial should be organized and given a solid form, and the marginal forms of the mourning of the trenches should be reintegrated into society. In practice, this meant forcing wartime grief into a single, homogeneous expression. Thus, Jagielski notes that French politicians vied with each other to condemn factionalism, expressed, for example, in arguments concerning the Treaty of Versailles or reparations, because this would 'deprive the ceremony proposed by the government of its unanimous character.' Any right versus left dissent, then, was seen to betray 'the greatness of his anonymous sacrifice' that 'surpasses our little quarrels and disagreements.' Vito Labita stresses that the Italian parliament focused on the Unknown Soldier's 'absolute suppression of his individuality' in order to establish maximum state power. Once again the Unknown Soldier was presented as above left and right factions, and his sacrifice was evoked to demand unanimity from politicians; in fact, the published *Atti parlamentari* show that in Italy the majority of politicians, in favour of the Memorial, pressured the few dissenters to vote in favour as well, so that the Memorial could go down in history as 'unanimous.' I will explore political dissent more as I discuss the Memorial's actual construction; but what is remarkable as regards the 'original' idea is that political unanimity is seen as a necessary response to a choral unanimity inherent in the war experience. Though politicians, intellectuals, and veterans did in reality argue about the Unknown Soldier Memorial, they felt beholden to a putative first proposal, to a deeper unanimity, that no one would argue against, even as, or precisely because, it was implicit: Simon's speech, Railton's bare cross, Douhet's humble soldier, d'Annunzio's Randaccio. In this way the deviant mourning of the trenches was a powerful undercurrent that the Unknown Soldier Memorial was supposed to contain, to reintegrate into society; at the same time, as will be demonstrated, such mourning, like trauma, with its contained violence, remained latent and ready to erupt anew. This was, one might say, the primal scene, which the Memorial could not fully represent, but which it evoked, even as it also sought to exorcise it from the haunted imagination of survivors and civilians.[23]

The Primal Scene

Among the many literary evocations of the primal scene where choral unanimity is created, Jean Giono's *Le Grand troupeau* stands out because it connects death on the battlefield to a missing body at home. Giono was a veteran, but would not be able to write about the war until this novel, published in 1931. In a first moment, one of the novel's protagonists, Joseph, begins his experience of the war by encountering the dying before he even experiences battle:

> The stiffness had risen up Jules' body and forced back his head until he looked as if he had been tied to a post. He seemed to be looking up at his hair. His eyes stayed open while he slept. For a long time he chewed at the air, sleeping. [...] Finally, he set out for those great countries that lie inside the head and he started talking. [...]
>
> Joseph puts his arm under Jules' head, but Jules' head cannot be lifted. It is no longer possible to hold him tight against oneself, with his soft manly thing that is life and heat, and the truth of the living body. He is as hard and stiff as dead wood. Joseph lay down next to Jules and could finally hold him. [...]
>
> It was only when night came that Joseph tired of holding this stiff, cold, dead thing in his arms. He got up. He looked at Jules lying down. Then he took his identification tag off [...] and started walking on the road.[24]

Joseph does not know Jules, and they come from different regions in France, though they are able to share a typically Gionoesque vision of country life and its everyday pleasures, which Joseph speaks for Jules who can no longer speak himself. More important, however, Giono emphasizes that in the journey to death, towards 'those great countries that lie inside the head,' Joseph develops a symbiotic relationship with Jules, which is manifested in their physical intimacy. Choral unanimity is represented here by the way Joseph throughout the novel continues to carry with him, not literally but metaphorically, 'this stiff [...] thing,' like 'dead wood' ('piled up' in his body). Notable also is the fact that Joseph does not even attempt to bury Jules: a few pages earlier we watch him try to bury another dead soldier, whose death he also presided over, and we see him realize that this is beyond his own strength to accomplish, alone; he has to settle for protecting the body from the flies by hiding it under a hedge. Thus, Joseph departs, with two sets of identification tags, hoping both to notify families, and to ask the battalion he is supposed to join to retrieve and bury the bodies.[25]

We do not find out whether this happens. But a chapter or so later we are back in Joseph's home town (the novel alternates his story with goings on at his home), where his wife attends the wake for Arthur, whose family has received news of his death but no body:

> They were coming, they were all there in the farm's big room with the cold fireplace.
>
> They are there stiff and silent keeping watch over the absent body [...]
>
> 'We are gathered here,' Marthe declaimed, 'to keep watch over the absent body of Arthur Amalric who was killed in the war. Let us all treasure in our thoughts our friendship for one who was the salt of the earth ...' She dipped her hand into the pot, took out a handful of salt and put it on the centre of the table. They made a little pile of it. [...]
>
> The heavy silence returned.
>
> 'Oh, my Arthur!' Félicie cried out. She was stiff as wood. [...]
>
> 'Poor one, if only I'd been there to close your eyes for you! Oh, my poor Arthur, dead all alone against the earth like an animal!'[26]

The implication is clear that Jules and the other dead soldiers helped by Joseph are likely to be mourned in the same way as Arthur. The two passages are linked, moreover, by the image of 'stiff [...] wood.' In the second, the wood of the table metonymically comes to represent the missing body, referred to as 'dead wood' in the first. Moreover, Félicie, like Jules, becomes wooden in her identification with the dead man. I will address in detail in Part II the notion that to die unburied is to die like an animal, for the moment what is important is how Giono connects the primal scene of choral identification to the need for an improvised ritual, a reinvention of traditional mourning. This is still rudimentary, as the pile of salt on the table seems to represent the missing body, as though reducing it to its mineral and vegetal components; this recalls d'Annunzio's image of '*carniccio carbonizzato*' (scorched flesh). It is precisely this sort of identification to a minimal humanity that is expressed in the Unknown Soldier Memorial. As Douhet notes in his proposal, which like Giono's novel dramatizes the humility of women, 'from this tomb the greatest consolation will come to those who are most sorrowful: those who never found out anything further about the one they bore, who seemed to vanish into the storm.' Moreover, this is linked to bones by a commentator in the *Corriere della Sera* who honours 'the unknown soldier more than the hero [...] the nameless dead man more than any military courage [...] this symbol of bare bones.'[27]

In contrast with Giono's pastoral and ultimately optimistic if pitiful tone, Blaise Cendrars' novel *La vie et la mort du soldat inconnu* refuses monumentality and evokes a deeper unanimity via satire. Cendrars, too, was a veteran, who gave up writing poetry after the war and was never fully able to write about his war experience, as this novel, begun in 1928, remains unfinished. In the novel's frame, we find its author aboard a cruise ship, drunkenly courting Caralina, and getting into arguments with the cosmopolitan couples he meets at the bar:

> – I am sorry, Monsieur de Mesle, I was daydreaming, what is this all about?
> – It's about the Unknown Soldier, said de Mesle.
> – Ah! l'Poilu, I said.
> – Yes, continued de Mesle, Monsieur de Tang claims that 'our' (de Mesle insisted on 'our' because he was a deserter) that 'our' Unknown Soldier that rests under the Arc de Triomphe in Paris is a German soldier!
> – And why not? I said, rather annoyed.
>
> Everyone protested indignantly, but Mr de Tang was yelling louder than the others, standing up, triumphantly:
>
> – You see, you see, it is possible! And Monsieur Cendrars, the only one among us to have fought in the war, sees no objection, in principle, to the fact that, or rather the possibility that, to even the probability I would say, that, if the removal of six anonymous bodies on six battlefields [...] was done in a proper fashion, nothing can guarantee, theoretically, that this body, to which the entire world today renders homage, nothing can guarantee that this body is not that of a German soldier. And I would dare to affirm, even, that here lies all the greatness of France, its nobility, its detachment, and that this gesture is very much in the chivalric tradition of this beautiful country, to have the daring to stage the apotheosis of one of its

heroic children, but also, but even, perhaps (were it not a chance out of ten thousand), of one of its heroic enemies.

– And why not a Jew? I added.

Mr de Tang was dumbstruck. He favoured the Germanic race, and [...] my remark had dented his enthusiasm. [...] Then he smiled ironically, bent his round body, broke it in two, dove down, saluted me, raised his glass, and said, smiling at me once again:

– Ah, these Frenchmen! They always want to have a laugh. Nothing is sacred for them and, subjectively, yes, they are all of the purest nobility. Ladies and gentlemen, I drink to the health of France! *Hoch! Prosit!*

Everyone was quiet. Everyone was embarrassed. [...] No I didn't want to fight with von Tang, nor, to please a bunch of nouveaux riches or noble cosmopolitan travellers, get myself into the defence and the history of the Poilu. I would not have been capable of it, anyway, as they all professed to venerate him, to love him, and to have adopted him, him, the Unknown Soldier, just as they had adopted the races, the theatres, the boulevards of Paris, Montmartre and Montparnasse, Poincaré and Mistinguett, the *Folies Bergères*, the *Chabanais*, Coco at the *Boeuf sur le Toit*, dresses from chez Patou, perfumes from chez Chanel, and, with enthusiasm, the basement gigolo, the gigolo from down below, from the low tones, from the la-di-dah, from the *Lido*. (I was looking at Caralina and wanted to laugh, as some humour is proper to the art of love, and I was her *petit poilu chéri* [...] .) [...]

– We the Americans (he pronounced it *Hammm'ricanne*), Ten Cate was addressing von Tang, [...], we cannot accept this, Monsieur von Tang. The Poilu (he pronounced *la Poilioue*), the Unknown Soldier (he pronounced *la Soldatte Inconnioue*) he (he pronounced *alle*) he is also in New York, he (he was still pronouncing *alle*) is American (he again pronounced *Hamm'ricanne*), I am sure ...

– And why not a Negro? Said I, for though I have never liked to cry wolf, I do enjoy provoking them.

This time my repartee failed, for no one was enraged, and Ten Cate's total surprise made everyone laugh until they cried. He was just standing there, his mouth hanging open ...

[...]

Isn't life beautiful!

I made love until dawn.

[...]

O World!

I stick my head out of the porthole. The ocean is like steel. The sky is the deepest black. To the East, the wind begins before the day, and everything shivers.

[...]

And now that I am alone, I have finally written it, this book. It is not the beautiful book Caralina dreamed of, but it is not a nightmare either.

It is not a saint's life.

It is not a legend.

It's just a book.

A book that teaches in turn to bless and to curse life.

In other words to love it.

Nothing more.[28]

Initially, it would seem that nothing can be taken seriously here, and certainly not the comfortable and consumable heroism identically invoked by these three French, German, and American bourgeois. But we must not underestimate the shock value of Cendrars' humour: his frame is dated 1 March 1929, and thus follows closely the Kellogg-Briand Pact against military aggression (signed on 27 August 1928) and is during discussions of the Lateran Pacts (ratified on 7 June 1929). As a result his humour is a rebellion against the unholy alliance of pacifism, economic interest, and established religion, given the lie by the rise of militarism in Germany (evoked by Mr de Tang's '*pangerma[nisme]*') and Italy, as well as the capitalist ambitions of America and France (evoked by the long list of French 'goods,' from Mistinguett to Poincaré). If the Unknown Soldier represents any unity still, Cendrars implies, it is not this one. Furthermore, the list of multiple identities that might be ascribed to the Unknown Soldier according to Cendrars is not casual, not aimed merely at multiplication. As a German, of course, he would decry the economic starvation of Germany that would come to haunt France later on. As a Jew, however, he not only would critique anti-Semitism in Germany (and elsewhere), but would also evoke a major crisis in French culture, where anti-Semitism became an excuse for factionalism and personal gain: the Dreyfus affair, which was hotly discussed again when the French authorities decided not to bury the Unknown Soldier in the Pantheon, where Zola, Dreyfus' great defender, had been buried not too long before.[29] Ten Cate's comical French pronunciation then makes the Unknown Soldier a woman and Cendrars finally adds '*Nègre*' to the list. Here, political divisions and Europe's failure to address them are evoked (women's suffrage, racism, colonialism, and use of segregated colonial troops in the

Great War), but perhaps even more important, an embodied identity is called for that would transcend the biological (the Unknown Soldier is a body at once male and female, at once white and black).[30] Also crucial is the parenthetical comment that de Mesle is a 'deserter,' for as we will see, much ink was spilled to assert that however anonymous, the Unknown Soldier could not in any way be anything but obedient and loyal. Cendrars' is thus a direct attack on constructed unanimity, as it reveals not only its falseness, but points to many of the political and cultural interests that motivate it.[31]

But beyond satire, as Cendrars makes clear in the rest of his book, is once again the issue of chorality. We see it here already in the frame, as he refuses to 'get [him]self into the defence and the history of the Poilu,' and yet identifies with him, albeit again in a somewhat comical vein, as he is Caralina's '*petit poilu chéri*' (dear little *poilu*). If we recall Mr de Tang's earlier admission that Cendrars is the only one among them to have fought in the war (and if we try to picture his physical presence, with his mutilated right arm), his refusal to address his fellow passengers with anything but distancing humour, his silence about that experience, become eloquent. No words are adequate to the war experience, and all serious speech is punctured not just by provocative questions but by embodied identity, as much of Cendrars' humour here derives from the jarring presence of a mutilated body gesticulating in semi-serious drunken revelry. Faced with this presence, de Tang's description of the concretely gory and traumatic exhumation of bodies from the battlefield gets lost in tangles of bureaucratic syntax (I have cut his sentence down to about a third of its original length in my citation above); and his concluding praise of the French chivalric tradition begins to look like a childish fable. More important, at the very end of the frame, Cendrars' tone acquires a far quieter, pared-down power. He speaks of the silence of the night, and of oceanic identification with the living world, where the love of life is felt yet is also inseparable from suffering. Against any kind of closure or ultimate meaning, be it heroic or opprobrious ('*beau livre*' or '*cauchemar*'), he demands that life be lived in its flux and mystery (as represented by making love until dawn, but also by that dawn itself, as Cendrars' description veers to a non-canonical, non-denominational pantheism). The phrase '*que la vie est belle!*' (isn't life beautiful!), as well as the contemplation of the living night, recur a few pages later as part of the experience of the Unknown Soldier himself, where they are associated with a 'primitive,' non-Western or 'Oriental' connection with life's primal sacredness, presented as a cycle of life and death, of celebration and mourning, of peace and war, entirely at odds with the ideology of progress and linear time of the West.[32]

If there is a truth to the Unknown Soldier Memorial, then, for Cendrars, it is that it cannot contain but can merely point to multiple paradoxes, or indeed, aporias: wartime death demands a unanimity deeper than national, religious, racial, or gender divisions, but we have no language to speak it; similarly survival forces on the veteran the simultaneous experience of 'cursing' and 'blessing,' of dying with one's comrades yet also having to live fully for them – an experience that is sacred in the etymological sense of outside the norms of culture, yet foundational for it.[33] The rest of Cendrars' novel is, as we will see later, a meditation on the new ritual understanding of life and death required by the Great War and symbolized by the Unknown Soldier; that it remains unfinished seems to imply pessimistically the ultimate failure of all forms of commemoration or memory. Nonetheless, already in its frame, Cendrars' novel succeeds in insisting that there is a primal scene of choral identification with the dead, even if we can never be true to it, and perhaps we should not even try, for that could lead only to the political farce Cendrars critiques. Moreover, in comparison with Giono, Cendrars seizes the traumatic quality of this origin of the Unknown Soldier Memorial. For Cendrars the issue is not only that the soldier's body is missing or 'scorched,' but also that, more deeply, the link between embodiment, on the one hand, and social, racial, gender, or political identities, on the other, has been radically severed. Such loss of embodiment may be repeated but, like originary trauma, can never be seen directly. Since true unanimity, in contrast with a constructed one, is tied to this embodiment, we see that for Cendrars it can only be expressed beyond any socially determined discourse, pertaining, if it can enter language at all, to the realm of the poetic and the archetypal. Thus, the unanimity at the core of the Unknown Soldier Memorial is not only about reduced humanity, pared down to 'bare bones,' but about the 'primordial life' (Eliade, above) and the oceanic moment such bones represent. As one commentator notes, the day after the inauguration of the Unknown Soldier Memorial, more important even than the 'saintly relics' was the 'organic crowd' (*massa organica*) of men honouring them. This image is telling for it indicates that the choral unanimity at the origin of the Unknown Soldier is linked to physical symbiosis, and at the same time it points to the dangers of such chorality. The crowd as organism is an attempt to reassert the value of organic life in its heterogeneity, yet it runs the risk of becoming homogeneous matter to be shaped by totalitarian ideology. As we are about to see in the coming chapters, the ethical distinction between the two – heterogeneous life and homogeneous matter – underlies debates about the Unknown Soldier and, in particular, about the individuality of his anonymous body and the nature of his sacrifice.[34]

2 Identification and Chorality

Bones Manifest Themselves

As it became clear that the First World War was not going to end in a few months, the number of dead bodies that had to be dealt with quickly became a 'crowd' or 'mass' that defied yet demanded individual attention. In this chapter, we will see how choosing one body from among this mass to become *the* Unknown Soldier involved contradictory imperatives that prolong the tension between choral and constructed unanimity. These are reflected in the different ways that gathering and recomposing a skeleton were presented and imagined. On the one hand, broken bones could metonymically stand for a whole body just as, in turn, the Unknown Soldier could stand for the bodies of all the missing. This implies a homogeneous continuity between all of them, and also with the body of the nation. On the other hand, the Unknown Soldier's bones could be seen to express an individual story in their particular brokenness, a story that might be analogical to that of other missing soldiers, but never identical. In this case, discontinuity is foregrounded as we are faced with our own inability to comprehend the experience commemorated by the Memorial.

From a practical standpoint, as mentioned, before the war even ended, countless unidentified bodies remained in improvised battlefield graves, and many became part of the mud of the trenches without an actual burial. All combatant nations in the First World War had promised their soldiers the proper burial and official recognition that in the past had been reserved for decorated officers: this was perhaps the last war in which burial was such a pressing public concern, yet at the same time the first war in which all combatants expected to be treated equally. How were governments to respond to the need for countless ceremonial burials?

How were they to put an end to spontaneous illegal exhumations? How was the choice of one body to represent all the missing going to avoid singling him out and making him special, if not heroic?

The overall strategy, common to Italy, France, England, and the United States, was to enforce some kind of normalization on the diversity of individual desires. Thus, Daniel Ingersoll and James Nickell, comparing a number of countries, conclude that 'unknowns' (all the missing in action) were depicted either monumentally (via ossuaries, cenotaphs, or other collective monuments) or with standard stones (countless identical crosses, for example), but very rarely in a differentiated fashion. British policy was to leave all bodies where they lay, but to rebury each and every one (known or unknown) under an identical marker. French policy was mixed: some bodies were repatriated, but most remained in battlefield ossuaries, commemorated by lists of names either on location (if known) or in their home towns. Though the United States promised to repatriate all the bodies of those whose families requested repatriation, in reality this resulted in something more like France's mixed policy. Italian policy, not formulated until the Unknown Soldier Memorial was also approved into law, in August 1921, was theoretically identical to the American one, and had the same mixed result. In practice, many remained in battlefield ossuaries, commemorated by a common list of names, either on the field or in their home town; moreover, under Fascism separate grave markers that had accreted during the war and in its immediate aftermath were replaced with identical spots in monumental constructions. In sum, all these policies tended towards abstraction because they deliberately separated bodies from names, allowing the persistence and homogeneity of the latter eventually to replace the – often gruesome – passing away of the former.[1]

In this context, the Unknown Soldier comes as a strange reversal: one body persists, without a name, as though indeed to compensate for so many names that have lost their bodies. For the most part, historians have seen the Unknown Soldier Memorial as part of the more general tendency towards abstraction because, even if it reverses other practices, it still maintains the separation between bodies and names. Thus, both Thomas Laqueur and Daniel Sherman, exploring the British and French contexts respectively, have argued for a convergence of opposites whereby lists of names, which had the advantage of being 'spatially indeterminate,' balanced the accumulation in ossuaries of bodies 'whose meaning ar[ose] from their being forever, impenetrably nameless.' For Laqueur, the British Unknown Warrior, the only repatriated body, became the 'universal' 'distributive singular term,' while for Sherman the French one was seen 'not as a personality but as an idealization.' Labita, in turn, insists on the suppression of any sign of individuality or heroism in the Italian Unknown Soldier, while Miniero concludes quoting the affirmation that for Italians he signified a rare 'unity of passions and wills; in other words, of true conscience.' Both Labita and Miniero stress how unity and lack of differentiation were a response to the fear of dissent, caused by political divisions (stressed by Miniero), and even more by the chaos of bodies that no authority seemed to be able to put back in order (Labita, referring to Leed).[2]

While state policies did emphasize abstraction in the creation of the Unknown Soldier Memorial, as we are about to see, I will argue once again that in the popular imagination, and in significant literary and cinematic examples, the Unknown Soldier was a more genuine reversal of the divergence of bodies and names. Specifically, though each government took great pains to ensure that the body of the Unknown Soldier was as de-individualized as possible, in practice the actual choice of a body, and even more the depictions of that choice, sought to endow him with a new, embodied identity. This reflected an obsession with corporeality evident both in the writings of veterans and in related discussions of commemoration in the popular press. It was also evident in the behaviour of survivors. Some families went so far as to hire private contractors or to perform in person illegal exhumations; in one example, a woman witnessed one hundred and fifty exhumations in the hope of recognizing her son; in another case, builders of battlefield memorials assured mourners that they would there be physically close to at least 'a little bit of him.' In turn, in Gance's film *J'Accuse* (1919, 1922) the corporeality of dead soldiers rising up from their graves, covered with mud and with open wounds, is contrasted with the abstraction of military formations of living soldiers. This obsession is what state scripts sought to hide, yet it was also the very aspect of the Unknown Soldier Memorial that made it so popular.[3]

As we saw in the previous chapter, a crucial image for the choral experience of death during the war was that of confused bones, tied to a minimalist yet primordial vitality that, it was felt, needed to be commemorated. When it came to choosing a body for the Unknown Soldier Memorial – and thus also to establishing its relation to all the other dead bodies – the image of bones became even more important. Specifically, the issue was how to separate the bones of a single soldier from the rest without severing the bond of battlefield chorality; another way of putting it was, the Unknown Soldier had to be special, in the sense that he was singled out to represent the many, yet he must not become a hero, or superior to the many or, in effect, have any specific identity that might undercut his being 'everyman.' As a result, state policies emphasized absolute randomness in the choice of the body – or, in actuality, of the bones – that would become the Unknown Soldier: such bones had to be 'blank' or 'pure,' in order to spiritualize suffering. In contrast, we find that the press, but also veterans and writers, wished to see these bones as exceptionally expressive, even expressionistic, in their ability to embody suffering, and thus to commemorate it. For them, the choice of the body was not random, but dictated by the body itself.

In Italy, France, England, and the United States, the procedure for choosing a body was very similar. Bodies were dug up from each of the major battlefields of each country, at times from marked yet anonymous graves, at times from the trenches themselves: there were eleven Italian bodies, eight French, four British, and four American. In the next section, I will explain how in each country one body was chosen in a carefully scripted ceremony; here I will focus on this first phase. In each case, numerous bodies had to be dug up from each battlefield in order to come up with one body that would qualify at once as anonymous, yet clearly of a specific nation. Thus, Jagielski quotes a rare piece of testimony regarding the exhumations at Verdun, where ten bodies where rejected (because of dog tags but also of items such as a lighter) before an appropriate body could be found, in an identifiable French uniform, but with no traces of rank or other particularity remaining. Furthermore, the French ended up with only eight bodies as candidates for the Unknown Soldier, as it turned out to be impossible to find a suitable body on one of the nine originally chosen battlefields. Overall, exhumations were kept out of the public eye, as Jagielski notes, to ensure that the person who would later choose the one to become *the* unknown from among the candidates would know nothing about them, not even which battlefield they came from, and certainly not what state the body had been in. This was true in Italy and England as well, as Miniero points out.[4] To this I would add that such exhumations were a lengthy, grisly, and dangerous affair (they took place in 1920–21, whereas many of the men had been dead since 1916; moreover, the battlefields were full of unexploded ammunition), and it is clear that the authorities wished to keep the public away from horrific visions that would run counter to the sense of spiritualized abstraction they were seeking. In many ways, the authorities did not want people to see the bones until they had been cleaned up.

In part, the newspapers accepted this decision, and did not question the fact that exhumation was the single element in the creation of the Unknown Soldier Memorial that they were being kept in the dark about. A notable example in *L'illustrazione Italiana*, which covered every phase of the ceremony, is the juxtaposition of photographs of the Unknown Soldier medals, in which he is depicted with an intact body, with a text that emphasizes 'the sad bones [...] recomposed upon a shroud.' Similarly, the *Corriere della Sera* describes the 'nameless son of the nation [who] arose from his burial, up there, and began his long march,' evoking him as 'this symbol of bare bones [...] by which, within competing parties, and fractious envy, [...] we learn to accept what must be accepted.' In both instances, bones lead directly to resurrection and ideal unity, rising up from the earth on their own, setting aside the reality of decaying flesh and the divisions, social and psychic, that it evokes. In contrast, this is what the Socialist newspaper *Avanti!* would emphasize, arguing that 'shredded bones' do not make a hero, but reveal only 'a human body, shredded by other humans, like him in shreds.' Similarly, the Socialist newspaper *Ordine nuovo* imagines the returning Unknown Soldier worried about his family's poverty, and about a Fascist attack on his mother's home; yet, insisting that he will not live again, it still depicts his acceptance of all the ceremonies in his honour as one more suffered sacrifice; moreover *Ordine nuovo* feels obliged to invent a name for him (Giorgio C.). In France, the pacifist Madeleine Vernet writes a letter 'To the Unknown Mother of the Unknown Soldier,' imagining a son's rejection of patriotism in mourning, and appealing to women's bodily repugnance towards the horrors of war (purportedly due to their procreative capacity). She is echoed in Italy by Maria Sulli who imagines Italy as a wounded mother that her sons must console. In these latter examples, abstraction gives way to a need to make suffering visible and personal in some way, in order to draw a lesson from it. The bones are not blank but in shreds.[5]

In the Italian case, another important element that runs counter to state policies of abstraction is the role of the local press, and of longstanding divisions between north and south such that Italian national unity was, from a cultural standpoint, very precarious. In contrast with the situation in other countries, Italy's dead were concentrated in the north, and especially in territories Italy fought over with Austria and still felt deprived of in the war's aftermath. As Miniero points out, Italy's feeling of 'mutilated victory' and its association with the north meant that dissent over the meaning of the Unknown Soldier Memorial was greater in that country. Italy is also the only nation in which photographs of exhumations were published immediately after the First World War, albeit in a rare book printed in Milan for the Association of Mutilated Veterans, to benefit the ossuary on Monte San Michele in the Carso. *Ignoto Militi* by Augusto Tognasso consists mainly of photographs of the many anonymous 'bones' that were not chosen to be *the* Unknown Soldier, and insists that they must not be overshadowed by too much pomp and circumstance (see figures 6 and 7). Similarly, local northern newspapers such as *La voce dell'Isonzo*, the *Giornale di Udine*, and the *Patria del Friuli* followed the work of the commission in charge of exhumations, going so far as to publish maps, even though such publicity was discouraged by the government in Rome. In those regions entire communities were able to follow the transportation of bodies from the battlefields into the towns; in one case, at Ortigara, a crowd even tried to stop the transportation arguing that the soldier had died to defend that land and should remain on it. In general, as Miniero notes, such communities wanted to identify the unknown as somehow 'theirs'; this meant accompanying the body, and it also meant imagining its suffering.[6]

As I already intimated, d'Annunzio's identification with the Unknown Soldier highlighted such dissent. His takeover of Fiume in 1919 took place just as debates about commemoration began, and polarized opinions about Italy's 'mutilated victory' – the phrase itself was invented by d'Annunzio. As we saw, d'Annunzio emphasized corporeality, 'shredded bones' like those mentioned above, and the deeper individuality of the nameless soldier. As d'Annunzio was a charismatic war hero, some newspapers, and especially veterans' ones, followed suit, undermining state emphasis on a more abstract anonymity.[7] Moreover, throughout Italy, popular sentiment was attached to a dead soldier who had a local and individual identity, in clear contrast with state scripts that insisted on randomness. The link to a region and to the land was further seen as an expression of visceral loyalty. As Labita notes, readers' polls revealed a clear desire to 'determine the heroic status of the unknown combatant,' by choosing a body that was visibly and symbolically martyred even as it remained anonymous: for example, they called for a 'fallen [soldier] with a bullet hole in the forehead.' Even among those who wished to emphasize not heroism but the 'suffering and pain' of 'all the Human victims represented by that one unknown yet "so familiar" victim' the desire to see suffering dominated. At the extreme, Labita cites an oft-quoted example: 'The last [body], gathered from near the Timavo, was truly the expression of martyrdom; with its broken legs and its head punctured by

Figure 6. Skeleton with Cross (Augusto Tognasso, *Ignoto Militi* [1922]).
© Courtesy of Department of Special Collections and University Archives, Stanford University Libraries

Figure 7. Skeleton on Shroud (Augusto Tognasso, *Ignoto Militi* [1922]). © Courtesy of Department of Special Collections and University Archives, Stanford University Libraries

projectiles, it inspired so much pity to the Research Commission that its members fell to their knees.'[8] Though we cannot know if this body ended up being chosen, the implication is clear: shredded bones become the expression of martyrdom, and they are associated with the land (the area of the Timavo, in this case). This is what testifies to this particular body's capacity to become the Unknown Soldier.

One last notable example of wishing to have the Unknown Soldier's bones speak for themselves is the poem by Ada Negri, 'The Unknown Hero Returns,' published on 1 November 1921 in *Il secolo illustrato.* It expresses the wish to put an end to mourning, and invokes a collective elevation reminiscent of official abstraction. Yet at the same time Negri emphasizes 'at the centre of his forehead the great and bloody stigmata' and the 'shrapnel holes and burns' in his uniform.[9] The 'bullet hole in the forehead' cited above, which was a recurrent image, now becomes more fleshy, and is associated with the stigmata of saints, which are themselves visually equated with the shrapnel holes in the uniform and, presumably, at one point, in the flesh. The overall effect is to insist even more on the expressivity of the Unknown Soldier's body, which now has its spiritual authority literally written upon it. Negri may be revisiting the famous example of the warrior Er who returns from the dead and is the authority for the conclusion of Plato's *Republic.* He is first notable because of his intact, though apparently dead body beneath the corpses on the battlefield; as with the bodies of many Christian saints, lack of decay is a sign of otherworldly perfection and authority. In the case of the Unknown Soldier, the body is not quite intact, yet neither is it randomly mangled: the marks on it become intelligible and signify nobility or saintliness. Moreover, it too, does not decay, since bloody stigmata are still visible. Hence, by implication, the Unknown Soldier, too, in his anonymity has otherworldly authority, given him not by the state or any external recognition that can separate names and bodies, but by his body itself.[10]

Recognition, or Reaching across the Divide of Living and Dead

The contrast of blank bones, and absolute anonymity, with expressive bones, and self-generated anonymity is even stronger as we move from exhumations to the moment of recognition proper, when each nation chooses among its various candidates the one to become *the* Unknown Soldier. But first, it is remarkable that all four nations followed the same pattern of dividing the exhumations of multiple candidates from the later moment of choosing one. As I suggested above, this surely was a way to keep the grisly side of exhumation out of the way; but it was also, as we will see next, a way to foreground the dead's reintegration into society. That is, exhumation, for all that the authorities tried to keep it clean, revealed the fascinating yet horrific power of dead bodies in their radical alterity, deviant from social norms and from established authority. The choice of one coffin among eleven (or eight, or four) which took place indoors, according to far more scripted ceremonies, was intended to counter this by stressing the familial relationship between the living and the dead, which would re-establish social structures. Thus, in all cases, though with telling differences, recognition of *the* Unknown Soldier was staged as a family member recognizing their loved one. Nonetheless, as I will show, a different sort of recognition was also at play, once again countering reintegration into past structures of authority. This alternative recognition took a number of forms, but what they share is that *identification of* a family member turns into an *identification with* that establishes a symbiosis between the living and the dead that goes deeper than family ties. This symbiosis, in the end, is what makes the anonymity of the Unknown Soldier existential, because self-generated, rather than accidental, and imposed by external authority.[11]

In the case of Britain and America, the separation between exhumations and choosing the Unknown Soldier was less marked, and reintegration was not emphasized as much; I would submit this was because the dead were not on national soil, and both parts of the ceremony took place in France. On the night of 7 November 1920, Brigadier General L.J. Wyatt, at the time supreme commander of the British forces, chose the Unknown Soldier from among four identical plain caskets, in a small ceremony held at the chapel of Saint Pol in France. The American Unknown Soldier was chosen by Army Sergeant Edward F. Younger, who had been awarded the Distinguished Service Medal in the First World War, from among four identical caskets laid out in the city hall of Chalons-sur-Marne. In both cases, generic brotherhood was emphasized as the link between the chooser and the Unknown Soldier; both ceremonies stressed the randomness of choice via identical caskets whose plainness reflected the minimalism of the Unknown Soldier Memorial. Perhaps the only remarkable element here was the myth that immediately appeared in the British media according to which Wyatt made his choice blindfolded: a story he always denied, but which implied that without such protection he might have 'seen' something through or on the casket. This kind of 'seeing' or visceral identification is what popular reactions of mourners emphasized when they saw the coffin at Westminster; in contrast, in the United States, the Unknown Soldier was buried at Arlington Cemetery, and identification with him did not involve mass mourning as it did in Europe. Notably, issues of visceral identification did not arise in the American case until the Vietnam War; this confirms that they are a reaction to social divisions, and explains why they were so pronounced in Italy after the First World War.[12]

The stories in France, and even more Italy thus make the issue of 'seeing' into the coffin far more explicit, and involve ceremonies that were much larger and more choreographed. In France, eight caskets were gathered in a military building at Verdun, in an underground gallery decorated like a crypt. The entire city and various authorities came to pay homage to the eight unknowns on 9 November 1920 until late into the night, and on 10 November *the* Unknown was finally chosen in a public ceremony. The chooser, it had been agreed, would be a 'poor soldier,' who had lost either a father or a brother in the war, whose remains had not been found. At the last minute, the first man chosen fell ill, and the role was given to Auguste Thin, whose father was a *disparu.* Thin was given a bouquet of flowers from the battlefield, which he was to put on the chosen casket, as he indeed did, showing great emotion but also quick decisiveness, as reporters wrote. Ideally then, the Unknown Soldier was a father figure, an ancestor that the living could share, becoming brothers and overcoming class and social divisions: though this may seem surprising, given the youthfulness of many First World War dead, it reflected the notion that suffering gave them a new authority. Yet some remarkable elements in the French case included the request, in previous days, that Thin or his predecessor be replaced by a blind veteran; and the story, apparently verified by Thin, that he used the number of his 'regiment' (132, which adds up to 6), and his 'corps' (6), to choose the sixth casket. Once again, blindness is invoked in order to guarantee randomness against some kind of vision that now even a blindfold could not prevent; more important, a blind chooser evokes a poetic visionary, and we begin to see how in the popular imagination, the Unknown Soldier had to be recognized less as a father and more through a deeper spiritual or symbolic affinity. Thin's use of numbers reflects the general fascination with occult spirituality at the time, and once again establishes an arcane link between Thin's destiny and that of the Unknown Soldier, tied to the number six, not to a father-son relationship. At its most extreme, this relationship was such that the living became the 'offspring [...] of corpses,' united by the rather strange brotherhood of being haunted.[13]

The Italian story is by far the most dramatic and deviant from scripts; I believe this once again reflects greater divisions among Italians, but also the later date, which meant more time had passed for public discussion and, therefore, dissent. We thus find both more measures to guarantee a random choice and more elements that undo it. On 27 October 1921 eleven identical caskets were transported to Aquileia and carried into the basilica there; the first, by mothers and widows of lost soldiers, the others by veterans and mutilated soldiers. Even the greater number of caskets in Italy – ostensibly one from each major battlefield – betrays the heterogeneity of Italians. After crowds were able to visit the caskets – and they included at least one mother who suddenly threw herself onto one of them, screaming her son's name, 'Giovannino! Giovannino!' – on the night of the 27th, their order was switched, to prevent anyone from knowing which casket came from which battlefield. But Lieutenant Tognasso, who oversaw this task, claimed that he and his men could still distinguish their provenance 'because of the different grains and knots in the wood, and the different distribution of the nails, and certain nuances of colour.' While the mother's 'seeing,' though against scripts, is still familial if overly dramatic, it is Tognasso's 'seeing' which brings us back to the primal scene in which living soldiers carry the dead within them like 'stiff [...] wood.' Moreover, Tognasso's attention to detail once again emphasized a kind of organic anonymity that was not blank as the state wished.[14]

The chooser in this case was a Triestine 'woman of the people,' who had lost her son (and his body) as well as her husband in the war, Maria Bergamas. I will return later to the extraordinary presence of women in the Italian case, but what we can already note is that a very different familial relationship is established here than in France. First, the choice of a son in relationship to a mother – unique to Italy – emphasized matriarchy rather than patriarchy, which was associated with a more visceral, less abstract, and less civic bond. Though we may see here a regressive pull of the political towards the irrational and even the abject, this was, as we will see in more detail as we proceed, an attempt to acknowledge a basic anthropological link between death and birth, that was felt to be essential to mourning. In Italy, traditional forms of mourning still included women who were hired to wail, dance, and otherwise express grief. Second, Maria Bergamas was also presented as the personification of Italy, who was saved by her children's deaths, the deaths that the Unknown Soldier represented. This shows, far more clearly than in France, that brotherhood is established by dying together, and that it is not just the dead father who haunts the living, but the mother who exposes how both the living and the dead are subject to the same power of death. The *patria* as mother, in other words, reveals that mortality, the embodied self, and the presence of death in the unconscious enter the political sphere, whence they had been banished, at least ostensibly, in the post-Enlightenment era. Third, considering how the newspapers emphasized the absence of authority figures (such as d'Annunzio, but also Cadorna, the general most often blamed for the defeat at Caporetto), and bemoaned the weakness of Italy's leaders (who were, for example, unable to resolve the Fiume crisis and also incapable of defending Italy's interests during peace negotiations), the choice of Maria Bergamas implies a missing father. The spectre here is Italy's failure to unite as nation, its failure to establish patriarchy, and the related fear of illegitimate children, which the war had exacerbated.[15]

As a representative of the *patria* Bergamas outdid herself. She was meant, like Thin, to place a white flower on the chosen casket, expressing sombre contemplation. Instead, against the script, she halted before the second among the eleven coffins, overwhelmed with sorrow as though seeing the broken body within and then, wrenching off her widow's veil, threw it over that coffin: this dramatic gesture refused the abstract transcendence represented by the white flower (whose purity was also that of blank bones), in favour of a more spectacular manifestation of the darker chthonic mourning that confuses the dead and the living under a single shadow. Bergamas' gesture, moreover, recalls the reactions to exhumations that I discussed above, as her black veil evokes the 'broken legs and [...] head punctured by projectiles' imagined by the newspapers, and thus it establishes a relationship more viscerally embodied than the familial one. Or, to be more accurate, instead of reasserting the sort of family that is the cornerstone of the nation, the rational family, as it were, Bergamas' gesture brings us back to the reality of generation, of the physical bond between mother and son, and thus also of the 'primordial life' of the womb (to borrow Eliade's term again), as a darker and more unpredictable power. Bergamas does not so much identify the Unknown Soldier *as* her son, as she exposes the power of death and birth by which we are all united, forcing us into a deeper identification *with*.[16]

To understand how all these deviant gestures end up suggesting symbiosis with the Unknown Soldier's existential anonymity, rather than reintegration into standard family structures, three further evocations of the moment of recognition are salient. These are, first, in Gance's *J'Accuse*, second, in d'Annunzio's *Notturno* and related texts, and third,

Figure 8. Maria Bergamas (in *L'illustrazione Italiana* [1921]). © University of California Northern Regional Library Facility

in Ungaretti's *L'Allegria*. Before I even get into their content, a reflection on their style and circumstances of composition is important. Abel Gance's *J'Accuse* has been recently discussed in Winter's *Sites of Memory, Sites of Mourning*, and has also been released for the first time in DVD in 2008, in a restored original 1919 version with 1922 additions. As Winter notes, the film is a complex mixture of loyalty to French heroes and deep pacifism, which I will get into in a moment. In terms of its composition, Winter explains that it has two contexts, which we tend wrongly to divide: that of popular culture, 'images d'Épinal,' and other images used by the French army as well as soldiers to revive their flagging spirits, on the one hand, and on the other, avant-garde and Surrealist experimentation, as well as interest in spiritualism, the occult, and the collective unconscious. Thus, the film was financed in part by Pathé, a major film company, yet it also received the blessing, and help, of the French army's cinematographic service. The French army lent Gance numerous soldiers to play in the film, and only the major roles went to professional stage actors. There are, accordingly, a number of highly realistic scenes in the trenches and of battles and also, accordingly, a great romanticization of 'martial virtues.' At the same time, in his depiction of the 'return of the dead,' who demand to know whether they have died in vain, 'Gance found a visionary Surrealism, a romantic language of nightmarish quality' that Winter rightly associates with Breton's claim that 'the imaginary is that which tends to become real,' and with Surrealism's emphasis on dreams and visions, of which 'Gance was a pioneer [as regards their] expression on film.'[17]

One key element in the creation of Gance's film, which we may never know much about, is the role of Blaise Cendrars, who was a member of the Surrealist group in those years. As I mentioned, Cendrars was a veteran, and he lost his right forearm in battle in 1915: he appears in Gance's film, his bandages unravelling, alongside other soldiers, many of whom went back to the front after filming and died before the film came out. We know that Cendrars became Gance's helper in making the film, going to Paris in the fall of 1918 to 'arrange for clips and other items.' Perhaps most important for the atmosphere of the film, as Winter argues, he was as a result in Paris for the burial of the poet Apollinaire a few days later, and he said in a 1950 interview that for him 'Apollinaire inhabited not the kingdom of the dead but the kingdom of the shadows.' In particular, mourners saw 'a grave with a bit of frozen earth exactly in the shape of Apollinaire's head,' in a 'surrealistic scene [that] took place near the grave of Allan Kardec, the founder of French spiritualism.'[18] In the 1922 version of *J'Accuse*, Cendrars chose to play a returning dead soldier as well, and appears in the famous scene in which crosses turn into bodies that rise up from the earth. Whatever Cendrars' specific influence on the style of the film may have been, it is clear that Gance draws from the models of automatic writing and mystical delirium, as well as wartime amnesia, to give voice to those who have lost their lives but have not yet been properly recognized; a primordial power of life and death is evoked here, and associated with the feminine principle as well as with symbiosis between the living and the dead.

The crucial moment of recognition in Gance's film takes place when the two male protagonists are in a battlefield hospital. Jean Diaz, a poet and author of 'Les Pacifiques' has been presented to us as nobly and unhappily in love with the married Edith Laurin. Jean is genteel, cultured, and lives with his mother. In contrast, Edith's husband François is described as a 'brute' and shown hitting her as well as relishing the blood of a successful hunt. That François is of a lower social class is made evident in that he is conscripted as a foot soldier, whereas Jean is an officer. Thus, everything divides these two men, and most of all the apparently unsolvable issue of competing for the same woman. Scenes of anger, jealousy, and contained violence abound, until the two become companions in the trenches. After Jean risks his life for François, we see them crying over the fate of the now missing Edith, heads together, embracing, a picture reminiscent of Giono's primal scene quoted above. After more sparring, the extreme moment comes as Jean begins to go mad and both are wounded: first François agrees to keep mailing the many letters Jean has prepared for Edith, to spare her from knowing about his death; but ultimately it is François who dies, and Jean who survives, though 'mad' now, as he is about to lead the army of the returning dead back to his home town. Here, Gance's camera zooms onto the clasped hands of Jean and François, across the aisle between their hospital beds, as the latter lies there dead and is covered by a sheet and the former, still alive, refuses to let go. They have become as one, and for the rest of the film Jean speaks for the dead and not the living.

Edith's role is important in the symbiosis between Jean and François, and is not unlike that of Maria Bergamas because she represents the darker forces at work in war. Of course, Gance is even more provocative in his depiction of these forces, because Edith is a lover and not a mother, and violence is sexualized in the film via Edith's rape by German soldiers. Gance depicts Edith as a Christ figure, noting that she bears all the crosses borne by women during war and thus shows that war is just as hard on women as on men: she is tempted by the absence of her husband to love another man; she is raped; she loses her father who dies in battle; she has to defend the illegitimate child the rape results in. In Gance's truly feminist defence of her situation (given the times), I think we see his allegiance to the tolerance of a Zola, alluded to in his film's title (and also in the Jewish Menorah of Jean's mother). Yet what is relevant in terms of the recognition that the dead demand from the living, and thus in terms of the Unknown Soldier Memorial, is how the boundaries between Jean and François break down, and they almost exchange places, because of Edith's suffering. François creates a shrine to her after she disappears (modelled after her father's shrine to France's lost land in Alsace-Lorraine), and it is when Jean sees the shrine that he begins to respect François; later, we see the 'brute' François, who would normally kill both wife and illegitimate child, learn gentleness from Jean; finally, at the very end, he agrees to send Jean's letters, taking over for him. In turn, Jean will end up shredding his pacifist and idyllic poetry in order to raise up the army of the dead, thus voicing François' violent 'accusation' against the injustice of war. Crucially, Jean becomes both insane (like Mangin, the living unknown soldier) and dead (like the actual Unknown Soldier) as he chooses to identify with all the soldiers who return from the grave to decry not only war, but a violence that now seems to have become all-pervasive. Barbed wire tearing up living flesh, bodies becoming indistinguishable from mud, and the growing rapaciousness of the soldiers as they struggle to survive, repeat visually a crescendo accusation against technology's inhumanity, against nature's 'indifference to human suffering,' and finally against humans' own inhumanity.[19]

D'Annunzio wrote the first draft of the *Notturno* between February and April 1916, when he was confined to bed in a failed attempt to cure the progressive detachment of the retina in his right eye, which had been caused by an accident during the testing of a new military seaplane in January of that year. Though in significant physical pain, and in complete darkness, he managed to write by using long strips of paper that accommodated only a few lines at a time. The book, whose first sections were published in 1916, with a 1921 final version, has been heralded as a turning point in d'Annunzio's work because of its introspective tone and its modern fragmentary style, even though some argue that it does not overcome his earlier aestheticism and triumphalism. In the context of First World War discussions of death and memorialization, however, the *Notturno* is a deeply felt and prescient meditation on identity and mourning, whose explicit links to the Unknown Soldier have been overlooked.

Structured as a litany for deceased comrades, it evokes their return in the modulations of a few stark images: the moment of death, witnessed or imagined not only as heroic but as a complete breakdown of paradigms (hence, 'beyond heroism,' and incomprehensible, particularly in that it inexplicably leaves one comrade – d'Annunzio – still alive); the dissolution of each comrade's body as it is mirrored in d'Annunzio's own physical sensations of metamorphosis (the writer's sensory deprivation breaks down normal boundaries as dead comrades seem to invade his body and imagination); the attempt to halt this decay by associating the purity of bones with the 'sibylline' words written on white strips of paper in the dark ('Now I seem to perceive as they fall the resplendent white of my bones').[20]

These same images recur in a less known yet perhaps even more revealing text by d'Annunzio, the 'Comento meditato a un discorso improvviso' (Meditated commentary on a sudden/improvised speech), written and published in the fall of 1922, just before the March on Rome. Even more so than the *Notturno*, this text presents itself as a 'reliv[ing] of death,' or indeed as a writing that has traversed death like an 'illness' whose 'essence is magical'; it also explicitly critiques the monumentalization of the Unknown Soldier. The genesis of 'Comento meditato' makes clear what was latent in the *Notturno*, and what Leed has identified as a widespread '"fixation" upon the dead,' a 'cult of suffering and self-sacrifice' that seeks 'to affirm social [and psychic] dislocation as a permanent state' by re-enacting it. Thus, where the *Notturno* derives its authority from associating the writer's temporary blindness with wartime deaths, the 'Comento meditato' goes a step further, drawing its voice from the writer's actual repetition of those deaths. For the 'Comento' is based on the 'sudden speech' constituted by the images and illuminations voiced by d'Annunzio and recorded by his doctors while he was semi-conscious for two weeks, his life in danger, after he fell out of a second-floor window in his home, the (not yet named) Vittoriale on Lake Garda, on 13 August 1922. Though this procedure recalls contemporary Surrealist automatic writing (more than Futurist 'words-in-freedom,' given its psychological component), it also imitates the transcription of the mystical delirium of saints: cut off from individual intentionality and social identity, a voice speaks here from a place that is outside of time, beyond history, and beyond death – d'Annunzio's famous 'third place': 'beyond life, beneath death.' This is, of course, also the place inhabited by the amnesiac Mangin, the living unknown soldier. While it seems most likely that d'Annunzio's fall was an accident, even though some thought it a murder attempt, he immediately presented it as a providential 'mystical fall of the exiled and mutilated archangel,' and later even dubbed it a suicide attempt, to be understood as the ultimate manifestation of his physical symbiosis with Italy's 'mutilated victory,' which his takeover of Fiume had so miserably failed to heal. Thus, exposed more starkly to death and speaking from it, through delirium, even more directly than in the *Notturno*, d'Annunzio prolongs the 'radical discontinuity' of war experiences, and remains suspended between life and death, but also between sacrifice and the disappointing return to a home that was 'no longer a community,' not (or not yet) a *patria*. In many ways, then, d'Annunzio speaks from beneath the veil thrown down by Maria Bergamas, and as we will see, his speech emphasizes the feminine primordial power of death and life.[21]

In the 'Comento meditato,' this power is contrasted with the terms used in official descriptions of the Memorial. Thus, in Douhet's proposal, in its less contentious and often paraphrased sections, we read:

> Our Soldier withstood and conquered all. From the gratuitous insults of journalists and politicians and scandal sheets that [...] expressed bewilderment at his valour [...], to the terrible calumnies spread around the world in order to shirk a terrifying responsibility. [...] He withstood and conquered all, alone, nonetheless. [...] Therefore to our Soldier we owe the highest honour [...]
>
> Nothing must be known of Him. He was a soldier killed as he did his duty. This is all that must be said of Him. And this is immense. This Humble One is an innumerable multitude. [...]
>
> Every citizen must line up for his triumphal return, uniting in a unanimous sense of ideal elevation, in a common act of reverence towards the Son and Brother of all, gone to his death to defend our Common Mother.

Abstraction dominates here in that the 'multitude' united in the 'One' is undifferentiated, even if it is contrasted with politicians and newspapermen. The mother, too, in this passage, is a symbol, and this highlights negatively the parts of Douhet's proposal that were rarely reproduced, such as his paragraph describing the mourning of actual mothers, 'those so unfortunate as to lose the one they bore,' in which flesh (*carne*) is emphasized.[22] This flesh, as common yet not abstract, is what d'Annunzio's text stresses, as the mother is associated with 'darkness' and not 'elevation,' and instead of lining up to honour the Unknown Soldier (identification *of*) we 'confuse [...] our bones' (identification *with*):

> Mother, Mother, from what darkness is my rebirth? [...] Who spoke thus? Who speaks thus?
>
> He has no name. He is nameless. He is faceless. [...]
>
> We are all nameless. We are faceless. [...] But [...] we will recognize and as Italians confuse once again our worn or broken bones.[23]

The addition of being 'faceless' or 'disfigured' (*senza figura*) to being 'nameless' transforms social anonymity into a visceral mutilation, which is, in turn, linked to ineffability, since *senza figura* can also mean 'without figuration.' Dislodging the bourgeois family, this mother, like Maria Bergamas, exposes in her darkness a continuity of life and death that is manifested in brokenness and loss, not in a cyclic return to plenitude.[24] Similar images occur in d'Annunzio's 1916 'Licenza' to the *Leda senza cigno*, based on his war diaries, where d'Annunzio writes of the soldiers he is fighting with:

> In some, the whole skull is showing. It makes one think of the skeleton, waiting inside the flesh, mimicking its gestures, following its attitudes, trapped. These heads are already touched by death, the relentless Worker. A mass of flesh for the butcher, a nicely prepared slaughterhouse. [...] I hear the song of the earth, I hear the steady beat of hearts pumping sacrificial blood. I hear the silence beneath the ground; and the silence out there, beyond the blue. [...] Someone bends down and recognizes me, closes my eyes. The tide goes out under the dome of my head. Two men lift my body onto the stretcher.[25]

The more pure, transcendent, or ascetic version of the skeleton is here undercut by the flesh that continues to imprison it, even in death, as the images of the butcher's display suggests. The earth anticipates the symbiosis between soldier and land that northern Italian newspapers especially emphasized, as an identity more visceral than the name. Finally, an 'oceanic' experience of death as maternal, reminiscent of Cendrars and Gance, is implied by the 'tide' retreating beneath d'Annunzio's own head as he is 'recognized' and put on a stretcher which also suggests a coffin (as *barella* echoes *bara* in Italian), and someone closes his eyes as though he were already dead. All of this creates a sense of symbiotic identification like that of Jean and François, which is not at all a blank anonymity that a return to the family could undo, but an identity that soldiers wish to keep, disfigured, unstable, and broken as it may be. Finally, in the *Notturno*'s 1921 version, d'Annunzio writes again about soldiers' fleshy symbiotic suffering and adds, 'I no longer feel the borders of my skeleton.'[26] The most solid, enduring element of the self now becomes also the most porous, so that anonymity becomes, like Jean's madness or Mangin's amnesia, a symbiotic and visionary state.

Giuseppe Ungaretti tells the story of how he wrote the poems in *Il Porto Sepolto*, published in Udine in 1916 in eighty copies by Ettore Serra, in the following terms:

> I started *Il Porto Sepolto* on the first day of my life in the trenches, and that was Christmas day of 1915, and I was on the Carso, on Monte San Michele. I spent that night lying in the mud, facing the enemy [...] I was in the presence of death, in the presence of nature, a nature that I was learning to know in a new way, a terrible way. [...] In my poetry there is no trace of hate for the enemy, or for anyone else: there is awareness of the human condition, of the brotherhood of men in suffering, in the extreme precariousness of their condition. There is the will to express, [...] there is exaltation, in *Il Porto Sepolto*, that almost savage exaltation of the *élan vital*, of the appetite for life [...]
>
> We live a contradiction.

The poems were reprinted in the volume *Allegria di naufragi* in 1919, and found a final version in the 1931 volume *L'Allegria*, which comprises 'Il Porto Sepolto,' followed by 'Naufragi' and 'Girovago,' framed by 'Ultime' and 'Prime.' While this frame consists of post- and pre-war poems, the three central sections were all written in the trenches, and each poem contains a date and specific location. Ungaretti considered this book a 'diary' and also a rite of passage, as he wrote, 'I discover myself and know myself, in the horrors of war, as *man of sorrows [uomo di pena]*.' Ungaretti's war poems are famous for their minimalist immediacy – the collection contains some of the shortest poems ever written, such as the fugitive respite from wartime chaos of 'Mattina' (Morning), written at 'Santa Maria La Longa on 26 January 1917':

> M'illumino
> d'immenso
>
> [I am illuminated / I illuminate myself
> with immensity][27]

The precariousness of life and joy, and the discovery of existential brotherhood are Ungaretti's main themes, which he develops with a lack of pomp, heroic, nationalist, or religious, that made him beloved of following generations, in contrast with d'Annunzio. In terms of style, more than Gance or d'Annunzio, Ungaretti embraces the fragment and endows it with the stark restrained power we associate with high Modernism and the best of avant-garde experimentation.

Twentieth-century criticism has thus tended to see Ungaretti not only as a superior poet or artist, but also as the only true pacifist of the three. Yet to oppose him as progressive to d'Annunzio as regressive is simplistic, just as it would be simplistic to see Gance as Romantic and not also avant-garde. In fact, in works written concurrently during the war, both of the poets and the filmmaker develop similar imagery to express the loss of identity and the symbiosis of soldiers in the trenches. As Valesio has shown, d'Annunzio's peasant-soldier who disappears into the river is a strong influence on Ungaretti's *girovago* or soldier-as-wanderer (to be discussed in more detail in the next section). Also, while there is no doubt that d'Annunzio endowed wartime symbiosis with mystical significance, at times using imagery of cleansing violence that would later inspire Fascism, we cannot forget that Ungaretti, too, was an interventionist in 1915, who supported Fascism in 1925, for populist reasons that stemmed from his experience of brotherhood in the trenches. Thus, in her article on 'Ambiguous Joy' in *L'Allegria*, Vivienne Hand argues that the contradictions within this work are not only stylistic, as critics have pointed out, but also thematic, as patriotism vies with pacifism, and as Orientalizing images contrast with a subtler exploration of Ungaretti's mixed Italian and Egyptian origins.[28]

Ungaretti acknowledges this in the quote above as the contradiction between *élan vital* and 'precariousness,' and he writes about it also in his famous statement of poetics, the 1926 essay 'Innocenza e memoria,' which returns to the war as a foundational moment:

> We know what innocence is made of. It appeared to us, and held us under its great wings, in the turmoil of these years. It occupied all our thoughts. Memory's eyes were bandaged, it was as though abolished. Even our notion of time was new. Time seemed eternal, and this was not just a phrase. We could not avoid the horror of eternity. Only instinct mattered. We were so familiar with death that our mind saw the entire film of shipwreck in each instant, and every object reflected it; our whole life from beginning to end was the very object our gaze fell upon. In such extremes, I understood why the Black man uses mirrors for eyes on his idol. [...]

> We no longer have a philosophical aspiration to innocence, as was the case in the nineteenth century, we have a direct experience of it. We have a mystical knowledge of reality. I believe the art of tomorrow will be felicitous. Little by little, the knot will be loosed. The attempts to reduce memory to a puppet and to transform innocence into an oracle will all go up in smoke. And no flattery will reach frightful and maternal innocence, returned to her obscure place deep inside memory.[29]

In a vocabulary reminiscent more of Modernist fascination with the primitive and less of Nietzsche, like d'Annunzio, Ungaretti depicts the experience of mortality, reflected concretely in every object and moment of the war, as mystical and as a form of chthonic and collective rebirth or rediscovery of 'innocence.' We sense Ungaretti beginning to fear that wartime 'innocence' will be turned into an 'oracle,' that is, reified and politicized, his concluding optimism notwithstanding. His later poetry will thus emphasize 'memory,' and imperfection of the human condition, which can only see through the mystical mirror darkly. Yet when we look back to Ungaretti's war poetry, we are surprised to find that his moment of 'innocence' converges with d'Annunzio's losing the borders of his skeleton, or with Jean's 'mad' transformation into a living ghost, along with the rising ghosts of returning soldiers, in Gance.

In three related poems in *L'Allegria,* Ungaretti compares soldiers to leaves that are about to fall from the tree in autumn. Here is the first:

Soldiers
Bosco di Courton, July 1918

They stand
like leaves
on the trees
in autumn[30]

The impersonal collective '*si sta*' (they stand) asserts a common experience of mortality, which is nonetheless, as Robert Harrison writes, lyrical, and thus 'a universalized first-person-singular voice, [...] an individuation of the universal,' in contrast with the epic, which contains 'the prehistory of the poem's simile,' and 'synthesizes a plurality of voices,' in 'the appurtenance of the individual to his family, race, class, or species.' Looking back to the simile of leaves in Homer, Virgil, and Dante, Harrison shows how this tension between the lyric and the epic is also that between 'the individual and the community.' In Homer, it is soldiers who, their heroism notwithstanding, come and go in generational cycles like leaves; in Virgil and Dante, leaves caught in the wind are used to describe the souls of the dead awaiting passage to the underworld, but in the second case, 'Christ has liberated a space of interiority for the individual,' and leaves become 'infinite singularities,' as Harrison writes, like snowflakes. In this process of individuation, Harrison concludes, 'personal identity seeks its absolute and impossible absolution,' and can only return to the contradiction between 'personal death' and something greater and more enduring yet no longer individual. In terms of our previous discussion, this means that an imperfect, partial, memory-bound identity seeks an impossible return to a collective, mystical, oceanic, maternal 'primordial life.' Significantly, Harrison at the very end of his discussion sees this contradiction at play in the Vietnam Veterans Memorial wall, because in it 'the excess of names is uncontainable,' and no clear universal meaning emerges. As I suggested in the introduction to this book, though they are different in that the Unknown Soldier Memorial has a stronger official component of abstract universality, what these two types of monuments have in common is an underlying evocation of physical particularity – as that which 'exceeds' the wall, as that which 'breaks open' the Monument.[31]

In Ungaretti's two other leaf poems, we find an increasing sense of this contradiction, as the collective 'Soldati' gives way to the singular voice:

Brothers
Mariano, 15 July 1916

What regiment are you from
brothers?

Word shuddering
in the night

Leaf just born

In the anguishing air
involuntary rebellion
of man present to his own
frailty

Brothers

Night Gully
Naples, 26 December 1916

Tonight's
face
is dry
like
parchment

This nomad
Bent upon himself
soft with snow
forsaken
like a crumbled
leaf

The never ending
time
wears me down
like
a rustle[32]

In the first case, the plural 'brothers' (title and last line) surrounds a single leaf which symbolizes at once the individual soldier lost in the night, his word of hope, and his simultaneous recognition of and rebellion at his own finitude. That precariousness is represented by a leaf 'just born' collapses time, and again associates motherhood or generation to the search for a common mortality. '*Tremante*' (shuddering) and '*spasimante*' (or anguishing) suggest the fragility of human flesh, making mortality concrete. In the second poem, 'Night Gully,' plurality is almost lost, as the image of the dry face compared to a leaf and associated with snow evokes the unique veins or wrinkles of a singular, unrepeatable individual and moment. Moreover, the 'rustle' (*fruscio*), reminiscent of the wind in the leaves in Giacomo Leopardi's famous poem, 'L'Infinito,' brings back the world of memory, of the minuteness of the individual with respect to time and eternity. Most important, eternity or 'never ending / time' and thus also the collective – associated with the night as in d'Annunzio – is now recognized as that which may 'absolve' individuality, but only, contradictorily, by wearing it down. Only when the leaf is finally dust, we may infer, will the struggle between infinite singularities and the community be done.[33]

One last poem by Ungaretti brings back the contradiction between personal identity and its absolution:

Italy
Locvizza, 1 October 1916

I am a poet
a unanimous outcry
I am a clot of dreams

I am a fruit
of countless contrasting grafts
matured in a hothouse

But your people
Italy
sprang from the same earth
I sprang from

And in this your
soldier's uniform
I am at peace
as if it were my father's
cradle[34]

This time unanimity is evoked along patriotic lines, and the image of 'my father's / cradle' seeks to synthesize the maternal, dark loss of individuality associated with Maria Bergamas' deviant gesture, or with d'Annunzio's emphasis on '*carne da macello*' (flesh for the butcher) with the more serene abstraction of a 'father' who founds a more transcendent brotherhood. Nonetheless, though Hand sees this poem as an apex of Ungaretti's 'pro-war' sentiment, the two images describing the poet – as a '*grumo di sogni*,' a term, like 'clot,' usually used for the coagulation of blood, and as 'fruit / of [...] grafts,' that is, with multiple genealogies rather than a single one – returns us to the sort of visceral identification with comrades that is not familial, orderly, and homogeneous, and once again undercuts the power of the state to 'recognize' its soldiers.

Sculpted Water

Insistence on existential, primordial anonymity, in contrast with a more abstract loss of identity compensated for a by a familial relationship to the nation continues as we move from the moment of identification of the Unknown Soldier to the design of the tomb itself. At first glance, what is remarkable here is that very little remains in archival documents regarding the aesthetics of the tomb; in the case of Italy, France, and to a lesser degree Britain, this seems to have been overshadowed by heated argument about the tomb's location. In Italy, the choice of the Vittoriano was debated also because the Vittoriano itself had not yet been completed, and as a monument to Italy's unification, its ultimate meaning was still in flux. In France, the decision to bury the Unknown Soldier under the Arc de Triomphe was so controversial that it was taken at the last minute and made permanent only after the 11 November 1920 inauguration, on 28 January 1921. In England, Westminster Abbey was somewhat contested, but the decision to also keep Luytens' cenotaph as a permanent memorial seems to have helped. In all three cases, however, discussion was about the location within the city and its symbolic meanings, and was more concerned with creating a spectacular stage for the Unknown Soldier's body; thus it will be addressed in Part II. Here I will concentrate on the less overtly debated aesthetics of the tomb, which was more focused on representing anonymity. Once again, we find that state insistence on blankness is undercut by a more organic and self-generative image of the Unknown Soldier's loss of identity.

To a great degree, the tomb was presented as having no aesthetic program at all. It was to be the simplest flat horizontal stone marker. State archives show politicians to be rather unconcerned with this 'detail' that could be left to architectural engineers. Yet the choice of a horizontal slab, and the near-elimination of any vertical element, as well as the type and origin of the stone used all turn out to have great symbolic value. The British Unknown Soldier is buried under a stone of black Belgian marble; the Italian, under white marble from Monte Grappa, and the French, under a slab of grey granite known as the 'Dalle Sacrée.' These stones recall the reality of the battlefield and thus remind us of where the Unknown Soldier fell. But more important, they pay homage to the fact that many veterans felt that the Unknown Soldier Memorial should have remained on the battlefield, for to bring it into the City was to allow politics and civil society to appropriate it. Similarly, they emphasize an organic connection between the Unknown Soldier's body and the stone he died on (which, as we will see when we discuss embodiment in Part II, was emphasized by veterans as well). Westminster makes clear the genealogy of the horizontal slab, which was typical of medieval churches, and was generally associated with humility and the passing of time: unlike vertical tomb markers, which often included three-dimensional representations of the deceased as they lived anew beyond death, and thus indicated transcendence, horizontal markers, most often on the floor where they would be worn away by time, indicated transience, the passing away of the body, but also, eventually, of any memory of the name and deeds of the deceased. Furthermore, the stone floors of Christian churches were associated with water – the water of baptism, but also the abyss in which all human efforts disappear.[35]

The symbolism of the Unknown Soldier's Tomb is not non-existent or blank, but minimalist, in its emphasis on transience. In all three cases this comes in contrast with the verticality and symbolism of the Monument surrounding the tomb: Westminster speaks more clearly of Christian transcendence (though not without patriotic implications), whereas the Arc de Triomphe is more secular, and the Vittoriano, not to be completed until 1931, is poised to bring together national and religious transcendence. But beyond these important differences, to be discussed later, we see here a tension typical of First World War memorials in general. Winter has shown that minimalism in memorials was often the result of a desire to preserve them '*from* art,' in contrast with what was seen to be the excessively decorated and rhetorical monuments of the past. At their best, such memorials could thus be an 'extraordinary statement [...] about mass death and the impossibility of triumphalism.' At the same time, Winter argues, it is important to remember that, if popular reactions to minimalist memorials such as Luytens' cenotaph or the Trench of the Bayonets were highly positive, this was unusual, since the 'austere avoidance of allegory, figurative art, or ornamentation' was more often rejected by the large majority of people, who still looked to romantic and sentimental models of mourning. While we cannot equate verticality with decoration and horizontality with minimalism for all First World War memorials (in fact, Modernist minimalism is often vertical), we can in the case of the Unknown Soldier, where it is clear that the tomb itself is anti-rhetorical and intends to counter the louder, older, more decorated structure that surrounds it. From this point of view, it is telling that parliamentary instructions insisted on how important it was to make the Unknown Soldier's Tomb 'visible' to all, yet what is, in fact, most visible is its frame. Similarly, the eternal flame was added to the Unknown Soldier Memorial by most countries a few years after inauguration: said to represent the resurgent soul of the dead soldier, it is a comforting if contradictory complement to the bare slab on which it rests. This may be compared with the addition of the sculpture of three soldiers to the wall at the Vietnam Memorial, which also serves to soften an uncomfortably stark aesthetic. A great part of the success of the Unknown Soldier Memorial, therefore, is its mixed message: an austere core surrounded by an allegorical frame.[36]

Hence, to interpret the Unknown Soldier Memorial, we must ask: what relationship is established between this austere core and the allegorical frame surrounding it? Is this mixed message an ambivalent failure to decide, or is there something more going on here? On the one hand, the minimalist Modernist aesthetic can be seen as a critique of traditional commemorative forms, and at the extreme, Modernism's emphasis on speed and change would be irreconcilable with mourning in any of its traditional forms. Moreover, in Italy, we have come to associate the Modernist aesthetic with Fascism, which it adopted for its own war memorials, in which abstraction became more of a denial of death than a meditation on our incapacity to understand it fully. As Jeffrey Schnapp writes, 'the modern monument must fulfill what for Freud is a successful completion of mourning: it must keep the dead alive while burying them for all eternity; it must remember them, to be sure, but also sever them from present reality; they must be consecrated as that which is closed, completed, finished, enveloped and absorbed into the dynamism of the present.' From this perspective, older types of memorials do not quite reach closure, but that is largely because closure means a total absorption of the past into the present. Similarly, from this perspective, the Unknown Soldier Memorial in its various executions is a failure, for its critique of older monuments is too mild, as some traces of the past still remain. On the other hand, Schnapp also notes that Modernist 'anti-monumental monumentality' incorporates 'a peculiar archaeological imagination,' so that its works 'close the door to the past only to reopen it again to an archaeology of archaic forms.' In a similar vein, Winter writes that in the immediate aftermath of the First World War Modernism – not yet an established aesthetic – could hark back in its minimalism to the 'eternal monuments of Brittany, of that primitive age when man fought against the savage beast and the chaotic forces of nature.' More generally, as Mia Fuller sums it up, 'Modern' described a number of architectural styles in postwar Italy, evoking at once international Modernist work, the rationalist aesthetic, as well as more historical Italian or 'Mediterranean' attempts to reinvent the past. From this perspective, older forms of mourning are not all irrelevant to modernity; rather, the 'archaic' and the 'primitive,' though no doubt a projection of what modernity is missing, become a way to bypass a specific nineteenth-century vision of death, linked both to Christian dogma (rather than to popular Christianity) and to the secular capitalist nation. From this second perspective, the Unknown Soldier Memorial does not fail, but rather stages a confrontation between two types of transcendence, or better, of rebirth. Instead of erasing the past, or eschewing any understanding of death, the Unknown Soldier Memorial makes us experience the increasing distance, the gap itself, between a Romantic pre-war vision and the new inchoate one brought on by the shock of the First World War. Both elements – the minimalist and the allegorical – are therefore necessary to the Memorial's message.[37]

In the case of the Unknown Soldier Memorial, we find that state ceremonies tended to emphasize verticality, and the frame, as the locus of meaning, whereas, as noted, the tomb itself was seen as blank. Thus, while Labita finds it difficult to reconcile with a 'funeral rite,' verticality was multiplied, during the Italian ceremony, by low-flying warplanes which futuristically suggested the transfiguration of flesh into mechanical purity, or as d'Annunzio wrote, the fusion of 'wing' and 'bone.' In France, flight was represented by the departure of carrier pigeons at the opening of the ceremonies. In Italy, without the metallic funeral wreath placed over the slab to symbolize victory over death, from an architectural standpoint, the tomb became almost invisible; both the tomb under the Arc de Triomphe and the one at Westminster were even more invisible were it not for soldiers or barriers that prevented people from walking on them. In contrast, veterans saw the tomb as exalted by its very humility, a view echoed in popular reactions reported by the press: they associated the '*soldat obscur*,' the '*caduto*,' '*oscuro*,' and '*ignaro*' with the higher vision brought on by blindness and suffering. Here we see that the disappearance of the tomb is not so much an acceptance of its ideological frame as it is a representation of how ideology can overwhelm the individual's experience and existential anonymity itself. Hence, the contention between ideological interpretations of death and rebirth, be they religious, political, or both, and a more minimalist and existential interpretation is at the very heart of the Monument.[38]

The importance of rebirth as primitive descent, and existential confrontation with transience, not transcendence, is seen in the role that images of water play in various evocations of the war experience. Such images are connected to the Unknown Soldier by d'Annunzio, who as we saw, associated 'the nameless peasant who entered the ford dropping to his knees in the midst of the current, and, sacrificing himself, sculpted his sacrifice into the water' with 'the one to be buried in the monument.'[39] Writing on water is a recurrent image in the *Notturno*, where it is associated with the symbiotic relationship between the poet and his dead comrades; this, in turn, is what causes the poet to lose, once again, a sense of the limits of his skeleton, his body, and to experience an ideality or infinity that is not absolute but '*in cammino*,' constantly shifting and 'in motion':

> Thus are his death and my life one. [...]
>
> Like the wind's message, which is the speed of infinity in motion, with a spirit without borders, with a body without form, with a joy that feels like terror, I sense the ideality of the world. [...]
>
> I write not on sand, but on water.[40]

Terming it *miles patiens*, (the suffering soldier), Valesio has analysed d'Annunzio's expression of 'an elemental and existential suffering' in the image of a nameless peasant. He rightly insists on 'the idea of the Italian soldier as victim, as witness to a sacrifice,' which goes against triumphalism, and later Fascist optimism. Valesio also is the first to point out the genealogical relationship between d'Annunzio's image and one of Ungaretti's most famous war poems, 'I fiumi': for Valesio, the soldier as 'acrobata / [...] sull'acqua' (acrobat / [...] / over the water) in the latter recalls d'Annunzio's peasant, and thus the soldier as victim, yet also adds to this his identification with the artist and the juggler, and thus the ability to 'rediscover [...] joy in the midst of misery' – making him a Franciscan 'fool' figure, also of d'Annunzian descent. I have already mentioned the importance of such joy in connection to the primal scene at the core of the Unknown Soldier Memorial, what I would like to emphasize here is the role of water imagery itself, as it links the flow of 'infinity in motion' to the flow of the dead into the living and vice versa, that is, to their symbiosis rather than their division.[41]

Ungaretti's poem echoes images recurrent in d'Annunzio's *Notturno*, as he begins 'I Fiumi' with images of mutilation, abandonment, and water, going on to evoke '*le mie quattr'ossa*' (my carcass of bones), which have to be gathered up, like those of dead soldiers awaiting their 'return,' before he describes the four rivers of his life (the Isonzo representing the war, the Serchio his Italian origins, the Nile his Egyptian ones, the Seine his French education), only to conclude on a 'corolla / of shadows.'

> I cling to this mutilated tree
> forsaken in this sinkhole
> that has the weariness
> of a circus
> before or after the show
> [...]
> This morning I lay down
> in an urn of water
> as a relic
> in repose
>
> The flowing Isonzo
> polished me
> as one of its stones
>
> I raised
> my carcass of bones
> and like an acrobat
> I moved
> over the water
> [...]
> This is my longing for home
> in each of them
> shining through
> now that night has come
> and my life seems to me
> a corolla
> of shadows[42]

The way his personal, but also familial and national genealogy flows into these rivers and ends up in indistinct darkness anticipates a later image, from 'La pietà':

The road of the dead passes through the living,

We are the spate of shadows,

They are the grain that splits open in us in dream,

Theirs the distance that is left to us,

And theirs the shadow that gives weight to names.[43]

As Robert Harrison has shown, Ungaretti's is a delicate philosophical meditation on how the dead inhabit the very words with which we give meaning to our lives; once again, what I wish to emphasize is the image of the '*fiumana*' (river in spate) by which the living and the dead flow into each other – an image that connects writing on water to a rebirth that is not transcendent but 'infinity in motion.'[44]

Another poem by Ungaretti, 'Annientamento,' makes the link between these water images and the sepulchral stone explicit:

Today
like the blue asphalt
Isonzo
I settle down
in stone-bed ashes
laid bare by the sun
and I transmute myself
into a flight of clouds
[...]
I feel the marble kiss
upon my lips[45]

The image of the river that is a bridge between earth and sky not only recalls again d'Annunzio's 'spirit without borders,' 'body without form,' associating it with a stone slab or 'marble kiss' such as the one at the Unknown Soldier Memorial, it also anticipates a key image in Gance's *J'Accuse*. Water imagery is very important in this film too, and in its more overt aspect serves to emphasize both the misery and unanimity of life in the trenches. As the film progresses, soldiers are seen to inhabit a less and less orderly world, as the trenches first fill up and become rivers of water and then rivers of mud where the living mix with the dead. This progression is parallel to the increasing closeness between François and Jean, which leads Jean to later speak for the dead François. But even closer to d'Annunzio's and Ungaretti's imagery is Gance's depiction of the return of the dead, who as I already mentioned rise up from their battlefield graves. They are seen as a 'river in spate' of people, in obvious contrast with the military parade of living soldiers, which has clear rectangular formations; moreover, in one of the more famous scenes of the film, this river of the returning dead is seen flowing above the Arc de Triomphe. As this part of the film was added between 1919 and 1922, it is apparent that it is a reference to, but also a critique of the Unknown Soldier Memorial below. I will discuss this in more detail in the context of debates about sacrifice, but the point here is that, visually, Gance contrasts a rebirth that has to do with flow and, in particular, the flow that unites the dead and the living in the trenches, to a rebirth that has the rigidity of military formations or of the Arc de Triomphe itself.[46]

A Silent Inscription

If the Unknown Soldier's skeleton becomes 'infinity in motion' where the dead flow into the living and vice versa, if his sacrifice is indeed 'written on water' and not transcendent, what sort of inscription can be true to this? What sort of words might be appropriate? Both the contemporary press and historians who have written about the inauguration of the Unknown Soldier Memorial have marvelled at the silence maintained by such immense crowds. It seems the British came up with the idea of simultaneous silence throughout the country at the height of the ceremony, a procedure that was adopted in France and Italy as well. Though various pieces of music were played to accompany the Unknown Soldier's burial, these, too, were in all three countries reduced to a minimum. In Italy, the negation of sound was also expressed in the Savoy tradition of playing the drums off-key to indicate mourning. In France, the only speech was a half-hour one by President Millerand, whereas in Italy and in England there were no speeches, but only the traditional mass for the dead; also, the latter two used sirens, drums, and bells to evoke the sounds of war and to emphasize silence itself. I will look in detail at this choreography in Part II, but I mention it here because it helps us to understand the discussions that took place over what was to be inscribed upon the tomb itself.[47]

To a great degree everyone agreed that 'the only acceptable attitude [...] is "impeccable silence." [...] A father has no moral right to use the heroic death of his son to add bluster to his personal prestige [...] or to support a politico-commercial doctrine which, consciously or not, would provoke new massacres.' Winter here cites Henri Barbusse (author, among others, of the famous *Le feu*) to evoke how people felt about war memorials in general, and this was no doubt true of the Unknown Soldier Memorial. Crucially, as Winter notes, Barbusse also felt that it was necessary to 'prevent the forgetting of the light of moral beauty and the *perfect holocaust which burned in you* during the monstrous and disgusting horror of the war.' We see here contradictory imperatives typical of the era: the rejection of 'bluster' and ideology becomes an almost complete rejection of monuments (as d'Annunzio insisted, 'we do not want monuments'); at the same time, however, sacrifice must not be forgotten, and its sacred character must be recognized (as 'holocaust' implies). In terms of the inscription on the Unknown Soldier's Tomb, the question thus became, how can words be at once anti-monumental yet permanent and even sacred? Or, how can words be at once silent and eloquent?[48]

In comparing the inscriptions for various Unknown Soldiers, we find a number of significant variations. Some of the most salient ones are:

> Ignoto militi.
> [Unknown Soldier; entire Italian inscription]
>
> Ici repose un soldat français mort pour la Patrie, 1914–1918.
> [Here rests a French soldier who died for the *Patrie*, 1914–1918; entire French inscription]
>
> Beneath this stone rests the body of a British Warrior unknown by name or rank.
> [part of the British inscription]
>
> Here rests in honored glory an American soldier known but to God.
> [entire American inscription]

Where the British and the Americans mention honour and glory explicitly, the French let the Arc de Triomphe surrounding the tomb speak for itself, as do the Italians with the Vittoriano. However, the French do mention he was '*mort pour la Patrie*,' whereas the British are far more prolix in writing that he 'gave the most that / man can give life itself / for God / for King and country / for loved ones home and empire / for the sacred cause of justice and / the freedom of the world'; the Americans and the Italians do not say what he died for. Also, the British describe him as 'unknown by name or rank' 'yet well known,' and further mention the 'many multitudes' he represents; for the Americans he is 'known but to God'; in contrast, the Italians and the French do not mention another identity at all.[49] Overall, historians have claimed that Unknown Soldier becomes more and more of an abstraction, and that aside from the British case where 'Warrior' evokes a feudal hero, the details of his heroism fade in the face of his obedience, making him more a type than an individual. Yet, in the Italian case, the choice of Latin participates in a rhetoric of empire that would become overwhelming under Fascism, and even in 1921 already implies ancient notions of heroism revisited in the Risorgimento; in the French case the word '*Patrie*' carries with it references to revolutionary heroism and the ancient antecedents Enlightenment thought found for it; even the American 'known but to God' refers to the religious freedom underlying the American Revolution. Further, as Ken Inglis points out, the French shift from '*le soldat*' in the initial ceremony to '*un soldat*' in the permanent inscription takes him away from being a type and thus underscores how the indefinite article in all four cases implies one among many and not an example. In Italy, the indefinite '*un soldato*' was used in Douhet's first proposal and early discussions, but then shifted more of to the definite '*il soldato.*' And the definite article was used overwhelmingly in the press, and in other writings, where the goal was to specify discussion was about *the* Unknown Soldier in the Unknown Soldier Memorial as opposed to all the other unknown soldiers who died in the war. Once again, we can see that in the end the Memorial sends out a mixed message, which recalls Barbusse's claims: the Unknown Soldier's heroism, like the definite article, is present, but under erasure, and like revolutionary heroism, refers to old models but only to claim that they are made radically new.[50]

What we see here, I would like to propose, are two types of silence not dissimilar to the ones distinguished by Valesio in his discussion of rhetoric and anti-rhetoric in *Ascoltare il Silenzio.* One is a silence of plenitude, in which anti-rhetoric, and in this case anti-monumentality, conceals a fullness of meaning that is still present, if hidden. Like traditional ineffability, such silence indicates a truth beyond words, yet a truth that is at least potentially or in the eyes of God knowable. The other sort of silence, in contrast, is one of interruption: in this case meaning, monumentality, and the possibility of knowledge itself are broken by something that withdraws and may never be comprehensible or visible. I would like to say that here we have not ineffability but vacancy, or as Valesio notes, aporia, an empty space that may or may not be filled. That these two types of silence vie with each other in the Unknown Soldier Memorial

Figure 9. Ignoto Militi inscription (1921 photo). © Courtesy of Department of Special Collections and University Archives, Stanford University Libraries

is made startlingly clear by another inscription, which in Italy accompanied the Unknown Soldier's coffin as it travelled from Aquileia (where he was chosen from among the eleven) to Rome. This was a verse from Dante's *Inferno,* 'L'ombra sua torna ch'era dipartita' (His shade, which had departed, now returns), which refers to Virgil and suggests that the Unknown Soldier is a guide. On the one hand, this may be read as yet another Christian reference which ties the Unknown Soldier's journey to Dante's and associates his return to a plenitude at once religious and patriotic. Yet once again, on the other hand, what we are surprised to see so overtly here is the recognition that this traditional symbolic vocabulary is lacking, for Virgil cannot guide Dante all the way to salvation, and in the cited passage returns to Limbo: a place where bodies have only a ghostly consistency, just as their actions can never embody either damnation or salvation, it is most of all the place of interruption, where death is not or not yet 'meaningful.' Remarkably, in his 2006 novel, *Io sono il milite ignoto* (to be discussed in Chapter 6), Gian Pietro Testa refers to the Catholic Church's recent abolition of Limbo as a move towards increasing the fear of authority, by removing any alternative to the dualism of Heaven and Hell.[51]

In contrast, images from d'Annunzio, Ungaretti, and Gance take up this alternative and help us to understand what is at stake. Thus, d'Annunzio had imagined a marker that would correspond to the silence of the battlefield, writing from Fiume in 1919 already: 'An Italian sends you this silence as one would send an uninscribed grave marker, where a single word has yet to be inscribed: the greatest one.'[52] Later he would go back to this silence and claim that 'Resurgo, [...] [was] the only word that should have been inscribed on the tomb of the Unknown Soldier.'[53] D'Annunzio's first comment underscores what is unusual about the second: the present tense. Unlike the traditional Christian *resurrexit* (he is resurrected), *resurgo* is closer to 'I am resurrecting' and thus evokes a liminal state, still in the making, corresponding, then, to a stone where something is about to be inscribed, but we know not yet what. Interruption and Limbo, we understand now, are also potentiality or emergence: instead of emphasizing transcendence as an end-point, or focusing on what the Unknown Soldier returns *to*, what is important here is the process itself, and philosophically speaking, the phenomenon instead of its putative essence. A similar emphasis is found in Ungaretti's 'I Fiumi,' where the poet's meditation on existential flow takes place in the pause 'before or after the show,' in a 'circus' that feels empty of its normal function, awaiting movement that is not yet there. In d'Annunzio's imagination, then, as in Ungaretti's, through the Unknown Soldier Memorial, the soldier has not returned to any definite 'Patria,' but rather is returning, still not knowing to where. Just as his bones cannot be limited to a single blank skeleton but evoke a *fiumana* of living and dead always in motion, they cannot be limited to a hidden plenitude behind silence but evoke a new expressivity in the making. This recalls Leed's observation that, according to the letters of British soldiers, 'If the war was initiation, it remained forever problematical what state, condition, or station the soldier was being initiated into.' Moreover, Leed notes that 'mutism and speech disorders were the most common symptoms of war neurosis,' and he ties this to the image of the war as a 'Labyrinth' that soldiers could only find their way out of if they could develop a 'correct reading' of its 'text.' Though Leed is referring mainly to social reintegration, his reference to the event as 'text' and his use of the anthropological category of 'initiation' show that what was still lacking and in the making, as d'Annunzio and Ungaretti claimed, was also a symbolic vocabulary and a ritual that might define a different sort of resurrection or return, one that would be tied more to descent and to transience, as we have already seen.[54]

In Gance's *J'Accuse*, the issue of inscriptions is evoked in the title itself, which is repeated at key moments in the film and, most important, dominates its last scene, when Edith's illegitimate daughter Angèle (recall that Edith was raped by a German soldier) helps Jean write it on an improvised blackboard. For Winter, Edith is trying to console Jean, who has 'gone mad,' and who after accusing the Germans, the war, and the French villagers who profited from it, can only accuse nature itself now, for its 'indifference to human suffering.' He writes that this more avant-garde element of Gance's film contrasts with his 'Christology,' which is 'unsubtle' yet 'compelling' and 'romantic' as it endows soldiers' suffering with a transcendence drawn far more from popular religion than from strict Catholic theology. In light of d'Annunzio's *resurgo*, I believe these two elements are less in contrast than they seem (as Winter acknowledges when he sees Gance marrying the experimental and the popular), and that in particular the accusation against nature itself, which is a refusal to understand the ubiquity of evil in the world, can be read as part of a demand for a new Christology that has not yet been invented (but that, as I pointed out earlier, for Gance includes women in a way that was quite progressive for the time). In this respect, Gance's reference to Zola's famous letter in defence of Dreyfus and against anti-Semitism is important, for it is clear that whatever Christology may persist after the First World War, it will have to be non-sectarian and tolerant. More generally, while I see Winter's reading of Jean's accusations as legitimate, as a viewer I was struck by how much power Gance can instil in them precisely by leaving their object unsaid. The implication is that Jean's immediate rebellion may be caused by present pain (such as his mother's death) but that in a more philosophical sense, his rebellion is a refusal to make evil intelligible, to the point of even refusing to name it: his accusation is more like interruption than plenitude, in that it decries death at war, but refuses to ascribe *any* meaning to violence. From this perspective, the last moment when Jean tears up his collection of poems, *Les Pacifiques*, is not just one of tragic madness, but one of poignant warning. By finally finding a target for his accusations – even if just his own earlier pacifism – Jean runs the danger of becoming violent himself, and losing his poet's vocation, which is to refuse any political appropriation of suffering. At the same time, Gance does not suggest there is any alternative: Jean has learned from François that to withdraw from action is not acceptable.[55]

Thus, where d'Annunzio and Ungaretti (in very different ways) see a certain silence about the war experience as a sign that a new language is needed, Gance fears that *any* language drawn from the war experience will be contaminated by its violence. Another way of putting this is that if war is a sort of failed initiation, an interruption of normal categories, to make that failure into a new beginning is always in danger of transforming interruption into plenitude, and thus of betraying the experience itself. From the point of view of the image of bones, we could say also that if in the trenches the borders between one skeleton and the other dissolve, making individuality porous, in the making, and fleshy, it may not be possible in the war's aftermath to dwell in this instability, to be forever true to battlefield chorality. Once this moment of horror but also of illumination and 'innocence' (to borrow Ungaretti's terms) enters the world of 'memory,' memorials, and monuments, is it possible for it to be anything but reified, anonymity turned into a new sort of identity? This is what is at stake in the Memorial's ambivalence between two types of silence, or in its contradictory attempts to inscribe silence in its structure, be it even as an ironic acknowledgment of the impossibility of silence. As we are about to see, we cannot evaluate the quality of the Memorial's silence without exploring the debate around sacrifice that was part of its creation and that continues to shape our experience of it. In particular, the Christology present in Gance, d'Annunzio, and many others poses crucial questions to be addressed in the next chapter: can sacrifice be reinvented to be 'holocaust' without 'bluster' (as Barbusse wished)? Since existential anonymity is self-generative and in flux, who is to be the subject of sacrifice, and whose authority will it uphold?

3 Sacrifice and the *Non finito*

Taking Up Anonymity

As the First World War came to an end, dealing with irretrievable dead bodies and the extension of wartime estrangement to survivors became a question of how the war would permanently transform the cultures and the people that had endured it. Gathering the Unknown Soldier's skeleton, and drawing out its minimalist expression of the shared mortality of the trenches was thus only a first phase. The second one, which this chapter focuses on, involved giving estrangement or existential anonymity an interpretive frame, that is, creating a home, conceptually, emotionally, in which to house the Unknown Soldier. A core debate involved sacrifice, as soldiers were said to have sacrificed not only their lives but also their identities in the war. But what was the purpose of their sacrifice? What home did it found? At the time, the term had obvious Christian meanings, linked to Christ's sacrifice on the cross, yet as we will see, transposing these to the political sphere was not without difficulties, though it was not a new idea. These difficulties were compounded by the crisis in Christianity known as religious Modernism, which questioned dogmatic definitions. Moreover, to Christian notions of sacrifice nineteenth-century anthropology had added cross-cultural, cross-temporal comparisons that, for all that they were Eurocentric, sought to unveil the social and psychological mechanisms at play in sacrifice. Within the more overt debate regarding sacrifice *to* – to the nation, to God, to one's family? to which version of the nation, the Socialists', the Communists', the right-wing nationalists'? to which God, the Vatican's, the Protestants', the one within each one of us? – there is a more covert debate about the very structure of sacrifice. The issue became whether it was possible to have a sacrifice *with* that retained chorality and shared meaning without defining its object, or more accurately, while considering its object to be in the making, emergent, and debatable, rather than a priori. At stake was the very possibility of openly discussing, and thus reinventing, the meaning of the war, rather than seeing it as untouchable, sacred, or absolute. Of course, as will be demonstrated, such reinvention can be dangerous, as it puts an emotionally powerful collective experience up for reframing; yet it is more dangerous, our texts and history suggest, to cling to an immobile past.[1]

As we saw, contesting the idea of loss of identity as blank, the Unknown Soldier Memorial also depicted it as choral and organic, self-manifesting, expressionistic, fluid, and resistant to reification. These characteristics will continue to be important as we consider the nature of sacrifice. Specifically, existential anonymity caused there to be great anxiety about *who* had sacrificed, and this underlay the question of sacrifice *to* versus *with*. The main problem was that the Unknown Soldier had at once to be a passive victim of the forces of war and at the same time the agent of a sacrifice he freely chose. Or also: alienation was something he suffered (in the Christian sense of *passio*), and yet in order for it to become sacrifice such suffering had to become also active and authentic (in the existential sense of taking up one's burden, confronting it, or accepting it). It is thus telling that while there was, of course, some speculation about the Unknown Soldier's personal or civic identity (and, as we saw, some irony about his religion, race, nationality, and even gender), there was far more emotional urgency to discussions of his moral, ethical, and existential qualities. This was a question of courage, but not, as shown, of heroism in the traditional sense, as it was felt that the Great War had erased the very possibility of heroism. Was there a difference, and if so how could it be understood, between taking up wartime alienation with courage, and fleeing from it with cowardice? As seen in the work of William James, but also of religious Modernist thinkers such as Maurice Blondel, Alfred Loisy, or Ernesto Buonaiuti, a morality based on dogmatic religion was at the time increasingly being contested by a more experiential, phenomenological approach. From the latter perspective, courage was less about being true to an external authority than about acting in accordance with one's inner spiritual truth. As the existentialist philosophy that grew in the interwar years insisted, this meant each that individual had to confront the inherent absurdity of the world, and the arbitrary nature of morality, and thus take responsibility for the values he or she chose to live by and to contribute to society. This returns us to the contention between the individual and the community mentioned in the last chapter, for we find here a confrontation between inward, individual understandings of sacrifice and a collective, institutional one. The latter was felt to betray each soldier's experience of the war, and yet if this experience had no collective expression, courage would be left to become an empty marker, a blank value that any political or religious party might appropriate.[2]

These questions shape the two aspects of sacrifice that I will discuss here, after a few theoretical reflections and three stories about sacrifice: first, absolution, and second, initiation and the establishment of an authoritative voice. In most of the countries I am considering, a ritual benediction or absolution of the Unknown Soldier was a central part of the inaugural ceremony. This was, on the one hand, a reflection of the Church's desire to retain power over First World War mourning and to assert the importance of spiritual salvation in contrast with state authority. On the other hand, it also reflected awareness of a more anthropological view of sacrifice, which saw it as the foundation of the social cohesion Church, state, and many others wished to reaffirm. In this light, the First World War was an extreme breakout of the wild forces of violence that threaten society, and to put a genuine end to it required a martyr that everyone could honour. At stake were conflicting views of what it meant to die *pro patria*. Second, in all the countries I am considering, but most heatedly in Italy, Christ's sacrifice was taken as model and yet also reinvented. The Unknown Soldier's authority was seen to be both like and unlike Christ's, and his sacrifice was compared to an initiation. Even though not all writers and mourners would have been aware of the religious Modernist debates that reached their apex with the 1907 Papal encyclical *Pascendi dominici gregis*, they did discuss Christology along similar lines. Specifically, Modernist emphasis on the human Jesus and on incarnation instead of transcendence was echoed in the image of the Unknown Soldier as *alter Christus* but without resurrection, that is, as a liminal Christ. Again, even though not all writers and mourners would have been aware of it, this coincided with the influential work by Arnold van Gennep, *Les rites de passage* (1909), in which he emphasized that initiation had a phase of 'separation' that preceded 'incorporation' – a liminal phase, in which frames, collective understandings, and dogmatic meaning are up for reinvention. Also important was the discussion by Henri Hubert and Marcel Mauss, influenced by Émile Durkheim, in their 1898 *Sacrifice: Its Nature and Its Function*, for while they insisted on a relationship between morality and sacrifice, they also criticized the idea of seeing Jesus as a model for sacrifice, arguing, in particular, that 'sacrifice is *prior* to the god, not, as it were, the other (i.e., generally Christian orthodox) way around.' In other words, Jesus was not sacrificed to a prior sacred cause but rather invented a new sacredness via his sacrifice: once again, the 'purpose' of sacrifice comes after a period of openness and liminality. In Italy, such liminality was overdetermined by the notion that Italy itself was still incomplete, as political unification struggled with social, linguistic, ethnic, and religious differences. The liminal was associated with 'bare life' – the individual's embodied uniqueness that escapes categorization and socialization – and thus with the Unknown Soldier as a specific body that could not be subsumed to any ideology.[3]

In his seminal *La violence et le sacré*, René Girard argues that to understand sacrifice we must grasp its hidden connection to violence. Most relevantly for the First World War, Girard associates the threat of violence to a breakdown of boundaries between the living and the dead: this is for him a typical moment of crisis, which requires sacrifice to restore order. Hence, 'the crisis manifests itself as a loss of differentiation between the dead and the living, as the mixture of two realms that are normally separate. This shows how much the dead embody violence [...] The chosen victim dies, it seems, so that the entire community, that feared dying with it, might be reborn to a fecund and new or renewed cultural order.' Notwithstanding all sorts of putative justifications, the sacrificial victim is arbitrarily chosen to take upon himself a violence that belongs to all, so that his death can also ensure its expulsion. The sacrificial victim, then, is a scapegoat, whose innocence is necessary even as its purpose can never be overt: for, if the victim deserved punishment, then he would only be receiving his due, not paying for all of society ills; at the same time, the fact that all of society's ills are projected onto him must remain hidden for the projection to be successful. Thus, Girard argues, 'men can only be reconciled at the expense of a third [person]'; hence, 'there is always a human death at the origin of cultural order,' even as the foundational role of this death is repressed and returns only obliquely in stories about sacrifice. This, in Girard's analysis, is the essential 'scapegoat structure' of sacrifice. A typical example, for Girard, is Oedipus, who appears to be punished for killing his father and marrying his mother, but who in reality was fated to do so, attempted not to, and cannot be held individually responsible. The actual purpose of his death, then, is to exorcise the violence personified by the Sphinx, and to found a new cultural order. Christ is also a scapegoat for Girard, but what makes him unique is his awareness of his role, which he chooses to accept. Christ, therefore, undoes the scapegoat structure of sacrifice by forcing us to see that the victim is blameless and, more important, by challenging us to confront the violence we project onto him as our own. This vision of Christ is, not unlike the Modernist one, liminal with respect to the Church as a temporal authority and a political power.[4]

Girard's theory is very helpful in interpreting arguments regarding the Unknown Soldier's sacrifice. In particular, it suggests that the state's commitment to blankness, its preference for a soldier who was truly 'no one,' is not a rather strange acknowledgment of the soldier as 'cog in the machine' but rather an attempt to ensure that he is 'innocent' in the sense of passive, a blank slate upon which, indeed, all the violence of war can be projected, and exorcised. Moreover, discussions concerning sacrifice *to* (nation, God, family, etc.), however divisive, serve to reinforce this scapegoat structure, for they all imply that the soldier's death had a real purpose, and thus hide the way society has projected its violence onto him. In contrast, the notion of sacrifice *with* that emerges from the choral anonymity discussed in previous chapters comes dangerously close to revealing that violence circulates throughout culture and cannot be projected onto a single victim. Further, attempts to imagine the Unknown Soldier's courage or authenticity in the face of death – or even to imagine his fear or his cowardice – are ways to question the scapegoat structure by making overt that he is 'innocent' in an active, not a passive sense. Both imply he had a choice. In this he is like the human Christ emphasized by Girard, who is a willing victim, but also an authentically vulnerable one, who has to confront fear and has a choice. In this case innocence, we will see, is not externally given by the Church or the state via absolution, but is an inner quality.[5]

The possibility of choice is notable for its absence in government documents about the Unknown Soldier. Just as governments preferred blankness to bones that made suffering visceral, they imagined the soldier's passive obedience rather than examining his moral qualities. Or, at least, they tried to avoid such examination, which abounded in the newspapers and in fiction, by emphasizing 'duty' as the only form of courage. Hence, in its official approval of the Unknown Soldier Memorial, in June and July 1921, the Italian Senate established his role as 'the symbol of the collective virtue of our people,' while the Chamber of Deputies insisted on how 'the absolute suppression of individuality, [...] demonstrates, in the end, that to have been Italian and to have fallen for Italy is a great enough title for supreme honours and eternal veneration, above and beyond any other sign of identification.' In more colourful language, Minister of War Rodinò addressed the following description of the Unknown Soldier to the Pariamentary Commission in charge of the Memorial in June 1921:

> an homage to all our dead [...] who fell with supreme endurance, and supreme sacrifice, according to the humble duty that illuminates, thus, without individuality, bringing down such strength of light on our race. [...]
>
> I, *ardito,* foot soldier, bombardier, artilleryman, I well remember that there was no gun more accurate than a dead soldier's. The person who picked it up never missed. [...]
>
> The raised gun of the fallen man signalled a fated victory.
>
> The raised bones of the fallen man will signal the beginning of true peace.
>
> Thus it shall be if we are able to gather together and meditate upon the highest expression of death's majesty, passing over a funeral procession, from the silent trenches of war to the immortal greatness of the City under the Italian sun.
>
> We believe a supremely austere ceremony must take place.

'Humble duty' is associated with 'lack of individuality' and the image of light emphasizes abstraction. Yet this is undermined by the uncanny power to direct bullets from beyond the grave, which implies that the Unknown Soldier would not have been either a coward or a traitor. The accuracy of a dead soldier's gun was legend already during the war, and is mentioned in d'Annunzio's work more than once; it is not unlike imagining the Unknown Soldier's skeleton 'with a bullet hole in the forehead' (as we saw in Chapter 2), for both indicate that he met death frontally, and was not running away. This trace of individual choice within official narratives confirms Girard's point that scapegoat stories obliquely reveal that the victim did not 'deserve' his death, and covertly imply he did more than was his 'duty.' Nonetheless, overall, the very issue of courage is what these narratives wished to avoid by focusing on 'duty' alone.[6]

This makes it all the more remarkable that this topic, at its most uncomfortable, is mentioned in Giulio Douhet's newspaper *Il Dovere,* about a year before he published his proposal for the Unknown Soldier Memorial in it. In fact, the title of his newspaper itself, which means *duty,* was an overt attempt to reappropriate the term from the Italian government, as Douhet explained in his inaugural issue that he was referring to the 'duty' the state had towards all soldiers, dead and alive. As explained in Chapter 1, Douhet himself was court-martialled for criticizing General

Cadorna's strategy, and was later vindicated – along with many others accused of, and often executed for, desertion – by Italy's traumatic defeat at Caporetto. His newspaper followed the activities of the Commissione d'Inchiesta su Caporetto (Investigative Commission for Caporetto), which was created to investigate, and exonerated many who had been initially blamed. This Commission was still at work doing the government's 'duty' while the Unknown Soldier ceremony was being planned, and it is clear that at least for Douhet, and for many veterans, if not for the government, one important function of the Unknown Soldier Memorial would be to honour those who had been unjustly accused of being deserters. Discussions of the Unknown Soldier thus brought up not only Caporetto, but a broader worry over traitors in general, and the vexed issue of when a soldier might rightly rebel against his commanders. For instance, *Il Dovere* approved of d'Annunzio's defence of those taken prisoner by the Germans 'after an adventurous foray,' even if they were acting against orders or should have fought to the death. That is, against a climate of suspicion of which he was personally the victim, Douhet's paper approved of imagining a story of daring to explain why soldiers were taken prisoner in mysterious circumstances, rather than one of disobedience or defection. Even more radically, a few issues later, *Il Dovere* demanded that those summarily executed as traitors be given the title '*caduti per la Patria*' (fallen for the *Patria*). At the same time, the article emphasized that it was even more difficult, yet more urgent, to oppose the widespread idea that large numbers of soldiers were deserters. In other words, not only for prisoners, but for all the fallen, an understanding of courage that married loyalty and independent thought was needed.[7]

In this light we can understand better what seems at first to be a shocking story in *Il Dovere*, 4–5 September 1919. Almost a year before what historians consider to be the first proposal in print for an Unknown Soldier Memorial in Italy (*Il Dovere*, 24 August 1920), we find Douhet's newspaper is already asking for 'a monument to this humble and unknown hero':

> here the powerful genius of the race forms in the person of a humble corporal an immense symbol of unique virtue. He is the innocent one, condemned to death at the hands of a firing squad, who renewed the classical splendour of civic devotion, showing the soldiers who were to shoot him how salvation and the honour of the *Patria* can only be nourished by a serene and proud self-sacrifice. An officer who witnessed the events nar-

> rates to the Investigative Commission [for Caporetto]: '[...] as the [death] sentence was read, the corporal remained impassive, serene, like a soldier in battle. As soon as the reading was finished, he said: "I believe the judges have voted with their conscience and I go to my death with pride." In front of the firing squad he refused to have his eyes bandaged and, addressing the soldiers, said: "Soldiers, I have been with you, I have fought with you; aim carefully, aim to the chest and always serve your country. *Evviva l'Italia.* This is supreme beauty."' [...] we propose to begin a subscription in order to erect a monument to this humble and unknown hero.[8]

Even though this soldier turns out to have been innocent, the very mention of the Unknown Soldier and treason in the same narrative was, to say the least, provocative. Far more than the dead soldier's 'accurate gun' or 'bullet hole in the forehead' stories, this one exhibits an ambivalence in which scapegoating and its undoing coexist. Thus, the initial thrust of Douhet's story identifies sacrifice with absolute obedience, to the point of accepting punishment and death without question. This soldier confronts his execution serenely 'like a soldier in battle' and identifies with those who shoot him in that all 'serve [their] country.' This echoes official versions of the Unknown Soldier's sacrifice which, as we just saw, celebrate the 'absolute suppression of his own individuality,' according to a model of passive obedience that resulted in the notion that 'in the exact observance of duty [...] sacrifice is celebrated – *however it is accomplished*' (emphasis added).[9] In this reading of the story, passivity itself is a guarantee of innocence and courage. In fact, we suspect, it matters little whether this soldier was in actuality innocent of treason or not, what matters is only that he accepted the state's judgment of his actions and, ultimately, the state's role in determining the purpose of his death. If that state shot him as a traitor, he accepted it; if later he became a hero, he accepted that as well. Read with a more critical eye, this implies that all who accept the random violence of war – death in its absurdity and anonymity – are innocent heroes and sacrificial victims, no matter what they actually do. While this may seem to forgive those who were afraid, or even those who retreated when they were supposed to advance, what it really does is establish the state's absolute power in determining the value of death, hence, also of life. In contrast as much with those who wished overtly to commemorate visible 'martyrdom' and 'warrior's courage' as with those who insisted, instead, on 'nameless dead' and the 'mystery of miserable and venerable remains,' what this story reveals between the lines is the total lack of connection between the soldier's intrinsic qualities and his becoming an exemplary victim. As Girard's elucidation of scapegoating helps us to see, 'innocence' and 'courage' are in this case conditions conferred upon the soldier by the violence he is subjected to. The true point of the story is to bring up the dangerous issue of treason and to render it nil by showing that the purpose of sacrifice is never debatable, ambiguous, or open-ended, because it is not up to the individual, in the moment of action; rather, sacrifice is part of a closed structure, in which its unavowed but true purpose is always the establishment of power over violence.[10]

There is, however, more to this story. The soldier's reaction is not merely passive but dramatic – his refusal of the eye bandage evokes by contrast the fear, anger, or self-justification we might expect – and this turns things around because he claims agency for himself. Even more, his gesture asserts his superiority over current military practice. In this vein, Douhet's newspaper concludes the story by noting that this soldier was 'too heroic' for such 'small times.' As in the first reading, the soldier's prior actions, be they treason or not, are irrelevant, and we are not told what they were. But this time, what becomes important is not the state's power to determine the meaning of his death, but the present moment: refusing the bandage, asserting his pride and solidarity with fellow soldiers, and speaking with assurance, this soldier *shows* his courage. Crucially, in this reading, courage has nothing to do with the purpose of his sacrifice: it is not about sacrifice *to* nor about claiming that wartime violence had a legitimate purpose; rather, courage comes from confronting an absurd, meaningless situation with dignity. Such dignity is individual, but as the soldier's faith in 'the judges vot[ing] with their conscience' suggests, it is also tied to broader human dignity, and thus to what I have termed sacrifice *with*. This reading of the story implies that perhaps all First World War deaths were equally unjust, but that soldiers could choose to confront this truth rather than to deny it. The story also implies that perhaps no one is entirely innocent, as we all even if indirectly abet killing, as interwar existentialism pointed out; yet, as existentialist philosophers such as Camus would claim, we can 'refuse to have our eyes bandaged' and confront the truth of violence around us, but also within us. Perhaps most frightening of all, we learn here that not all violence, not all suffering is redeemable. As Girard's analysis helps us to see, what makes this interpretation of the story different from the first one is that the soldier is aware of being a scapegoat, that is, aware of the ubiquity of violence and falseness of situating blame all in one place; and most important, against this closed, circular structure, he courageously lays claim to a dignity that is more open-ended, made in the present and without any certain external justification. As regards treason, finally, this interpretation of the story makes the anti-establishment claim that it may be at times justified, courageous, and moral, and that no external authority can save us from our own responsibility.

Two more stories about deserters will help to show how these different notions of sacrifice – *to* versus *with*, passive versus active, closed versus open, affirming the scapegoat structure versus confronting it – emerge in the context of the First World War and its aftermath. The first is historical, and is the life of Nazario Sauro, narrated in multiple editions of *L'illustrazione Italiana* (2 October to 20 November 1921), concurrently with preparations for, and the celebration of the Unknown Soldier, as well as with comments on d'Annunzio's *Notturno* and return from Fiume. Sauro was from Capodistria, and chose to fight for Italy; when captured by the Germans, he claimed to be Venetian, but was nonetheless executed as a traitor since he was an Austrian citizen; he was depicted as a great martyr to the Italian *patria*, and ceremonies to honour him took place at the same time as Unknown Soldier ceremonies. We can see how only clearly defined borders, the closed system of the nation, can make Sauro incontrovertibly a sacrificial victim (for the Italians) or a traitor (for the Austrians); I believe Sauro's status as a martyr was immensely popular in Italy because to assert it was also to assert the absoluteness of borders that were in reality contested and had been for centuries. A closed notion of sacrifice, then, where nothing is irredeemable, goes hand in hand with a vision of the nation as absolute, if not literally in terms of borders, at least ideally in terms of it being unchanging in some essential fashion. This recalls debates occasioned by d'Annunzio's takeover of Fiume, when the Italian government termed members of his paramilitary force 'deserters' as they were fighting for a city outside Italian borders against their own government's orders; this was, in turn, condemned in *L'Ardito*, a pro-Fiume and pro-d'Annunzio journal, who compared these paramilitary combatants to Russian revolutionaries, and by implication saw dissent over Fiume as civil war. Once more we see that debates about where the borders should be conceal the more difficult problem posed by the reality of their constant instability, which calls into question traditional notions of sacrifice. In Italy especially, the absolution of the Unknown Soldier was also symbolically an absolution from being an unstable, contested, 'borderless' nation.[11]

My second example is Giono's novella, 'Le Déserteur,' which tells the life story of a man who leaves France on foot in 1850, who is also 'a Hugolian character, who predates *Les Misérables*.' The man is both an artist and wanderer who at first 'deserts one form of society to go live in another' seeking 'a sort of *Patrie*' which is tied to the pastoral values most often present in Giono's work. He is depicted as a 'stranger in everything' who is at once 'wretched and aristocratic' according to an 'aristocracy of the heart' revealed in his saintlike white hands. We are told at the beginning that he is forced to leave, but left with the questions: 'If he has even committed a crime, what is it? Is it petty? political? professional?' Only at the end do we read, 'No law can pardon Charles-Frédéric Brun: his crime is the crime of misery, he is a *misérable*'; he dies still a 'deserter,' but leaves behind a last work, a crucifix that works miracles. In contrast with Sauro's life, Giono's story gives sacrifice an open structure and shows how this is linked to questioning the very notion of borders as that which defines 'Patrie.' In the end, Charles-Frédéric is a stranger to external social categories, which he deserts in favour of a poverty that has Christological aspects, but not doctrinal ones, as he represents all those who are marginalized and misunderstood, those who are *misérables*. Crucial to Giono's story is that we never know what event or 'crime' made his character into a 'deserter,' that is, what occasion forced him to leave. All we see is him taking up this role and transforming 'desertion' into a life that has spiritual dignity, that is, into an active rejection of external borders and definitions. In terms of our discussion of sacrifice, the implication is that to the degree that there is an initial event (a loss or a crime, a sacrifice or a betrayal), its value is determined by the life that is lived later and is neither absolute nor able to be contained in ideology or dogma. Another way of putting this is that courage and sacrifice, according to this new definition, are possible even if innocence is uncertain, as difficult to read as Charles-Frédéric's white hands, which he grows into, but which at first seem to conceal a crime. This is a sort of sacrifice that, like the Unknown Soldier's initiation I will discuss in the last part of this chapter, may not be 'redeemed' or 'recognized' within a system, and yet generates its own spiritual authority.[12]

Absolution

As a prerequisite to sacrifice, the absolution of the Unknown Soldier was almost absent in the British case, more visible in the French one, and central to the Italian one. From a Christian, and especially a Roman Catholic perspective, of course, absolution was necessary for a 'good death' and without it the soldier could not be a true martyr. Was an official absolution necessary, or even possible, though, given the chaos of the battlefield? And what would an absolution after death, post facto, mean? In practice, many soldiers died suddenly, or alone and without any accompaniment, let alone spiritual guidance. Thus, exhumations, and reburials with a ceremony, not only had the purpose of giving bodies a more permanent and individualized resting place, but also served to accompany the dead. Absolution was part of freeing them from the travails of life. At the same time, for many of the dead and especially for the Unknown Soldier, absolution also served to re-establish the authorities who conferred it: the Church, but also the state. Finally, given that the Unknown Soldier was, as we have seen, caught between being an abstraction, a type, and a unique embodied individual, the question arose as to which sort of identity was being absolved. Was a specific body being absolved? Or was it the soldier as a type for all soldiers? Who would 'all' include, in this case, given the issues of courage and desertion?

In Britain, the Unknown Soldier's casket was carried across London from Luytens' cenotaph at Whitehall, and then buried at Westminster, according to the traditional, spare, burial ceremony, on 11 November 1920. The only detail that speaks to the rhetoric of absolution was the claim made that the Unknown Soldier's coffin lay 'on sand that has apparently been untouched since that portion of the Abbey was built. The diggers of the grave found there no remnants of other humanity.' This 'uncorrupted soil' metonymically indicated the purity of the Unknown Soldier, but no ceremonial verification of it was included. This reflected the Protestant view that official or external sanction of spiritual truths was less important than the individual's conscience. At the same time, burial at Westminster, and the uncontroversial participation of the Royal family, corresponds to relations between church and state that were far less difficult than they were in France or Italy. In the French case, the casket was transported from Verdun, traversed Paris, and paused at the Pantheon, finally to arrive at the Arc de Triomphe on the Place de l'Étoile. Here, on 11 November 1920, it was blessed by the Bishop of Paris, who performed a double blessing, first addressing the Unknown Soldier himself, and then 'all those who died in the war.' For Jagielski, this double blessing indicates a successful assimilation of the Unknown Soldier to all the dead; however, I would argue that it draws attention to the difference between the single individual and the individual as a type, and suggests that the bishop may have felt some discomfort at giving such a generalized blessing. In Italy, the Unknown Soldier was given not only a burial ceremony but also the entire mass for the dead, which took place after the casket came off the train from Aquileia and paused at Santa Maria degli Angeli for two days before moving on to the Vittoriano. As the Italian Unknown Soldier was buried on 4 November (anniversary of victory rather than armistice), the mass took place on 2 November 1921, which was also the day of the dead (All Souls Day). The casket was welcomed by 'all the parish priests of Rome' lined up on the steps of Michelangelo's famous church. The ceremony ended with absolution, marked by Monsignor Bartolomasi's aspersion of the casket with holy water. This last element makes explicit what was only implicit in England and France, that the Church was willing to consider fighting in the war to be enough to earn salvation. This is confirmed by a previous absolution, which took place at Aquileia, as part of the choosing ceremony, when all eleven of the candidates to be *the* Unknown Soldier were sprinkled with holy water from the Timavo River, which had been the site of major battles. I have not been able to verify whether Timavo water was used again in Rome, but the assimilation of water from the battlefield and holy water indicates the equation between suffering in battle and salvation.[13]

Even though Church participation in the burial of the Unknown Soldier seems to have been little contested, as historians have noted, its meaning and the value or necessity of its absolution was far from clear. In other words, almost everyone wanted the Church to be present, but for very different reasons. Thus, for Miniero, in all three countries, this was the result of the Church's emphasis on 'expressions of piety that were so dear to popular sensibilities,' that is, its ability to capitalize on a certain nostalgia for tradition, as a remedy to the shock of the war. For Jagielski, in the French case, this was an indication that 'the Unknown seems to have rapidly entered the universal and appeasing dimension that his "inventors" so desired for him.' That is, the Church was present as a sort of generic indicator of salvation, which did not disturb the French state's commitment to *laïcité* or secularism.[14] However, as Luc Capdevila and Danièle Voldman show in *Nos morts: Les sociétés occidentales face aux tués de la guerre*, the Great War was seen by the Catholic Church, but also by other religious institutions, as a chance to participate in European societies that had increasingly excluded them from politics. Absolute separation of church and state was consolidated in 1905 in France with the law on secularism; in Italy, after the wars of unification resulted in taking over most of the Papal States in 1870, the Church prohibited Italian Catholics from voting, and signalled its separation from the new Italy by the Pope's never leaving the Vatican. *Pascendi dominici gregis*, Pope Pius X's

1907 encyclical condemning religious Modernism, cemented the Vatican's uncompromising stand against what it saw as the modern alliance of political and cultural liberalism. On the other side of the debate, the liberal-era *Illustrazione Italiana* approved of eliminating speeches and music as much as possible at the Unknown Soldier's inauguration, for this would also indicate 'no reconciliation of Church and State.' In both countries, priests nonetheless participated actively in the war, on the front lines, following far older traditions; and Catholicism remained the dominant religion, even as Protestant ministers and Jewish rabbis were increasingly present. This dovetailed with, but was not identical to, popular sentiment which, confronted with deaths shocking in their violence and numbers, sought comfort in forms of mourning that were religious, if not dogmatically correct. Capdevila and Voldman give the example of the Rabbi Abraham Bloch, known for bringing a cross and comfort to dying Christians, but also discuss how the '*veillée du corps absent*' (wake for the absent body) – the very scene of mourning we saw in Giono, in Chapter 1 – was adapted from Catholic rites for those who died at sea. In general, the Catholic confession and extreme unction, previously considered essential to gaining salvation, were replaced with the absolution that was already a traditional part of the mass for the dead (originally as an *addition* to confession and extreme unction), in order to 'soften' the horror of death at war. Also, recent work on the material culture of the First World War shows that in France, Italy, and other nations, soldiers went to war with a variety of religious objects – scapulars, medals, Bibles, and reproductions of religious art – whose origin and meaning lay far more in popular culture, and in its pagan roots, than in codified Catholicism. Keeping Girard's analysis in mind, it is clear that the extreme outbreak of the violence of the war made people turn to religion and gave the Catholic Church, in particular, an opportunity to assert its power, but that, at the same time, it was the occasion to question dogmatic notions of salvation and to seek a 'new or renewed cultural order.'[15]

Hence, though it would be easy today to consider absolution in generic terms, it is important to grasp how much anxiety underlay it in the culture of the time. Antonio Fogazzaro's 1905 novel *Il Santo* provides an excellent example of the fear of a 'bad death' that existed already before the First World War, and that wartime exacerbated. This novel tells the story of Piero Maironi who, after much spiritual travail, becomes the monk Benedetto. A charismatic speaker full of spirit, Benedetto becomes involved in religious Modernist circles, who wish to reform Catholicism, returning to its sources in early Christianity and modernizing its dogma.

These were the same circles Fogazzaro himself was a part of, and his novel is considered the main literary expression of religious Modernism; it was very successful and was immediately translated into many languages, even as it was condemned by the Catholic Church and put on the Index in 1906 – a decision its author accepted, suspending all further publication. Fogazzaro was influenced by the thought of Alfred Loisy and George Tyrrell, was friends with Giovanni Genocchi and Giovanni Semeria, and was associated with Ernesto Buonaiuti. These were some of the major thinkers of religious Modernism, and though they could hardly be said to agree on the reform they wanted, they defended two related ideas. First was the importance of the historical study of religion, and of Christianity in particular which, Loisy famously argued, showed the variability of dogma. Second, Tyrrell, but also Maurice Blondel – following nineteenth-century spiritualism and inspired by William James – insisted on the primacy of religious experience, and of the actions of the soul, over their external, institutional, codifications.[16]

These ideas were central to Fogazzaro's novel, which was condemned by the Church for one scene in particular, which it saw as encapsulating them. In it, one of Benedetto's followers, a devout young woman who works with the poor, anxiously asks for reassurance about her father. He was a good man who helped others all his life, she says, yet he died recently without confession and absolution for he could not reconcile himself with the Church. Is there any way he can be saved? Is there anything at all to be done? Benedetto tells her to pray for him, and when she reacts with disappointment responds, 'Does one pray for the forgiveness of someone who cannot be forgiven?' For 1905 readers, the implications were clear. First, even a non-believer, who died without the Church's blessing, could be saved: this might be a question of last-minute repentance, known only to God, but it might also be, more clearly following Modernist ideas, that God's decisions were greater than, and at times mysterious to, the Church. Second, prayer is more effective than dogma, and a daughter's love and forgiveness is more powerful than following the rules about repentance; also, it is implied, the daughter's prayers are no less valuable than those of a priest or other cleric. Overall, this scene shows that Church authority over who was eternally saved or eternally damned weighed heavy, even as it seemed unjust and difficult to understand. That this scene was singled out for condemnation reveals how much the Church felt under siege in its capacity to control salvation. It is thus unsurprising that when it came to death at war, both these elements resurfaced.[17]

Behind the convenient agreement of 'secular silence' and 'Catholic prayer' most historians emphasize, then, we find some voices of dissent. In particular, in Italy, Giovanni Semeria, who had been the chaplain of the Supreme Command during the war but before that had been associated with Catholic Modernism and Fogazzaro, reacted to the Unknown Soldier Memorial with some very clear distinctions. On the one hand, Semeria approved of the Unknown Soldier as a representation of Italian unity in dignity, duty, and sacrifice – values Christianity should uphold, in his view. On the other hand, Semeria disapproved of the ceremonies' bellicist side. This was an overt complaint about the conflation of the political and the religious, but it was also an implicit critique of the Church's encouragement of such a conflation, as seen, for example, in the influential work by Father Agostino Gemelli, *Il nostro soldato*. Like most Modernists, Semeria felt the Church needed to be reformed from within. Following a thinker such as Loisy, he also felt Church involvement in politics was not true to early Christianity and should be limited. He was a popular priest, already before the war, to the point that Pope Pius X, who promulgated the encyclical condemning Modernism in 1907, and created the anti-Modernist oath in 1910, allowed Semeria to take the oath while leaving out the parts he disagreed with. Semeria's capacity to move soldiers is attested to by d'Annunzio in his diaries, where he comments somewhat backhandedly on the power of his 'lack of eloquence' or minimalism.[18]

In the *Corriere d'Italia* of 2 November 1921, coinciding with the absolution ceremony, Semeria thus wondered whether the Unknown Soldier ceremonies would be 'a glorification of warriors or an apotheosis of war.' Insisting that war is 'a scourge' and 'ugly and sad,' he maintained that 'we will glorify the soldier. [...] through [...] the anonymous [one] we honour dignity not the person. [...] We will glorify him because he is a man who did his duty [...] a man who sacrificed. Go and kill is something you can tell an executioner; go and die can only be said to the soldier.' Here we find Semeria condemning bellicist nationalism, as he also derided 'Nietzschean supermen [who] preached the cult of violence. [...] overly zealous patriots [who] found a necessary evil beautiful.' This last may have been a critique of the Fascists who, as Labita notes, kept their distance from the 'liturgical union' around the Unknown Soldier because they were not at the centre of it; it may also have been a reference to Filippo Tommaso Marinetti's claim that war was a 'hygiene' for Italy's ills; finally, though Semeria expressed admiration for d'Annunzio's courage when they met in 1916, it may have been a complaint about his continued bellicism after the war, as Miniero suggests. In any case it is clear that for Semeria unity around the Unknown Soldier was desirable, provided it was linked to pacifism. Hence, he still defended a notion of heroism as obedience, sacrifice, and self-effacement, which sounded very much like the 'discipline' invoked by Gemelli as the necessary depersonalization of the soldier, but linked it now to peacetime values.[19]

This was overwhelmingly the state's wish as well. Thus, Rodinò (quoted above) saw the Unknown Soldier Memorial as 'the beginning of true peace.' But, as Winter rightly points out, pacifism after the First World War had a very different meaning than it might have now, as for most people it did not exclude the possibility of a just war, certainly in the past and perhaps even in the future. Insistence on the Great War as 'the last war' or '*la der des der*' (the last of the last), as Capdevila and Voldman argue, was a way to keep awareness of dissent at bay. In Italy, government insistence on peace often comes across as an attempt to avoid discussing the purpose of the war, so as not to reveal deep rifts over it. State archives reveal much worry over such dissent in the months prior to the inauguration of the Memorial. Interestingly, worry was focused just as much on possible agitation by the *ex-legionari fiumani* (legionnaires from Fiume), and the arditi, both Fascist and anti-Fascist, as it was on dissent by the socialists, the communists, or the anarchists. The Ministry of the Interior was, for instance, warned of a possible terrorist attack by anarchists, in a letter that decried the instrumentalization of mourning, and asked the government to rise above all factions. At the same time, the ministry received information that a group of legionnaires from Fiume was planning to have d'Annunzio lead them in a march towards Rome that would be anti-monarchical and would, of course, defend Italy's claims to Dalmatian territories; the file continues with relieved reports that d'Annunzio refused to participate and nothing came of it. In fact, overall positive reports came in from all over Italy, after the ceremony, showing that disruption of the Unknown Soldier Memorial's inauguration was almost non-existent, and thus far more rare than the government had feared. One purpose of absolution, which church and state could agree on, and which was at least temporarily successful, was to overcome political divisions, all equally seen as 'sins' of partisan passion.[20]

Disagreement surfaced, however, in the latter portion of Semeria's reflections. Here he noted that 'courage, duty and sacrifice are all great Christian words, great Christian realities.' This brought him to distinguish between 'glorification' and 'apotheosis,' which 'has a pagan odour'; the latter, Semeria argued, led people 'precisely and only for love of the patria' to 'return to the sacred ritual,' whereas 'We are led to the patria by God, and not to God by the patria, or worse yet, to God in the patria. Our dead are heroes to us, not Gods or something similar.' Reacting to many newspaper headlines that referred to the Unknown Soldier's apotheosis, Semeria here sought to separate religious from political sacrifice. In doing so he returned to a very old debate, as I will explain shortly; at the same time, his critique is one of the first lucid notations of the birth of a new religion of the nation, about to find its expression in Fascism, and quite different from past examples, as Emilio Gentile has argued. Specifically, Semeria attacked those who thought they might 'go up to [...] the altar of the patria to revive the Rome of Caesar within the Rome of Christ.' The Unknown Soldier, in other words, could be a great symbol of unity, and could be mourned with true religious fervour, but his sacrifice was not to be confused with that of Christ, nor could death at war be per se sufficient absolution to attain salvation. Most important, then, Semeria celebrated a 'dignity' that applied not to gods but to embodied individuals – to 'warriors' not 'war' – in terms that echoed Douhet's initial proposal, which asserted, 'the apotheosis of the soldier does not in any way mean the apotheosis of war.' This was very different not only from state insistence on the blankness of the Unknown Soldier, but also from a more conservative or authoritarian Church view.[21]

These latter two come together in a postcard conserved in the Ministry of the Interior's archives in Italy. It depicts the Unknown Soldier as an armed ardito standing up in his casket with an angel right above him, and it was accompanied by a poem, attributed to d'Annunzio, referring to the soldier as a saint. This image was similar to commemorative medals reproduced in *L'illustrazione Italiana*, which showed the Unknown Soldier getting up from a battlefield burial and rising straight to heaven. In all of these, we see that any passage through confession, absolution, penance, or purgatory – all spiritual travail, in sum – has been eliminated. Another purpose of absolution ceremonies, then, which Semeria decried but regarding which the Church might not disagree with the state, was to overcome the divisions that were internal to the individual himself. That is, the individual would be saved, or not, by a power that lay beyond him. While this appeared to take away much of the uncertainty of death at war, and in particular to 'forgive' all those who were too frightened to rush out of the trenches into no man's land, it also separated salvation from individual agency, and disembodied it. Thus, this approach acknowledged that people felt that the war was such horror that nothing further could possibly be needed to guarantee salvation but at the same time depersonalized the experience of that horror. Keeping in mind Fogazzaro's *Il Santo*, we also see that it relieved survivors of worry over the salvation of their loved ones, but at the same time deprived them of participating in it, as Fogazzaro's Benedetto implies is so important.[22]

Ernst Kantorowicz opens his famous essay on medieval political thought, 'Pro patria mori,' by citing two other First World War priests, one Belgian and one French, whose debate helps us understand what is at stake in Semeria's remarks. For Cardinal Mercier, death at war was the ultimate sacrifice, such that 'of a sinner instantly it makes a saint.' In contrast, Cardinal Billot responded, 'To say [...] that the mere fact of dying consciously for the just cause of the Fatherland "suffices to assure salvation" means to substitute the Fatherland for God [...], to forget what is God, what is sin, what is divine forgiveness.' Semeria's distinction between heroes and demi-gods echoed Billot's refusal to see death at war as sufficient for Christian salvation. Mercier's vocabulary of sainthood was also used extensively in descriptions of the Unknown Soldier, who was often linked to Christian martyrs as well, and seen as an *alter Christus* (discussed in more detail in the next section). One of Kantorowicz's important points, relevant to the Unknown Soldier also, is that the disagreement between Mercier and Billot, or also between the Unknown Soldier's apotheosis and Semeria's insistence on glorification, has to do with shifting understandings of the patria that go back to the Middle Ages.[23]

Specifically, Kantorowicz notes that in the ancient world the dead heroes of war were often deified, at times in response to fear regarding their return; however, he shows this was not a death for the *patria* as a territory but of for a *polis* and its ideals. In the Middle Ages, he argues, the true patria is seen as the one beyond this life, and sacrifice at war is linked, rather, to personal loyalty to one's 'Lord.' Also, such loyalty is not sufficient for religious salvation, though it can be its prelude. A turning point comes in the thirteenth and fourteenth centuries when, for the first time, patria is equated with territory and taxes are levied for it. Crucially, even though the Council of Clermont in 1095 had stated that those who died on the Crusades received complete remission from doing any penitence, but did not receive absolution from sin, over time they would be considered de facto saved. As the Middle Ages gave way to humanism, Kantorowicz writes, *caritas* as love of one's brothers would make the link between the love of God and the love of a territory that needed to be defended, ultimately justifying the association of soldiers and martyrs. This analysis confirms that, as intuited by Semeria, one important purpose of the Unknown Soldier's absolution was to cement this association.[24]

We find traces of the shift analysed by Kantorowicz in two early proposals for the Unknown Soldier Memorial in Italy, Douhet's, in August 1920, and an anonymous article in the *Corriere della Sera* in October 1920. In the latter, as Miniero writes, we find a 'mystical tone' that prefigures the 'religion of the patria,' as we read:

> The great war has stirred up names and faces: true and less true heroes emerge from our memories. But there is the Hero without a name and without a face, who took the road destiny willed and immersed himself in the shadows, [...] he laid his arms upon the earth as though it were his cross [...] on [mount] Grappa he marked with his body the sacred border between despair and hope, [...] he bent his knees on the mud of reconsecrated Friuli as he fell, dying, bowing to the victory that was advancing towards Trieste and Trento. Was his courage willing love? Or was it only discipline [...]? It matters not. [...] Names, faces? Nothing, nothing so small or so precise. [...] for this most obscure [man] [...] every particular is abolished in the solar incandescence of glory – the tomb in Rome, on the Altar of the Patria.

We can recognize here the association of blankness (true and less true heroes are equated in suffering) with a sacrifice explicitly linked to the borders of the nation, and equated with Christian salvation (individuality becomes irrelevant to making the tomb an altar, that is, to transforming the Unknown Soldier Memorial into a vision of the Passion). But the image of territory itself becoming the cross is especially notable, as it enacts a shift by which the *corpus mysticum*, traditionally the body of Christ, and the Eucharist, become the body of the nation. What is redeemed here, or indeed resurrected, is the patria itself, rather than the individual soldier, as the two grammatical readings of '*Altare della Patria*' (Altar *of* and Altar *to* the Patria) show. We can see this more clearly as we contrast this article from the *Corriere* with Douhet's vision, according to whom the Italian soldier 'thrown naked and unarmed against the reinforced concrete and the hard steel' won the war, against 'politicians,' and thus deserves an honour far greater and quite opposed to that of his 'leaders.' Douhet's image dissociates the body of the soldier from the body politic and emphasizes its vulnerability. This, too, evokes Christ, but not the one of resurrection (not a 'solar incandescence of glory'). Instead, the juxtaposition of a naked body to cement and barbed wire evokes the flagellation, the agony, and the deposition. Moreover, the enemy here is not another nation but war itself, and the technology that rendered the First World War so inhumane. To the degree that redemption is possible,

Douhet insists, it has not yet happened; moreover, were there to be such a salvation, the Memorial is not equated with it but is seen rather as calling for it.[25]

It would be wrong to see Douhet's vision as anti-mystical, however. For one thing, his admiration for d'Annunzio and his takeover of Fiume is all over the pages of *Il Dovere*, which even at one point claims that '*il santo volto*' of Italy's new martyr, the Unknown Soldier, is present there far more than in Rome (presumably a reference to d'Annunzio's commemorative rites at Fiume, and to his image of Randaccio's face as the face of the Unknown Soldier). Moreover, several cartoons in the newspaper depict soldiers (often mutilated ones) in Christ-like poses, suggesting that they are still being crucified (by 'dollars,' in one, that is, by postwar greed). Others suggest a mystical identification with that suffering (hence, the dead, it was proposed in the same issue, should be kept on electoral lists along with the living, instead of being relegated to 'cold monuments'). Douhet's emphasis on the face (in contrast with the *Corriere*'s 'Hero [...] without a face'), and identification with its particular suffering (in contrast with 'nothing so small or so precise') brings to mind Kantorowicz's discussion of 'mysterious materiality' in 'Pro patria mori.'[26]

Specifically, according to Kantorowicz, the meaning of the term *corpus mysticum* shifts drastically from the body of Christ to indicate the Church as political organism starting around 1150. Yet it is not until the seventeenth century that the metaphor of Church as political person becomes associated with the abstract, juridical person, and with the state as territory. At this point, for Kantorowicz, 'the mysterious materiality which the term *corpus mysticum* had still preserved was here abandoned and exchanged for the heuristic abstraction of the "mystical person," which was synonymous with the lawyers' "fictitious person."' Thus, even as sacrifice to the nation was being increasingly equated with salvation, as we saw above, that nation was being reconfigured as a mystical body. Crucially, while this made sacrifice at war as heroic as Christ's salvation of humankind, Kantorowicz also shows that it 'strangely devaluates the physical individual who alone, according to Genesis, was created in the likeness of God.' In other words, the mystical body is no longer the concrete and physically present *agent* of salvation (as was Christ, as would be the Unknown Soldier in Douhet's vision); rather, it is the political entity (at bottom, an abstraction or a 'fiction' in the sense that borders, too, are a fictional partition of the natural world) *to whom* bodies are sacrificed (*corpo* or 'body' is sacrificed to uphold 'confine' or 'border' in the *Corriere* article).[27]

Kantorowicz concludes that this evolution leads to a dilemma. Either we accept the devaluation of the physical individual in order to preserve a mystical heroism that now belongs to the state, and this leads to Fascism; or we run the risk of making death at war meaningless, or worse yet, something no different than a petty murder. As we have seen, debates about the Unknown Soldier Memorial express this dilemma clearly, as they refuse not only traditional heroism but also the celebration of a blank impersonal duty, yet have difficulty in defining the sort of courage or individuality that they can celebrate. What I would propose now, though, is that by making the dilemma unavoidable, the Memorial also seeks to recover something of that mysterious materiality that modernity had lost, which the First World War paradoxically or scandalously brought out in its absolute disdain for bodies. That is, Douhet may emphasize the horror of technology whereas the *Corriere* may stress the heroism of absolute borders, but it is in bringing them together that we can finally perceive the materiality of the soldier's body as the key element in their confrontation. This confrontation may be termed a contradiction or a scandal like the incarnation and the Passion that the Eucharist renews: in it the eternal and the transcendent enters the material – perhaps irrevocably as the more extreme Modernist theology would have it – but is never reducible to it. This is, I believe, what Kantorowicz means by his felicitous characterization of the materiality of the *corpus mysticum* as 'mysterious.' Far from being an abstraction, a 'fictitious person' and thus an institution, neither is it 'purely' material in the sense usually associated with the modern Western scientific world view. Rather, *corpus mysticum* in this sense is more than the self-enclosed individual body yet resists reifying that 'more' into an authoritative, institutional definition.[28]

Initiation

The Unknown Soldier's suffering was often associated with the Passion, be it in the popular press, in official documents, or in fiction and poetry. As we are about to see, this appeared to be self-evident but was in fact not so, just as Church absolution appeared uncontroversial but hid some real conflicts. What kind of resurrection awaited him? If it was not a canonical Catholic one, as many felt, what 'other state' was the soldier being 'initiated' into? What was the social and political role of this initiation? D'Annunzio's 1922 'Comento meditato' gives a good sense of the complexity of the new Christology that arose around the Unknown

Soldier. In one remarkable passage this becomes an overt critique of the transformation of the Unknown Soldier into an official monument:

> I want to help the dead express themselves. Even if I must carry them within me and one by one regenerate and inspire them, I want them to learn to express themselves. I want the unknown soldier to express himself and force the marble of his sepulchre, breaking open his monument and disappointing the bearers of useless balms. [...]
>
> For how many centuries had Italy been unable to express itself as a nation, to be heard only in single voices and signs?
>
> Had it expressed itself nationally in the Renaissance? Had it expressed itself nationally in the Risorgimento?
>
> Today I enclose in museums all of its masterpieces in order to vivify and recelebrate its one incomparable masterpiece of expression: its war.
>
> Its war is the effort for expression, unprecedented for the multitudes, be they armed or disarmed.
>
> This is why I put Michelangelo to work on the Carso's rocks, this is why I put Dante to work on the Carso's *bolge* [...]
>
> Desperately, hopefully, in the shadow of the Piave's banks, there was one who cried: 'What can we throw to the great flame? We will throw everything if need be, even the most sacred tablets.'[29]

Aesthetics, politics, and religion are here interwoven in a series of typically d'Annunzian polemical reversals: it is not the living who mourn but the dead who must express themselves; the First World War is not a military victory but an artistic effort; national unity is re-established not as a collective recovery from the war, but as the religious recelebration of the war. We recognize here the existential alienation discussed in earlier chapters, which the dead who return must teach to the living. Imagining that he carries the dead within him, d'Annunzio evokes again the flow that unites them to the living in choral unanimity. This chorality is presented as healing the dividedness of the individual (who is 'regenerated') but also of Italy (who failed to be united in the Renaissance or the Risorgimento). More important, we also find in this passage a nationalism that, as I am about to explain, acquires a mystical dimension. As Kantorowicz's analysis helps us to grasp, at stake is the relationship between the individual body and the body of the nation. Thus, the most important aspect of d'Annunzio's imagery here envisions the ideal Unknown Soldier Memorial as an abandoned, broken one. This is far more than a rejection of past memorials. D'Annunzio's makes a provocative claim about First World War mourning by associating the broken marble of the Unknown's tomb with the tablets of the Mosaic Law being hurled into the flames of First World War battle. The implication is not only that the Unknown Soldier, like Christ, signals the end of the old law, but that his death itself signals the limit of all laws, Catholic or otherwise: in Foucauldian terms, his death demystifies coercive ideological uses of the law. Further, and most remarkable of all, the Monument is broken by a concrete body. The Unknown Soldier is an *alter Christus,* for d'Annunzio, whose resurrection has nothing of the immateriality suggested by the Gospels, where the bearers of balms find Jesus' body gone, and the stone mysteriously rolled away, by an angel in one version. This is confirmed by the image proposed by d'Annunzio a few pages before the passage cited above. In it, he makes a parallel between Christ breaking out of his sepulchre and the peasant-soldier who is the prototype of the Unknown Soldier, depicting the latter as so poor that he must literally plough stones, breaking them into pieces that will become fertile soil by mixing with his blood. All these images show that it is the Unknown Soldier's body, with its flesh and blood, that shatters and does not merely roll away the stone of his sepulchre. By implication, his resurrection is a return to full embodied humanity, a rejection of transcendence in its abstraction.[30]

At the same time, however, this passage from d'Annunzio is an apt warning for how the rejection of transcendent purpose, sacrifice as initiation that breaks rather than makes a monument, can itself be monumentalized. First, as mentioned earlier, the 'Comento meditato' was written and published in the fall of 1922, a few months before the March on Rome, and is based on words uttered by d'Annunzio and transcribed by his doctors when he was in a coma after falling from a window in his home on 13 August 1922. Though we will probably never know whether he was pushed for purely accidental reasons (by a young woman refusing his advances) or for political ones (as that young woman's sister, d'Annunzio's lover, Luisa Baccara, was accused of being a Fascist informant by d'Annunzio's children), it is known that d'Annunzio was to meet with Mussolini and Nitti two days later, on 15 August, and that he saw his role as that of a 'pacifier from above,' a '*superiore pacificatore.*' As Renzo De Felice shows in his analysis of the correspondence and relations between Mussolini and d'Annunzio, this was the small window of time, after Fiume, when many veterans and especially the arditi (a portion, but not all, of whom were joining Mussolini's *Fasci*) still saw d'Annunzio as a potential leader, and as veteran and author of the *Notturno,* one who transcended party politics. His reaction was strangely indecisive, as De

Felice and others have pointed out, for on the one hand, he refused to take an active political role and yet in his writings and speeches continued to voice a dissatisfaction with current institutions that the arditi and many others shared. Perhaps the anarchic unlivable situation at Fiume, however gloriously symbolic, led d'Annunzio to be sceptical about his ability to govern, as opposed to leading a rebellion. Undoubtedly, he also underestimated Mussolini. But in any case, d'Annunzio's fall providentially put him out of commission at a moment when a decision might have been made, when Mussolini's control of the Fasci might have been questioned. The 'Comento meditato,' published a few months later, sounds in this context like a somewhat vain attempt by d'Annunzio to reiterate his vision of the war, and of commemoration as open initiation, against the growing power of Fascist rituals and monuments.[31]

Second, d'Annunzio's take on commemoration became a model for Fascist ritual, which superseded it by transforming its very emphasis on battlefield alienation, brokenness, and marginality, into something monumental, state-run, and permanent. Looking back at the quotation above, it is difficult not to read its nationalism as a preamble for the Fascist colonial enterprise. D'Annunzio's assertion that the First World War was 'an effort for expression' by 'the multitudes' 'armed or disarmed' reflects, especially in 1922, his experience of the war as undoing social divisions and promoting existential awakening. It is not necessarily bellicist as regards the future, but rather echoes the widespread need to feel that so many deaths were meaningful. But, as Antonio Bruers writes, as the postwar era wore on, and social and cultural renewal failed to materialize, d'Annunzio's re-evocations of the war became increasingly nostalgic and lost their concrete and provocative power. Fascism would appeal to this nostalgia, which was not d'Annunzio's alone, promoting both its inner colonization of the Italian south and its colonial wars in Africa as the just 'expression' or outlet demanded by Italian vitality. More important, d'Annunzio's image of breaking the Monument and its association with stones made fertile by blood and suffering prefigures Fascism's ideology of permanent revolution (as does the image of turning away from museums, which is a softening of the Futurist call to destroy them, also a prelude to Fascist revolution).[32]

Specifically, in its first phase, as presented in the philosophy of Giovanni Gentile's Actual Idealism, Fascism saw itself as synthesizing transcendence and immanence, or also ideal institutions and historically determined changeable ones, in the 'pure act' whereby self-consciousness coincides with cosmic consciousness. In less abstract terms, 'pure act' meant taking the contingent action of rebelling against past ideologies, and making it absolute: the point was now not to contest a specific political program, but to make permanent contestation *the program itself.* Going back to d'Annunzio's image, under Fascism, monuments would have to be permanently in the state of being broken – a gesture of rebellion frozen into abstraction. To be sure, both in terms of monuments and political action, this was extremely impractical, and Fascism in its second phase emphasized reconstruction. At the same time, however, the Unknown Soldier Memorial was central to Fascist rituals, starting on 4 November 1922, when Mussolini took great pains to identify his body with that in the tomb. Part II of this book focuses on this as a spectacle of embodiment (among others that took place); what is important for the moment is that the Unknown Soldier Memorial is at once a model for the rejection of transcendent sacrifice, *and* for how this, in turn, may become a modern form of transcendence. Going back to Kantorowicz's terms, the Memorial calls for a mysterious materiality that would loosen the dilemma of totalitarianism versus meaninglessness, and at the same time, it seems also capable of reifying that dilemma, transforming the lack of meaning of wartime deaths into something monumental.[33]

To understand the appeal of making meaninglessness monumental, or having broken monuments become the norm, it is important to remember how much veterans, and survivors in general, felt compelled to repeat an experience of wartime death that they could never fully understand or let go of. Though I will discuss this as a form of traumatic repetition in Chapter 5, my point now is that many felt as d'Annunzio did when he wrote, 'To be a "survivor" is for me, now, the worst calamity that could befall my body and my spirit' since 'death who was supposed to take two took one, against the pact, against the offering.' This sentiment was echoed in Douhet's newspaper, *Il Dovere,* where a headline insisted that soldiers both dead and alive were still 'awaiting their return,' which meant a place in society and appropriate honours, but also a recognition of their new, lacerated, identity, of a physical symbiosis between the living and the dead that could not be undone. This survivor's syndrome, as we now call it, was expressed most poignantly not just in the need of the living to repeat the death of their comrades, but in the anxious feeling that no such repetition was sufficient, that there could be no closure. We see this at the end of Gance's *J'Accuse* when the dead are still continuing to return, like an endless river. But even more telling from this perspective are Gance's multiple remakes of the film, which has 1919, 1922, and 1937 versions. We see it also in how d'Annunzio's *Notturno,* already an extremely popular re-enactment of the deaths of comrades by the author's blindness and convalescence, is repeated in the 'Comento meditato,' and then again in other texts. Further, d'Annunzio described Fiume as a 'state of immolation' meant to renew and undo the calamity of survival, and thus depicted the unsuccessful end of the Fiuman enterprise as a call for further attempts to repeat wartime alienation.[34] Leed documents the same feeling in the letters and diaries of British soldiers and, in particular, insists on how each 'death of a friend' would intensify 'the bonds to those who remained,' creating a crescendo of repeated loss whereby 'one's own demise could be welcomed as a resolution to an insupportable, continuous, mourning.' What we find here is a cycle: the impulse to renew sacrifice, to relive death, and finally to produce an ideal meaning and consummate mourning ends up exposing anew the lack of meaning, the breakdown of identity, the failure of chorality.[35]

What was at stake in this cycle is expressed, once again, by d'Annunzio, in a surprising image that connects sacrifice and resurrection to the issues of purification and traitors that we saw earlier in this chapter:

> Men of sorrows, workers, comrades, your faces are pale; and my face too is pale, perhaps even more than yours. Visit the sick man, comfort the convalescent. Is all blood gathered around fraternal hearts?
>
> So be it. [...]
>
> No doubt behind us we had all our dead, [...] but in front of us we had the unborn, more numerous than those who had been killed. [...]
>
> In our wretched bodies, in our humble souls, eternal forces dwell and work. [...]
>
> I have suffered the hardest test. I have been thrown from the Tarpeian rock, [...] I have felt my [skull] break. [...]
>
> I have felt the disaster of Caporetto pass upon me; I felt it upon all of me, as though it could trample and spatter and bury me alone. [...]
>
> It is true, men of sorrows, workers, sailors, peasants, what the visionary Fathers proclaim is true: there is only one mortal sin, only one; [...]
>
> And you ask what it is? Giving in to despair, despairing of oneself, as Gregory the Great would say.
>
> Of this sin I must accuse myself, in my unbent humility. [...]
>
> But I survived that will to die [...]
>
> Bleeding at the foot of the Tarpeian rock, I did not despair. I do not despair.[36]

Addressing his audience but also describing himself as a 'man of sorrows' (*uomo di pena*), d'Annunzio uses the same term as Ungaretti (see Chapter 2), to describe being permanently changed by the war. This expression is most often associated with images depicting Christ with the wounds of the Passion prominently displayed and thus emphasizes again his humanity and physicality. Extended also to the dead, the living, and the unborn, it brings them together in a choral suffering whose concreteness is stressed by expressions such as 'blood,' the 'wretched bodies,' 'trample,' and 'spatter.' But what is most remarkable is how d'Annunzio reconfigures his fall from the window in August 1922 as a fall from the Tarpeian rock, which is then presented as a reliving of the defeat at Caporetto, of death itself, of all defeats and all despair. In the 'Comento meditato,' d'Annunzio's fall was associated with the Unknown Soldier and, in turn, their similarly broken bones were linked to Italy's 'mutilated victory.' Here, the association is more allusive, as falling from the Tarpeian rock, which in ancient Rome was located on the southern tip of the Capitoline Hill, would land one quite close to the Vittoriano's present location. But first, in both texts, it is clear that the Unknown Soldier's sacrifice leads not to an ideal elevation, but to a fall. Once again, this recalls Eliade's distinction between an initiation that is a descent into the primordial in contrast with an ascetic and metaphysical ascent. Second, and far more important, the Tarpeian rock was used in ancient Rome for the execution of 'the most notorious traitors, and as a place of extra-legal, unofficial executions.' It 'was a fate worse than death, because it carried with it a stigma of shame.' This brings back the association of the Unknown Soldier with traitors, adding the important notion that even in death, he remains a dangerous outsider, and that his very 'execution' is beyond the law and society. Moreover, d'Annunzio suggests that his fall was a suicide attempt when he accuses himself of the sin of despair. This echoes Leed's assertion that re-enacting wartime deaths does not lead to reintegration into the community but to an intensification of alienation such that death seems welcome.[37]

Most surprising, however, is the association of commemoration and execution. Giorgio Agamben's discussion of the '*homo sacer*' helps us to understand it, and also brings in executions from the Tarpeian rock as a prime example. For Agamben, 'defenestration from the Tarpeian rock' is more of a 'purification rite' than a 'death penalty in the modern sense,' and this is because it brings into play a '*vita sacra*' (sacred life) or also '*nuda vita*' (bare life) that, initially, pertains neither to the religious nor to the political sphere, and is therefore dangerously wild and in need of containment. He shows that those executed there were considered '*uc-*

cidibile' (killable), that is. they could be killed without legal prosecution; at the same time, he argues, they were *insacrificabile* (unsacrificeable) in that their deaths did not or did not yet have meaning, that is, they were not already part of a sacrificial rhetoric. This makes them sound more and more like First World War soldiers, suspended between categories, subject to killing according to an '*offensive à outrance*' or 'attack to excess' that saw individual lives as expendable (killable), and at the same time unable to be repatriated or mourned properly for their meaning was still in the making (unsacrificeable). In fact, Agamben himself argues that being exposed to death without dying – and he gives the specific example of soldiers who were 'supposed' to die in battle, or survivor's syndrome – liberates an excess of 'sacred life,' an excess of embodied estrangement. This excess is 'living proof of his [the survivor's, bare life's] subjection to the power of death,' and is for Agamben 'the original political element.' In other words, bare life is precisely what exists before or beneath any broader, more abstract, ideological meaning, be it political or religious, that it might be given. This recalls d'Annunzio's image of the Unknown Soldier as a body who shatters any monument that might contain it.[38]

For Agamben, absolute power is established not by endowing bare life with a given meaning, but instead by banishing it, that is, by asserting its utter meaninglessness. This may seem paradoxical, but Agamben draws on the notion of a 'state of exception' to explain it. During a state of exception, according to the theory of Carl Schmitt, and others, the sovereign can transcend the rule of law in the name of the public good. This might seem like ending the rule of law, but it is not. Following a dynamic of inclusion via exclusion, rather, the state of exception is a rule of law that is absolute in its power to abolish all other rules. Similarly, for Agamben, by declaring bare life to be utterly meaningless (killable and unsacrificeable), power makes it into an exception. Following the dynamic of inclusion via exclusion, this makes power absolute, for it can declare what is and is not an exception to life itself. As he explains:

> *Sacratio* takes the form of a double exception, both from the *ius humanum* and from the *ius divinum*, both from the sphere of the profane and from that of the religious. The topological structure drawn by this double exception is that of a double exclusion and a double capture, which presents more than a mere analogy with the structure of sovereign exception. [...] Just as the law, in the sovereign exception, applies to the exceptional case in no longer applying and in withdrawing from it, so *homo sacer* belongs to

> God in the form of unsacrificeability and is included in the community in the form of being able to be killed. *Life that cannot be sacrificed and yet may be killed is sacred life.*[39]

To the degree that the Unknown Soldier does not merely break open his monument once, but becomes part of a permanently broken Memorial, he too becomes the exception transformed into an absolute. For a permanently broken Memorial does not just say that the meaning of wartime deaths is in the making, painfully incomplete and uncertain, caught in the contradictions and divisions of lived life; rather, it asserts that the meaning of wartime deaths can never be found and, according to the dynamic of inclusion via exclusion identified by Agamben, it therefore covertly affirms the state's absolute power to declare the meaning or meaninglessness of death. Commemoration and execution come together here because in both cases the act or the life being commemorated or condemned is deprived of intrinsic or internal meaning in order to assert the state's power to create meaning post facto. This recalls the structure of scapegoating discussed in the first part of this chapter, insofar as meaning is external, but adds an element that is very important to the Unknown Soldier Memorial: the scapegoat, in this instance, is not just an individual who is marginal, it is embodied individuality itself, it is bare life or embodiment insofar as it threatens the stability of social cohesion.

Let us return then to the cycle by which the impulse of survivors to relive wartime deaths ends up exposing their meaninglessness anew, and ask whether this inevitably plays into the hands of absolute power, or whether, conversely, something gets in the way of the dynamic of inclusion via exclusion. In other words, is there a difference between d'Annunzio's soldier who breaks his memorial, as a rejection of ideological meaning, and the later Fascist assertion that wartime deaths permanently await their redemption? Is it merely a question of time before rebellion becomes institutional, or before the war enters the museum, or is there a qualitative difference here? Agamben's discussion helps us to understand that central to this question is the role of embodiment: can absolute power succeed in banishing it, that is, in declaring it utterly and permanently beneath meaning, or does embodiment still manage to produce its own, more inward, meanings? To answer these questions and conclude, at least as regards the sort of sacrifice enacted by the Unknown Soldier Memorial, I will consider briefly how inclusion and exclusion, or also presence and absence, relate to the Unknown Soldier's body. This, in turn, will be connected to the play of absence and presence in Gance and in d'Annunzio, where we will find that the body of the soldier is increasingly connected to the unfinished and the emergent as positive values that undercut ideological appropriation.

The dynamic of inclusion via exclusion identified by Agamben depends on a particular play of absence and presence. When bare life is banished or declared utterly meaningless it is made absent from society. At the same time, since power asserts itself via this banishment, bare life is constituted as a negated concept. Crucially, if this concept replaces what is actually missing, we fail to see power asserting itself, and the process of abstraction that this entails. We may think here of how the bodies of those considered expendable by society (slaves, soldiers, minorities) become absent or invisible *as bodies*, yet are made present as concepts via negation (in permutations of *they are not like us*: not free, not individual, not integrated, etc.). Eventually the concept of 'slave,' 'soldier,' or 'minority' seems to replace actual bodies that are no longer seen in their individual concreteness. This entire process depends on what is absent becoming abstract so that it can be made present conceptually and we lose track of the reality that something is missing. Ralph Ellison's *Invisible Man* counters this sort of abstraction by insisting on the specificity of one individual's story as an element of political change. Making what is absent concrete again – and not merely discussing it abstractly – is what disturbs the power structure that had excluded it.

Absence and presence are also important to mourning, and this will help us to understand yet another aspect of inclusion via exclusion. As Robert Harrison writes, 'funeral rites serve to effect a ritual separation between the living and the dead [... and] first and foremost, they serve to separate the *image* of the deceased *from* the corpse to which it remains bound up at the moment of demise [...] the dead must be detached from their remains so that their images may find place in the afterlife of the imagination.' The image, then, is an abstraction which allows us to make the dead present in a way that is consoling because it helps us to forget the horror and sorrow provoked by the corpse, which we banish through burial or incineration. Though this is a necessary resolution to mourning, Ariès argues that Western modernity's increasing hurry to hide or otherwise make invisible and abstract corpses amounts to a failure to mourn and, eventually, a denial of the reality of death. Harrison makes a similar point when he discusses the 'imperative to dispose of the corpse so as to *liberate* the person from its tenacious embrace,' for this disposing is a ritual acknowledgment of and physical closeness to the dead body, which needs its time and place, and allows mourning to become a dialogue with rather than a denial of death. As Harrison points out, 'when the dead body goes missing' this work of mourning 'is all the more arduous' and, as was overwhelmingly the case in the First World War, this now '*undead*' body becomes an expression of 'the open-ended, unreconciled psychic state of the grievers.' As he points out, such an unreconciled state can be a failure to mourn, yet it can also, perhaps at the very same time, be a claim that the dead 'remain [...] politically active,' and thus a rejection of 'forgiving or premature forgetting.' To pass to the image or to abstraction too quickly, in sum, makes it impossible to grasp the existential and the political aspects of death; in contrast, a return to the physicality of what is absent, to dwelling with it, indicates a limit or critique of abstract understanding.[40]

How do absence and presence, concreteness and abstraction shape different visions of the Unknown Soldier's sacrifice? First, as we saw in Chapter 2, from exhumation to ceremonial inscription, an immense effort was made to associate one specific body with existential anonymity. However, an equally remarkable effort was made to keep the actual body invisible. In fact, even as legal documents insist on the importance of making the Unknown Soldier Memorial 'visible to all' in order to facilitate mass mourning, what was visible was only a coffin. No doubt, a decomposing body would have been too horrific (as we see, or rather fear to see, in Tavernier's *La vie et rien d'autre*, where people are shown fainting at exhumations). But it was also too concrete, too much of a reminder of individual physicality. In this respect, it is quite noticeable that even as the authorities and the press presented the Unknown Soldier as potentially the body of each mourner's loved one, and moments of such imaginary identification did occur, we find that *at the same time* his body was a reminder of all the bodies that were still unclaimed, and that, as one newspaper put it, were even greater in their suffering for not becoming exemplary. Thus, we read, 'the women of Aquileia seemed taken by even greater pity in front of the ten [soldiers] who were excluded from Roman apotheosis.' In England, the juxtaposition of the Unknown Soldier Memorial with Luytens' cenotaph makes this coexistence of presence and absence clear, as mourners saw the two monuments to be intimately connected. More generally, it is remarkable that the burial of the Unknown Soldier did not, in fact, stop exhumations, as might have been feared by veterans who wished to respect each and every one of their comrades (as is dramatized in Tavernier's film). Similarly, the commission investigating Caporetto did not stop working once 'all the dead' where honoured along with the Unknown Soldier, even as veterans like Douhet feared it might. What we see here is that for some, and especially for the authorities, the Unknown Soldier's body was present as an abstraction, and as such replaced the bodies of all the missing; for others, in contrast, it was a concrete evocation of the physical remains that could not be retrieved, and thus it pointed to innumerable absent bodies.[41]

Second, in Italy, absence and presence were also important to the relation between the Unknown Soldier's body and authority figures. As the full-page headline in *L'illustrazione Italiana* read, when the Unknown Soldier descended to his final resting place, *Gli Assenti* or 'The Absent' were there along with king and military and ecclesiastical authorities. These were, as the article explained at some length, for very different reasons, d'Annunzio and Cadorna, who did not attend the inaugural ceremony on 4 November 1921. As explained earlier, Cadorna was the general widely held responsible for Italy's defeat at Caporetto and, more traumatic, for the unjust execution of Italian soldiers as deserters. He was absent, then, from a practical standpoint, to keep unrest to a minimum; as the Ministry of the Interior archives show, the inauguration aimed to overcome such divisions. Yet Cadorna's absence was eloquent, and not unnoticed, as an admission of error and perhaps cruelty on the part of the authorities, as the *Illustrazione* implies. As for d'Annunzio, his absence was seen as even more eloquent than Cadorna's, in particular because the facing page for this article in the *Illustrazione* was an announcement for the publication of the *Notturno*, billed as the ultimate true description of the war experience. His being kept away, or keeping his distance, from the inauguration of the Unknown Soldier Memorial was thus seen as suspicious, a possible critique if not of the Unknown Soldier at least of government presentation of him, and in any case a sign that Italian institutions did not understand veterans. Ultimately, d'Annunzio's absence made manifest Italy's lack of a credible authority figure to heal the country after the war. Once again, the body of the Unknown Soldier, with whom the *Illustrazione* associated d'Annunzio, pointed to the concrete reality of what was missing.[42]

Between 1919 and 1922, Gance added the famous scene in *J'Accuse* where dead soldiers march *above* the Arc de Triomphe. They starkly contrast with the military display below, not only, as mentioned, because they are more of a river and less of a formation, but also because their bodies are emphasized in their difference from one another, and in their woundedness and mutilation. These soldiers may be resurrecting, but the image is of something unfinished that, it is implied, will never be finished. For Gance, it is clear, the Unknown Soldier is not one among many, one just like the others, as the military display below the Arc implies. On the contrary, as the march of 'men of sorrows' above indicates, the Unknown Soldier is a specific embodied man, and his presence reminds us that the 'countless dead' for all that they flow together, were each a specific individual. No generalized meaning for their deaths can be given, beyond the very struggle for meaning that their return makes palpable.

For d'Annunzio, the physicality of the Unknown Soldier, and bringing back concreteness to his memorial, was associated with the art of Michelangelo. When he wished the Unknown Soldier would break open his monument, he also saw this as putting 'Michelangelo to work on the Carso's rocks' (see full quote above). This was a recurrent theme, as when d'Annunzio evoked his fall from the Tarpeian rock, he also compared himself, and all the 'men of sorrows' he addressed, to 'heroes as yet imprisoned in silence like those of Michelangelo in the hewn stone.'[43] As mentioned earlier, the Unknown Soldier was given a mass for the dead and was on display for two days at Santa Maria degli Angeli in Rome. This church was built by Michelangelo who adapted the extensive ruins of the Baths of Diocletian. The massive concreteness of Michelangelo's art was seen in various newspapers as appropriate to the Unknown Soldier's suffering, to the point that one article suggested that he might have preferred this church to the Vittoriano as a final resting place.[44] But it is d'Annunzio who makes explicit the connection between Michelangelo's *non finito*, the concrete reality of the battlefield, and the new Christology of the Unknown Soldier Memorial. Thus, in the *Notturno* there is a passage that prefigures the image of the Unknown Soldier breaking open his monument, in which the suffering of soldiers at war is compared to '[the] sublime drawing [by] Michelangelo [...] the titanic Christ who in forcing the heavy covering of his tomb still has a foot in the concave stone. Yet with his erect head, with his raised arms, with all the impetus and the rage of his passion, he hurls himself at the sky.'[45] This Christ is caught in the present of his still unfinished resurrection, recalling d'Annunzio's proposal of '*resurgo*' as an inscription (discussed at the end of Chapter 2). More important, the image of the 'foot in the concave stone' conflates the idea of breaking open the tomb we saw above with the notion that the body itself emerges from the stone it is breaking, just as Michelangelo saw himself to be liberating his statues from the marble rather than imprisoning them. The physicality of the Unknown Soldier is thus shown not as the opposite of idealization, or abstraction, but rather as made up of the struggle between the material and the immaterial. It is this struggle, ultimately, that makes impossible the replacement of concrete by abstract or the dynamic of inclusion via exclusion that we have been discussing. Through this struggle, bare life in its uniqueness seeks a meaning for death that is greater than life but that would not overwhelm it.

This recalls Leed's comparison of the war experience to an incomplete initiation, for 'it remained forever problematical what state [...] the soldier was being initiated into,' but suggests that such openness, while painful, may also make possible a confrontation with death in its concreteness and mystery.[46] As Agamben sees it, especially after the First World War and the concentration camps of the Second World War, bare life has been entirely ejected from society. From an ontological perspective, he sums up, this also ensures that 'existence is nothing more than what essence excludes.' That is, existence is the negative mirror of essence, and this absolute negative is all that we are left with. He does, however, conclude that beyond this ontology there lies possible research on a 'life that, being its own form, remains inseparable from it.'[47] The unity of life and form is precisely what d'Annunzio's reference to the *non finito* expresses for it claims that if there is a resurrection for the Unknown Soldier, it lies in his struggle to recover a body that he has lost, a body that is unique, irreplaceable, yet also politically active (not an image) and expressive, emerging from stone and silence (not merely caught in it). Part II of this book thus focuses on how the body of the Unknown Soldier was imagined, made into a spectacle, and in a variety of ways brought back to life.

PART II

Embodiment and Spectacle

INTRODUCTION TO PART II: THE UNDEAD BODY, THE PHOTOGRAPHIC IMAGE, AND THE RELIGIOUS ICON

Was it even possible to be sure that 'the Unknown Soldier' was made up of body parts that came from the same person? And were they all there, was the body 'whole,' was there enough bone to make a whole body? First World War trench warfare and modern weaponry created this new problem that was even more haunting than the attempt symbolically to bury irrecoverable bodies. Significantly, this problem is never mentioned in discussions of the Unknown Soldier Memorial in the early part of the twentieth century, and even Cendrars' satirical questioning of his nationality, race, and gender does not bring it up. Yet it was the central if latent motivation for the most innovative element of the Unknown Soldier Memorial, namely, the attribution of symbolic power to *one* concrete body, to represent other lost bodies. This body responds to an anguish that goes even deeper than loss of identity and alienation, and has to do with breaking down the boundary between not only self and world but also between animate and inanimate matter. I ended my discussion of existential anonymity and competing notions of sacrifice by emphasizing the mysterious materiality, associated with the *non finito*, which undercuts ideological appropriations of the Unknown Soldier's suffering, even as it dangerously prolongs mourning. Such materiality, of course, reminds us that that there is an actual body in the Unknown Soldier Memorial, and more accurately, a corpse, rather than a resurrected body. In the next three chapters, on embodiment and its role in making mourning spectacular, I will seek to understand how the repressed materiality of the corpse re-emerges in the Memorial, especially in its inauguration and reception, where emphasis on alienation gives way to a more uncomfortable fascination with abjection. At stake is whether such materiality is only the source of anguish and endless mourning, or whether it can become the basis for a new understanding and acceptance of mortality.

Let me return to my opening question and draw out some of its main implications. Was it possible to ensure that the Unknown Soldier was a single, integral corpse? The first answer, on a practical level, is that this was very difficult, if not impossible. In Tavernier's 1989 film, *La vie et rien d'autre*, one of the main characters is Major Delaplane, the officer who runs the organization that identifies remains, as battlefields are slowly combed through after the First World War. When he is asked to produce some bodies that might be candidates to be the Unknown Soldier, he is doubly disgusted. On the one hand, his mission is to identify, and he intends to do so to the utmost of his ability, using even the most insignificant traces, such as the parts of a watch, scraps of letters, anything that survivors wait in long desperate lines to recognize. For him any body that is whole enough, and disinterred, will bear some trace that can potentially yield an identity, and to rebury it as the Unknown Soldier is to foreclose this possibility, depriving someone of their loved one's body. On the other hand, in a moment of great anger, the Major also admits that because the Unknown Soldier has to be unrecognizable, he might as well be made up of disparate body parts – and they have plenty of those. The comment is meant to be a grisly joke, but reflects the real difficulty of ensuring both anonymity and integrity. Moreover, Delaplane remarks to a young woman who has been pleading to be let into a mental hospital, to perhaps find her lost lover, that she will most likely find a man mutilated in mind and body, and that she would do better to abandon her search; we are then shown images of disfigured, mutilated soldiers. In Tavernier's film, survivors are kept away from both of these horrific truths – dismemberment and mutilation – which they are depicted as incapable of accepting, even as they doggedly seek traces of their loved ones.

As we saw in Chapter 2, the public was indeed kept away from exhumations, in part because they might also dig up live ammunition (and explosions were not uncommon), and in part because the state preferred to emphasize 'bare bones' rather than 'suffering flesh' and, as we see now, it wished also to gloss over the confusion and the mutilation of bodies. Exhumations in preparation for the Unknown Soldier Memorial were even more off-limits, in order to stress the randomness of the choice. Thus, even in Italy, where northern communities were able to follow the proceedings more closely than southern communities could, people were allowed to transport the individual corpses that were not chosen to be the Unknown Soldier from the battlefield to their cemeteries, but they did not disinter them themselves, and no mention is made of disparate pieces of remains, or of mutilated corpses. This was, of course, the traditional purpose of ossuaries, and they were used in the First World War. But what is remarkable from the perspective of the invention of the Unknown Soldier Memorial is the care that was taken to insist that his was a single body, and the emphasis on such singularity as the 'redemption' or 'honour' finally given to the soldier, in contrast with the 'humiliating' mass grave. Even when bodies had to be buried quickly as the front line moved, instructions were to bury them side by side rather than on top of each other, so that they could later be disinterred and kept separate. Of course, historians have shown that such instructions, created post facto after the war began, reflect the very impossibility of actually following them. Nonetheless, archaeological evidence shows that people did try to keep the bodies separate from each other as much as possible. What is revealed here, as in Tavernier's film, is not just the state's, but even more survivors' psychological block against seeing disparate body parts on the battlefield. As will be demonstrated in Chapter 4, 'Embodiment and *Imbestiamento*,' the symbolic journey of the Unknown Soldier's corpse from the battlefield back into the city was meant to ensure that the 'bestial' death he suffered, with its horrific dismemberment, which was also a dehumanization, would be brought back to a human dimension and wholeness. In Chapter 5, 'Mutilation and Spectacle,' I will focus on how the inaugural ceremonies sought to control what we might call the return of the repressed, that is, displaced but undeniable expressions of the body's breakdown. Finally, in Chapter 6, 'Mourning, Transcendence, and Re-enchanting the Flesh,' I will turn to four recent works that explicitly thematize the Unknown Soldier Memorial in order to demand that modernity develop more genuinely embodied and holistic forms of mourning.[1]

Hence, a second answer to the question of whether the Unknown Soldier was a single, integral corpse is yes, but such a corpse is a creation of the imagination as much as it is a concrete reality. When confronted with a corpse that we identify with, especially a horribly mutilated one, the imagination helps to process what is too shocking to be immediately assimilated. In *The Body in Pain*, Elaine Scarry contends that 'the objectlessness, the complete absence of referential content, almost prevents it [embodied pain – and she also gives the example of a mourning mother] from being rendered in language [...] But it is also its objectlessness that may give rise to imagining.' Imagining, according to Scarry, is a 'ground of last resort' but also a 'visionary solution' that makes up for unacceptable or incomprehensible objects in the natural world; as such, it is ambivalent, at once 'creation' and 'wounding,' at once alleviation of pain via its projection onto symbolic objects, yet also betrayal of the body's silent suffering via the overwhelming power of such symbolic objects. We will see in Chapters 4 and 5 that a similar ambivalence is at play in the Unknown Soldier Memorial, and that in particular as a 'visionary solution' it seeks to make embodied mortality spectacular, and to give this experience ritual power, and yet it also recognizes the voyeuristic side of such a spectacle, in which trauma is repeated without being healed. The recent works discussed in Chapter 6 seek to counter such voyeurism and thus explore the ethical and political implications of the visionary, aesthetic dimension of mourning.[2]

As Robert Harrison writes in *Dominion of the Dead*, all corpses by their very nature are in part imaginary vision:

> Heidegger makes a serious blunder in *Being and Time* when he states that the corpse is a mere 'thing' which, in its presence-at-hand, gives Dasein no access to its own death as such [...]. A corpse is indeed present-at-hand, yet its lifeless mass is what is left of what is no longer there, what is on its way elsewhere. The corpse is the site of something that has disappeared, that has forsaken the sphere of presence, that has passed from the body into ... into what? [...] Only something as resistant as the insensate body of a loved one – the enigma of its expired life and remnant thinghood – could give the anthropomorphic mind its first access to what we call death, and with it to the ethos of finitude. In its perfect likeness of the person who has passed away, the corpse withholds a presence at the same time as it renders present an absence.

To the degree that it withholds presence the corpse as thing is a traumatic embodiment of loss, and as such wordless; but to the degree that it renders present an absence, still evoking, giving form to what was lost, the corpse is imaginary in Scarry's sense, it provokes a vision of a passing away that is also the beginning of an understanding of mortality, as Harrison writes. In contrast with the dynamic of inclusion via exclusion discussed in Chapter 3, the corpse gives us access to death for it 'renders present an absence' without replacing it or disembodying it. The corpse, provided we do not hide it too quickly, or clean it up, transforming it into a two-dimensional, pristine, immortal image, does not provide us with a concept of mortality – but with an experience of it.[3]

The Unknown Soldier accentuates this 'disquieting' dual aspect of the corpse, discussed by Scarry and Harrison, stretching it to the limit. For, on the one hand, his body is more 'thing' than most, and it embodies a loss that is not only of life but of human integrity, both in the physical and the moral sense. On the other hand, in reaction to this, his body provokes powerful imaginings, gives form to shadowy and haunting images of finitude. That is, the absence that is made present is restless, angry, and even violent, to the extent that it decries the abjection of the 'thing' that is present. One of the main images of how the Unknown Soldier tremendously increases the disquieting nature of corpses is that bare bones grow new flesh just as mutilated bodies acquire prostheses; somehow the Unknown Soldier's body is reanimated, undead even as he is dead, and able to contaminate the living not just with mortality but with the abject mortality of the trenches.

As we saw in Chapter 3, ritual absolution sought to banish divisions among but also within individuals, and was the prelude to a rhetoric of sacrifice that rendered human bodies abstract while attributing agency to the nation. At the same time, we saw at the very end of the chapter, a different notion of sacrifice emerged when mourners and writers emphasized their physical symbiosis with the dead, refusing to bring mourning to its traditional closure, that is, refusing to separate the dead 'thing' from the 'image' that lives on, in Harrison's words, in 'the afterlife of the imagination.' What I propose to do in next three chapters is to reconsider this refusal as an attempt to articulate a new relationship between the absence and the presence ambiguously generated by the corpse. In particular, Harrison notes that the corpse is a 'perfect likeness of the person,' and this reflects traditional forms of mourning, in which the image that lives on is based on this likeness, which 'detaches' itself from the dead body. It is this very notion of 'likeness' that the next three chapters seek to put pressure on, for in the case of First World War mourning, it was far from perfect, and people reacted to corpses with recognition, to be sure, but also with a sort of fascinated revulsion at how much likeness had been defaced, betrayed, or perverted. What sort of image, if any, could such horrific or false likeness generate? Would dead soldiers remain defaced or mutilated for all eternity in the imagination of survivors? We see here that the image of the Unknown Soldier as a single integral corpse is undercut by the far more disturbing sense that this integrity is artificial or monstrous, Frankenstein-like in its badly hidden disparate parts and poorly replaced missing limbs. A prime example, present everywhere but most starkly in Gance's *J'Accuse*, is of the soldier who resurrects and departs for the afterlife with all his wounds, such that he appears to move and live again even as the state of his body would make that impossible. The flesh is resurrected here, but not redeemed from suffering.

Another way of putting this is that some of the dead body's 'thingness' is retained in the 'image.' The absence of life and the presence of an afterlife are no longer separate and symmetrical, the two sides, as it were, of a single corpse. Instead, they intermingle metamorphically, even exchanging places, so that the 'image' of the 'afterlife' often seems more dead than the purportedly 'lifeless mass' of decaying flesh. Moreover, representation itself does not seem to function in its traditional fashion, as the flesh is also the 'letter' which cannot be detached now from the 'spirit' or abstract meaning. This recalls an assertion by Leed that I already quoted in discussing alienation: 'the symbols of invisibility, death, burial, and pollution are particularly apt descriptions of individuals who are for a moment passing between categories [...] in war, these are not symbols but experiences.'[4] We see here that physical integrity, which has to do with separating outside from inside, and inanimate from animate, also affects the distinction between 'world' and 'representation.' Obviously 'invisibility, death, burial, and pollution' have always been experiences and, at the same time, the (traditionally religious) symbols by which humans become conscious of these experiences as fundamental. What happens, not so much in war in general, but in the First World War as the documents used by Leed show, is thus a reversal of the normal process of coming into consciousness and representation. Things that were thought of as representations turn out not to be, and this questions the very stability of the separation of appearance from reality. When it comes to the Unknown Soldier Memorial, it is a commonplace among historians to claim that the body within became 'distributive,' a 'type,' or an 'abstraction,' that is, a representation of the many irrecoverable bodies. Yet what anxiety about this body's integrity reveals is not just concern over 'which image' of death it projects, but more profoundly, fear that this body might be 'not symbol but experience.' In other words, if representation is no longer stable because of wartime trauma, the Unknown Soldier's body might not *re*present but instead more dangerously make present again wartime violence, causing mourners to experience it rather than to move it into consciousness and abstraction. Thus, I submit, the Unknown Soldier's body is not so much distributive as it is contagious.[5]

To understand better what was at stake in this confusion of reality and representation, before I get to the details of how the Unknown Soldier's corpse was imagined, a few words are needed on the role of photography and cinema in the First World War. I already mentioned that Gance was able to make *J'Accuse* during wartime in part thanks to the French army's cinematographic services and that he gave all but major roles to actual soldiers. One of the truly disquieting results was that by the time the film came out, in 1919, a number of the men present on screen were dead and buried. The film's representation of their deaths had thus become a reality. While we may be used to seeing people now dead as actors on screen, what is uncanny in this case is that the screen prefigures not only their death, but its specific manner. This kind of blurring between reality and fiction comes close to putting the viewer in the position of a voyeur, the witness to a real horror that she or he is actively part of, even while pretending to maintain distance from what is only a fiction. Gance's pacifist film challenges us not to remain voyeurs because it foregrounds the materiality of the war, from the mud of the trenches to the blood of mutilations, resisting aestheticization through his innovative mixture of avant-garde and popular visual references.

In Italy, the case of documentary footage by Luca Comerio is rather more uncanny, since it was purportedly the numerous mothers recognizing their dying sons on screen that inaugurated censorship, starting in 1916. Comerio's intent was not pacifist, and his films dwelled with real fascination on the speed and violence of modern warfare. In this context, deaths were heroic and anonymous, close to the Italian state's emphasis on the 'sacrifice [...] however it is accomplished' of the Unknown Soldier. Their materiality, which was linked to the corpse's uncanny capacity to make us experience mortality, was replaced with an image that was all surface and formal beauty. A prime example was Comerio's depiction of foot soldiers ascending the Alps in the snow, seen from a distance as spiralling organic forms growing over an almost entirely white abstract backdrop. Yet the public did not see them this way, and once again cinema's ability to blur the lines between representation and reality was a large part of it: deaths on screen suddenly became 'real' via mothers' recognition, and this undid Comerio's impersonal narrative. This time, we see the viewer refusing to become a voyeur. In both Gance and Comerio what emerges, however, is that cinema and photography, and more generally, representation, is not detached from action but involved in it. Thus, to Paul Fussell's famous argument that pre-war literature provided a symbolic vocabulary that shaped combatants' actual perceptions of the war, we can add that in wartime literature and film, the process of making a new symbolic vocabulary was accelerated, if also chaotic, as symbols and experiences morphed into one another. But even more important, Gance and Comerio suggest that the photographic image, and even more the moving image, bears a special relationship to death and mourning, for it may be used to heighten or conversely to deny the uncanny, disquieting aspect of the corpse. The new symbols of mourning inaugurated by the First World War and epitomized by the Unknown Soldier Memorial cannot be separated from the fact that symbolism itself was being transformed by modern technology.[6]

To understand better why real deaths on film are so disturbing, and how this relates to the role of the body in mourning, I would like to reflect briefly on the 1980s and 1990s films of Yervant Gianikian and Angela Ricci Lucchi. Drawing in large part from the archive of Luca Comerio (which includes his footage of the First World War, but also that of others, as well as earlier Comerio footage and Comerio photographs), which they salvaged from destruction, Gianikian and Ricci Lucchi refilmed all the footage frame by frame. Then, to make their films, the footage is not only reassembled in a complex montage (at times with later images as well) but it is also slowed down, and the focus and framing of the image are shifted. Hence, for example, footage of mutilated soldiers in *Oh! Uomo* puts their mutilation at the centre of the image, or pans over it slowly, whereas in the original a surgeon showing off his work or a maker of prostheses would have been the focus. In *Dal polo all'equatore,* images of men being shot in the First World War are brought closer and centred, undoing the homogenizing distance of the original, and repeating the gaze of the mothers who recognized their sons on screen. Also, this film deliberately juxtaposes images of the First World War with films of colonized countries, including Indochina and East Africa (Gondar), using reframing to make us conscious of a similar voyeuristic gaze of the camera in both instances. Finally, and perhaps most revealing, Comerio's filming of the famous hunt by Baron Franchetti in Uganda in 1910 is also added, and reframing is used to emphasize that both the war and the hunt are similar massacres. This comments on the popular notion that soldiers 'died like animals' whereas they deserved to die with dignity by suggesting that true bestiality lay with their leaders who, like Franchetti and Comerio, revelled in massacre. Gianikian and Ricci Lucchi, thus, subvert the aesthetics and intent of the original documentaries that they use.

Their films are disturbing in two very different ways. First, by juxtaposing their own understanding of photography and film with Comerio's, they propose a thoughtful critique of his depictions of violence and especially of death. It is not merely two aesthetics that are contrasted here, but their ethical and existential implications. This emerges from changes in point of view, proportion, and depth of field. In Comerio's images, the camera is detached from the action, and as unobtrusive as possible, establishing a neutral, Cartesian space in which the mind can abstractly reflect upon chaos and its impersonal forces. (This recalls more abstract Futurist depictions of speed and warfare, such as Gino Severini's *Armored Train* of 1915.) In Gianikian and Ricci Lucchi, the images are separately reframed, so that the camera is now closer, inside the action, constantly shifting and present, asserting the interconnected subjectivities of the directors, the viewers, and the supposed objects of the gaze. Comerio's proportions are those of immensity and awe: the Alps themselves, vast construction projects involving networks of tunnels, explosions that transform the landscape, and symmetrical military formations dominate the human form. By zooming in so persistently that no larger point of view is available, Gianikian and Ricci Lucchi emphasize the vulnerability but also the dignity of the human body, which is assailed and defaced yet never deprived of its centrality. Finally, Comerio's wide depth of field conflates the human and the machine, the organic and the abstract into a single homogeneous movement. (Again, he is close to Futurist notions of simultaneity and mechanized man.) Gianikian and Ricci Lucchi, by reframing and zooming in, create a very shallow depth of field that emphasizes situatedness, granularity, and heterogeneity: we see processes and transformations rather than movement as a vector or abstract force, and we are brought away from vision and as close to touch as film can render. (This recalls the *non finito* with which I ended Part I.)

These three types of changes have the overall effect of stressing that in Comerio beauty has become about analysis, unveiling, and even dissection: inseparable from a gaze that reifies, such beauty sees consciousness as an objective observer, separate from the body and, in turn, can only understand ethics as a set of rational laws rather than as dynamic relationships. In such a context, the depiction of death on film overdetermines what Harrison describes as the traditional separation of corpse from image, which is a dynamic and dialogical journey: indeed, within the aesthetics of Comerio, film makes an image that is unchanging, permanent, and abstract, reifying and freezing the essence of the person, almost as though stealing his or her soul before the body, and mourners close to it, have bid it goodbye. But what do Gianikian and Ricci Lucchi propose instead? I submit that their films gesture towards what Hans Ulrich Gumbrecht calls 'athletic beauty': this is presence before analysis, and '*focused* intensity' rather than observation; it is not 'interested' in the sense that the desire to control and, perhaps, even the anxiety of death have been momentarily lifted. It is, in short, grace in its ecstatic finitude. It is also that moment in which life and death mysteriously come into contact, a breach between two worlds that can be frightening or 'devastat[ing,' as Gumbrecht notes, but also healing, and a core part of the mourning process. But I say *gesture* because I do not believe the films of Gianikian and Ricci Lucchi quite capture such grace.[7]

Why? Because their films remain ambivalently poised between undoing and repeating Comerio's reifying gaze, which replaces embodiment with surface. We see this most of all in their decision to elide their own procedure of manipulating Comerio's images: we know that they have been refilmed and reframed if we read about Gianikian and Ricci Lucchi, but we never see this taking place in the films themselves, and thus Gianikian and Ricci Lucchi's pacifism seems weirdly to emanate from Comerio's films, as though it had been there all along. Thus, for example, in the scenes drawn from Franchetti's hunt in Uganda how are we to read the hunter's identification with his prey, and is it fair to see it also in Comerio's treatment of First World War soldiers, or is this parallel only an illusion of later perspective? We know that, from a standpoint of conscious intention, pacifism was absent in the original films; yet Gianikian and Ricci Lucchi imply that it was symbolically present, if repressed. But how much are we allowed to read into history? Can their films be called 'documentaries,' as they claim? Is this new image drawn from Comerio's unconscious the result of a journey and an engagement with his work or is it an imposition all the more unfair in that its authors remain invisible? These questions become especially salient when it comes to images of mutilation and death. To zoom in on them is, to be sure, to remind us that real bodies are at stake and, hence, to insist that grace must not be forgotten. But to elide the zooming is, rather uncomfortably, to render historical responsibility for violence latent, so that grace remains haunted by something that cannot be seen. This is the second reason the films of Gianikian and Ricci Lucchi are so disturbing: though they claim to uncover a hidden truth in Comerio's images, they end up destabilizing the relationship of truth to image. This, in turn, leads us to wonder whether intensely embodied grace as well as an authentic image of death are compatible with modernity and technology that increasingly erases process and time.[8]

To draw out this last point, I would like to reconsider the editing done by Gianikian and Ricci Lucchi in light of Michel Tournier's short story, 'Les suaires de Véronique.' This story centres around an innovative photographer named Véronique, who seeks to produce uniquely truthful images by imitating the Veil of the Veronica. Similar to the Shroud of Turin, this veil is said to have been used by Veronica to wipe Christ's face during the ascent to Calvary, when his image remained impressed upon it. The veil itself became known as 'the Veronica' or *vera eikon* (true icon), the model for all religious icons in that its veracity is based on physical continuity rather than resemblance. Accordingly, Tournier's

Véronique develops a special acid in which to bathe her model, Hector, before rolling him in a canvas that will retain his image, 'just like a cadaver in a shroud.' Eventually she produces a whole series of such images, displayed hauntingly on all the walls of a chapel, where they are 'reminiscent of a series of torn off human skins.' Tournier's narrator, also a photographer, is initially fascinated but increasing foreboding leads him and us to conclude at the end that Hector has died from 'too many samples taken every day from [his] very substance.' What is the moral of the story? Véronique's photography, much like Comerio's films, seeks to unveil, to wrest an essence from within the living body, to seize a fixed and absolute truth. Tournier explicitly associates Véronique's world view with that of scientific rationalism, metonymically described as 'vivisection': such a view prefers the *écorché* – the flayed cadaver with muscles and skeletal structure revealed – to the nude; and the nude itself it prefers without a gaze, pure flesh without subjectivity; everywhere, this view seeks to separate surface from depth, body from mind, appearance from truth. But is Tournier merely condemning photography, and somewhat nostalgically decrying the loss of aura caused by mechanical reproduction? By extension, is the broader point that Gianikian and Ricci Lucchi cannot use Comerio's medium to critique him without repeating the violence of his gaze?[9]

I believe not. Beyond Tournier's own passion for photography, the reference to the Veil of the Veronica highlights a philosophical point that recurs throughout his work, which concerns how we use – or misuse – this medium. It can be encapsulated in two images from Tournier's *Vendredi ou les limbes du Pacifique* (1967), a novel which rewrites *Robinson Crusoe* to explore what would happen if, instead of imposing rational order on his island, Robinson were to completely transform his understanding of identity, mind, body, and world. Robinson's initial world view is not unlike Véronique's, and Tournier's image for such knowledge is that consciousness is like a candle that projects light onto the things it moves towards; as such, it seeks to probe them or unveil them, not understanding that if the objects have shadows this is due to the candle, rather than something intrinsic to the objects. Consciousness is then compared to a stranger wandering around a dark house and only able to know a few objects, and only one side of them, at a time; hence, such consciousness forms an ego 'that exists only intermittently' and needs to appropriate the other's gaze to supplement itself. Ultimately, this creates a 'knowledge of things *through the other*' in which something is always lacking. Alone, however, Robinson is forced to discover a different knowledge, which he compares to that of 'objects that are phosphorescent in and of themselves, without any external source of light.' And, he adds, in this case, 'the consciousness I have of an object is this very object, the object is known, felt, etc., without a person who knows, feels, etc.' In this case, then, consciousness is at home in a world that it knows without need for a separable light or gaze, without instruments, and without self-consciousness. Body and soul, surface and depth, the veil and the essence are organically bound in radiant grace; such consciousness is closer to Buddhist awareness, or also to William James' definition of mystical experience, infinitely receptive, ego-less, and ecstatic. This recalls Gumbrecht's comment on grace as related to 'how distant a body and its movements appear to be from consciousness, subjectivity'; also, as he notes, 'grace and violence often go together,' and we see here that what feels violent is the disappearance of the 'person who knows,' the ejection of the self-conscious ego from the world.[10]

If we return now to the question of what makes Véronique's art so deadly, we can see that the issue is not photography per se, but rather that her notion of photography is part and parcel with her feeling of estrangement from the world, and her insistence on a self-conscious, objectifying, distancing consciousness. To be sure, photography facilitates this sort of distance, for it appears to be a separable, portable gaze that we can observe and aim at the world. (We can also see here that it denies time in that it denies the situatedness of the gaze.) As Robinson writes, existing pre-photography but anticipating its vocabulary, as soon as the gaze is separable as 'candle' (or camera), 'a click [*un déclic*] takes place. The subject is torn away from the object [...] something has cracked in the world [...] [the world] dies under my sceptical gaze. [...] An object has been suddenly degraded into a subject.' What kills, then, is the separating out of subjectivity from the world in which it was embedded. Unlike 'Les suaires de Véronique,' *Vendredi* makes explicit that what might seem to be an abstruse philosophical debate has not only ethical but political consequences. Robinson's initial sense of consciousness, as estranged and existing only through the other, is associated with his attitude of conquest and imposition of order on his purportedly desert island. The parallels with colonialism are evident and were topical in 1967, when the novel was published. Some forty years later, as we confront potential ecological disaster, what is more striking is Robinson's slow process of learning not to wage war against the island itself, which comes from discovering that it is not empty space, but a living organism. All this violence, Tournier implies, comes not from photography or any new medium, but from Véronique's transformation of the distance created by any tool into a permanent ontological truth. In other words, the camera, but also vivisection, allow us to think we see ourselves; but as with a mirror, we must not forget that what we actually see is just a reflection. Tournier's statement, 'an object has suddenly been degraded into a subject' is almost shocking because it reverses our common valuation of subjects over objects. His point is that what we call the subject is merely a reflection in the mirror: whereas the real is made up of objects, radiant and present irrespective of our gaze.[11]

How does this relate to the Veil of the Veronica, to the films of Gianikian and Ricci Lucchi, and to the Unknown Soldier Memorial? In the Eastern Christian tradition, icons were meant to be representations of the divine that functioned far less through visual accuracy and far more through physical transmission. Moreover, traditionally, like relics, they were thought to have a physically divine origin; hence, they broke the opposition of the physical and spiritual worlds, as does the Incarnation itself. At the same time, icons were not meant to be contemplated as we contemplate art objects, but were part of elaborate religious ceremonies meant to awaken the senses and provide an embodied, experiential, essentially mystical apprehension of the divine – an experience of the sacred, or indeed of the sacredness of the flesh, and thus also of the world. In many ways, nothing could be further from this than the art of Tournier's Véronique: surrounding the viewer panoptically, her shrouds freeze the movement of life and indeed seem to suck it out of the viewer, who is oppressed by a 'feeling [...] of suffocation.' She aims at a truth that is objective, external, and unmovable, epitomized by her gesture of unveiling, wrenching away the surface to expose what lies 'beneath the body,' yet she is left with nothing but flayed skin. This points out by contrast a key aspect of the traditional Veronica's gesture, which will be very important to keep in mind when I discuss in Chapter 5 how the Unknown Soldier Memorial was compared to the Veil of the Veronica: far more important to her than the unveiling, and the subsequent creation of an image or object, is the moment of veiling, a gesture of compassion and identification with Christ's suffering that has no ulterior goal. Tournier thus teaches us that the meaning of the Veronica lies in the alternative between anxious, ever-lacking 'knowledge [...] *through the other*,' and compassionate radiant consciousness that does not separate the world into subject and object.[12]

When it comes to the work of Gianikian and Ricci Lucchi, and to photography and cinema in their relation to death and violence more generally, this implies cultivating an awareness that the medium provides a distance that is only heuristic: it provides an image of truth that, like the icon, allows us to embark on a journey, a search for meaning, that is situated, embodied, and ever-changing, and should never be mistaken for an abstract eternal truth. The historical materials, it is implied here, do contain a trauma and a grace that can be lived anew, and cannot be limited to the framework in which it was, or is, presented. But such grace, and the trauma of loss, do not have a single true expression that we can wrest from the veils of history. Indeed, neither Comerio's vitalistic celebration of violence, nor the melancholy gaze of Gianikian and Ricci Lucchi exhaust grace or loss, as we can envision many further reframings. The films of Gianikian and Ricci Lucchi are deeply disturbing, not only because they reach back into the past to reawaken its traumatic nature, but also because they reveal, albeit rather unwillingly, that there is no unmediated vision of this past, of the grace that was and is no longer, and thus no final interpretation that will put trauma to rest once and for all.[13]

Returning to the body of the Unknown Soldier, which as 'not whole' retains in its 'image' some of the horror and the physicality of the 'thing,' we can see better that it functions rather like Comerio's images, but also like those of Gianikian and Ricci Lucchi. Like them, it contains a trauma to which we do not have unmediated access; like them, it provides such trauma with an initial framing which nonetheless, as we are about to see in detail, is such that more and more reframings are needed. Ultimately, in this case, too, we are led to wonder whether trauma can ever be put to rest, whether grace is possible, and what this would entail. Can the new forms of mourning that emerge in the First World War and in the Unknown Soldier Memorial, especially in their concreteness, and link to the body, become, in the words of Nicholas Saunders, a 'memory bridge,' in contrast with abstract history? Does this mean that at the same time there is always a haunting presence of the past in our midst, ready to reactivate traumatic pain? In Chapters 4 and 5, we will explore the descent into embodied imagination of the Unknown Soldier Memorial, its engagement with 'thingness' and refusal to allow the 'image' of the corpse to become fully abstract. Specifically, we will see that the Unknown Soldier was associated with mutilation and trauma, with women's 'hysterical' bodies, and, finally, with the appropriation and purported redemption of all three by Fascist ritual. This forces upon us the violence of the past in such a way that we become responsible for it and, like the viewers of Gance or Comerio, must decide either to be voyeurs or to recognize ourselves in it.[14]

This raises two crucial questions to be explored in Chapter 6, through recent works that associate the Unknown Soldier with a modern reinvention of the Veronica. The first concerns the focus on suffering as the basis for sharing, the related fascination with abjection, and the marginalization of those who bear its symbolic burden (women and soldiers, especially traumatized ones, here unexpectedly converge). Like the films of Gianikian and Ricci Lucchi, and like Tournier's short story, these contemporary works about the Unknown Soldier are aware that no measure of connection to ancient or non-Western holism is a guarantee of escape from reification: they know they have to invent new forms for compassion. Hence, a second question they raise concerns the projection of suffering onto the body alone, which tends then to be degraded and separated from the mind, as the latter attempts to transcend the physical. At stake here is whether it is possible, in the modern West, to envisage shared, embodied mortality as something other than horrific and animalistic, to express the sacredness of life without separating mind and body, and to enact an honest existential confrontation with death without falling into nihilism.[15]

In 2003 mourning in Italy for the dead of Nasiriyah, Iraq, some journalists claimed 'we call them heroes but they are martyrs,' while others made a point to insist we must be careful to 'not exalt the cult of martyrs and heroes.' In other words, while we must be wary of its reification and political manipulation, the pervasive reality of suffering in the modern world cannot be denied, especially as war in its historical forms gives way to terrorism, ecological disaster, and unprecedented economic inequality on a global scale. Suffering therefore – be it understood as 'bare life' in Agamben's terms or as 'body in pain' in Scarry's – becomes an extraordinarily powerful political force, whose links to religion cannot be ignored. As Edward Tick puts it in *War and the Soul* (2005), quoting a popular adage, 'there are no atheists in the foxhole'; yet when veterans return home, 'the quality of sacredness is usually missing,' because Western modernity lacks rites of passage to reintegrate soldiers, to transform them into 'mature warrior[s],' who 'exercise restraint' and become models for converting violent impulses into spiritual growth. Tick adapts traditional rituals in his psychotherapeutic practice (such as the Native American sweat lodge or *inipi*) yet also notes: 'veterans' stories need to be told in ways that transfer the moral weight of the event from the individual to the community,' that is, veterans need to be heard and to feel society has learned from their stories. But, recognizing the importance of embodied understanding, Tick adds that this is difficult when 'our entertainment-driven consumer society markets empty icons of the war experience.'

In such a context, Amy Hollywood concludes: 'What is still unclear is what forms of ritual might allow us to negotiate the dangers of irrationalism and emotionalism lurking, for many, within any public "return" to religion. If we disavow or repress these dangers through appeals to a purely rational political discourse and practice, however, they will inevitably return in unexamined and, thus, more politically and ethically dangerous forms.' The pressing question thus concerns the relationship between aesthetics, religion, and politics that is established by specific representations of suffering. Gianni Vattimo's reaction to Nasiriyah mourning, in his article entitled 'Ritrovare le parole' (To recover the power of speech) argues that we must 'stop worshipping martyrs,' but also decries the political appropriation of 'Christ,' recently 'reduced' to hanging in every government classroom, and he implies that a different spirituality is needed. Vattimo criticizes the homily of Cardinal Ruini, who blessed the dead and added, 'We will confront the assassins. But without hate.' Ruini was also criticized by Monsignor Nogaro, who insisted 'we cannot fight terrorism with armed combat.' Their polemic echoed the 1921 complaints by Semeria, discussed in Chapter 3, as both Semeria and Nogaro went against the Church's apparent willingness, in 1921 as in 2003, to equate death at war with Christian values and salvation. We are here warned that existential confrontation with death is never purely personal, or isolated in a separate ontological or religious sphere; nor, however, can it be subsumed to the social and the political. Thus, Tick rightly wants veterans' spiritual struggle with soul to become public, yet sees their potential role in politics as that of a critical moral conscience. In contrast with 'empty icons,' then, what new symbolic forms, beholden neither to traditional religion nor to political ideologies, can we develop for mourning and healing? What rituals, what kinds of representation, what sorts of icons are more, or less, politically and ethically dangerous, even as they are existentially powerful?[16]

4 Embodiment and *Imbestiamento*

Trauma and Animality

Now, the distant battle of the Mosa enters my fire. [D'Annunzio refers, initially at least, to the fire inside his wounded eye as he convalesces in the dark.] The battalions, drunk on ether, advance like those parts of a burning pine forest called 'counterfire' in my Marsh of exile, and they are like herds of animal flames [...] I see them through the posts and the barbed wire. [...] The slaughter turns into pyres, but they don't burn, they don't turn to ash. They smoulder for a long time without flames, like turf.

I lie all night against the barbed wire that blocks off the hill. I count the corpses.

They are all caught up in the iron thorns, compressed by the twisting and broken metal wires, they hang between posts like thieves poorly nailed to their crosses, they twist like animals caught in snares.

They have no eyelids, they have no lips. I see frozen naked eyes. I see frozen naked teeth.

I see the blood dripping down the wood and the iron, thickening, blackening, viscous like bird-trapping glue.

There is no more dew, there is no more dawn for the world.[1]

We enter the dark coffin. [...]

Suddenly, I am afraid. Someone is with me. [...]

As I write in the darkness, my thoughts break and my hand stops. [...]

I shiver in fear. And I stop moving, my whole body rigid, and I dare not trace a single sign in the darkness.

[...] Then only my bones exist, only my skeleton bandaged with flesh. And in my skeleton there is a sudden coagulation of life.

> Life precipitates, thickens like blood which is no longer fluid. It is a horrible weight. [...]
>
> Inner life, in its every movement, is abolished. I have no strength left to move and stir up the great incoherent masses of lyrical substance that form my melancholy. [...]
>
> My suffering is vile and powerless. [...] Before I sink into the horror of transformations, I feel my mouth turn to metal in the slower breathing of sleep.[2]

These two passages from d'Annunzio's *Notturno* provide a vivid image of how in wartime embodiment is experienced as a breakdown of normal boundaries. Thus, in the first the soldiers' bodies become both a herd of animals and a barrage of Dantesque flames (earlier, d'Annunzio described himself and his aviator companion as '"two inside a flame," but the flame is undivided'). Moreover, *carnaio* (slaughter) is a butcher's display, continuing the theme of animality that culminates in the image of cadavers caught in barbed wire like prey in traps, hanging lifeless from the steel. Then, without eyelids or lips, but with a glassy stare and a terrible grin, they are again not human, yet also the human defaced, the monster. Finally, their blood becomes viscous and saplike, still flowing as if they were alive but heavy with death. The liminal status of these bodies is further accentuated by two related paradoxes. First, they burn without being consumed, which implies that, as in Dante's *Inferno,* they are material yet also more than material. Second, they are like badly crucified thieves, but there is no Christ here, nor, as the last image of endless darkness and thirst indicates, is there any possible redemption. The second passage is linked to the first by the image of thickening blood, now d'Annunzio's own, as he identifies his convalescence in the dark with the death and burial of his comrades, both actually witnessed and imagined by him. His body, too, loses its boundaries as his dead companion seems to penetrate his darkness and contaminate it, making his body grow new flesh and precipitating his blood, forcing it into a liminality similar to that of the soldiers'. Finally, writing and thought are broken by this disquieting embodiment, just as the Unknown Soldier's body, in d'Annunzio's imagination, would break his memorial. All that we are left with is a melancholy that is an incoherent mass, a suffering that is shapeless, metamorphic, and 'vile.'[3]

Two important themes emerge here, which the coming chapter will explore in order to understand why the materiality of the Unknown Soldier's body was both so fascinating yet so repressed. The first has to do with the link between dehumanization and reanimation. 'They died like animals' could not be replaced with 'we remember them as men' because the 'image' or 'vision' associated with these cadavers was one of monstrosity coming back to life rather than of humanity transcending its physical shell. In other words, wartime suffering was such that even in death it remained beyond the norm, beyond words, beyond the human as culture defined it. The second comes from the association of such monstrosity with contamination, expressed in images of viscosity, dissolution and coagulation, falling apart or pouring out. As cadavers lost their humanity, in other words, they also lost their integrity, falling into pieces that the living could no longer keep separate from themselves. The inability to distinguish dead body parts from living ones in the trenches, therefore, continued in the mourning process, after the actual violence of war was over, so that living bodies now seemed monstrous as well.

In both cases, and as I already briefly mentioned, the body of the Unknown Soldier as representative of soldiers' bodies is not just disturbing or mysterious but associated with trauma. We see this in a basic way in that it is associated with an experience of radical loss and horror beyond words, and with the subsequent repetition or return of that experience. The concept itself, however, was fairly new during the First World War, and in particular, the link between 'traumatic lesions' that were only physical, and the emotional scars of soldiers, such as Mangin's amnesia (discussed in Part I), was still being explored. Thus, the work of Freud and Breuer in the 1890s opened the way for understanding hysteria not as a female and physically originated illness, but rather as an illness that at times questioned the very separation between mental and physical, and could morph into a variety of symptoms hiding its true or original cause. Yet for Freud, after 1896, that cause remained in the realm of the imaginary or of the unconscious, stemming from a violence that he saw as internal to the psyche and as specifically sexual. This made it difficult to connect the materiality of war with trauma and, as noted earlier, mental illness in soldiers was most often considered a weakness. Much of the Unknown Soldier Memorial's emphasis on anonymity and sacrifice, as we saw, sought to repress materiality, to keep it separate from a suffering that then could be made pure or 'cured.' But, as this and the next chapter will show, much of the Memorial's power came from a latent emphasis on embodiment as the locus of a suffering that is ambiguously mental and physical, and therefore also perhaps incurable, as it is not an illness but an act of accusation against war itself.[4]

More recent work on trauma and Post-Traumatic Stress Disorder (hereafter, PTSD) can help us understand what was at stake. In particular, Laura Brown in 'Not Outside the Range: One Feminist Perspective on Psychic Trauma' questions the definition of trauma given in the *Diagnostic and Statistical Manual of Mental Disorders* (hereafter DSM), version III-R (1987) and anticipates some of the changes made in DSM-IV-TR (2000). The former begins by stating that 'the person has experienced an event that is outside the range of human experience,' and much of Brown's argument is that such language is based on a normative notion of human experience which in practice equates it with what is normal for white heterosexual financially secure males in Western culture. As she points out repeatedly, sexual abuse of women is far from 'outside the range' in that it is tremendously common (worldwide, she states, perhaps as much as a third of all girls are abused before they are sixteen years old), yet it certainly deserves to be considered traumatic. DSM-IV-TR, in turn removes this statement and replaces the above definition of trauma with 'the person has experienced, witnessed, or been confronted with an event or events that involve actual or threatened death or serious injury, or a threat to the physical integrity of oneself or others,' and 'the person's response involved intense fear, helplessness, or horror,' noting that 'in children, it may be expressed instead by disorganized or agitated behavior.' However, in his introduction to *Haunted by Combat*, 'Private Traumas, Personal Mythologies: Post-Traumatic Stress Disorder among Combat Veterans,' Jeffrey Kirkwood stresses that trauma is defined far more by the unwanted recurrence of feelings of helplessness and horror than it is by any original event, and that it may come from an accumulation of events or, more importantly, from the unique interaction of an individual's world view and psyche (or also, personal mythology) with external circumstances (in other words, what is traumatic for one person is not *always* such for everyone). Kirkwood thus warns against allowing 'the quest for comprehension' to become an 'obstinacy of knowing.' By evoking the 'private' aspect of trauma, he also goes back to Brown's discussion, showing that even if DSM-IV-TR's language is less overtly normative, it still enacts the power of institutional definitions to reduce and codify experience.[5]

These issues are highly relevant to First World War experience, and were discussed in connection with the 'excess of bodies' this war produced and with the Unknown Soldier Memorial's attempt to express it, though not, of course, in the same language. Specifically, first, the question of whether trauma is outside of human experience can be connected to the problem of the dehumanization, reanimation, and monstrosity of soldiers' bodies. To the degree that such monstrosity can be reintegrated in culture, that is, to the degree that bodies 'on the battlefield' can be 'brought home,' wartime trauma is both acknowledged and reduced to a normative definition as that which is 'outside' with respect to an 'inside,' as the temporary animality or aberration of war. In both cases, as Brown's discussion makes clear, what is covertly upheld here is a specific set of distinctions between outside and inside, human and non-human, normal and aberrant; in the case of the First World War, these distinctions were not only those of the white patriarchal culture exposed by Brown, but even more sharply, those of a culture that wished to disavow its responsibility for wartime violence by claiming control over trauma. Thus, it is precisely when violence cannot be reintegrated but remains radically outside that trauma can be seen not as 'beyond human experience' but as 'beyond' a narrow and political definition of the human, and as, in reality, endemic. Countless soldiers experienced 'bestialization' or *imbestiamento* but this can only be understood as many sorts of dehumanization, as multiple and private as well as public losses of integrity, rather than as a codifiable breaking of a putative human norm, which can then be redeemed. As we are about to see, 'The Symbolic Journey,' the desire to bring the Unknown Soldier home was both an impulse to normalization and a realization that perhaps normalcy could not be re-established, and that in its liminal undead quality the Unknown Soldier's body brought the battlefield into the City.[6]

Second, the contrast between 'the quest for comprehension' and the 'obstinacy of knowing' can be tied to the question of contamination and repetition by which living bodies are linked indissolubly to dead ones, reliving their monstrosity. Such reliving is no doubt a symptom, an indication that the living are still affected by what happened during the war, and therefore in need of help. What very recent work on PTSD, such as *Haunted by Combat*, shows, however, is that we must not be too quick to erase these symptoms as undesirable lest we lose our sense that trauma is itself a part of healing. As Brown puts it, 'how, rather than desensitizing survivors to symptom triggers, [...] can we help them to reconstruct their world views with the knowledge that evil can and does happen?'[7] When it comes to the First World War survivors, reliving comrades' deaths may be traumatic, but desensitization, once again, enforces a normalcy that minimizes the cost of war. In contrast, treating not only symptoms, but a deeper world view means reassessing the costs of war. This may indeed be a long and painful process, since, as Jay Winter has shown so well, one of the strongest persistent reactions to the First World War was to seek reassurance that so many had not died so horribly in vain. Yet recent work shows that by not insisting on 'knowing' what trauma is about, by not reducing it to triggers and symptoms, we begin to 'comprehend' something new by listening to what trauma has to say. The realization that was looming in the case of the First World War, I believe, is one that ushers in our modern era, because the scale and disregard for human life of this war irreversibly changed the human. Though it was only in the Second World War with the atomic bomb that total annihilation became technically possible, we find already in the First World War the alarmed realization of humans' self-destructive power. Thus, the second part of this chapter, 'the darkness within,' examines the image of the human that haunts the Unknown Soldier Memorial, which is expressed through anxiety about his body's decay, and is echoed in fictional explorations of violent forces hidden within the body. These are associated with a shapeless and dangerous melancholy, with the 'horror of transformations,' in d'Annunzio's words, which cannot be repressed, as we will see in the next chapter, where I explore how the Unknown Soldier Memorial seeks to make it spectacular and yet ritually to contain it.[8]

The Symbolic Journey

As explained in Part I, in Italy as in other countries, the Unknown Soldier's body was progressively moved from the battlefield to his final resting place in the Unknown Soldier Memorial in the nation's capital. What I wish to emphasize now is how much this was depicted as a symbolic journey – a return home – that was both emotionally and politically fraught. To bring home physical remains came dangerously close to exposing survivors to the bestiality of war, unless these remains could be 'domesticated' and detached from no man's land. At the same time, what home could they be given that would not betray the war experience, transforming dead bodies into a political symbol? A new relationship between the land, the body, and the nation had to be negotiated. It is telling that France and Italy, where civilian populations could not be separated from landscapes strewn with bodies, are the countries in which the 'home' and the journey to it were most controversial; and Italy was the only one to stage this journey, reflecting a desire to enact an Italian unification that was still felt to be lacking. As we are about to see, the Vittoriano and the Arc de Triomphe were, of course, symbolic locations, but discussions show that it was an open question as to whether they transformed and spiritualized the Unknown Soldier Memorial or whether, vice versa, it brought the materiality and the liminal embodiment of war to the core of the nation.

It is notable that in both Italy and France a proposed burial in the Pantheon was rejected, though for very different reasons. In the French case, Zola had just been buried there in 1908, and further, the Unknown Soldier ceremonies coincided with the transfer to the Pantheon of the heart of Léon Gambetta, a controversial political leader who had resisted the Germans in the Franco-Prussian War and later defended a liberal stance against the monarchy. For the French, then, the Pantheon was too left-wing to represent all soldiers. More covert, yet visible in parliamentary debates, as discussed by Jagielski, was the French government's desire to avoid recalling the Dreyfus affair with which Zola was associated and, more broadly, to prevent communist or socialist pacifist agitation, which the authorities felt such a location might have invited. In the Italian case, the Pantheon went back to ancient Rome, was transformed into a church, and later also became the burial place of the Italian kings. For the Italians therefore, the Pantheon was too right-wing to represent all soldiers. What is evident in parliamentary debates, as discussed by Labita and Miniero, is nonetheless once again a desire to avoid dissent from the left, which might have opposed the religious and monarchic associations of the location. However, Douhet's proposal – rather left leaning in its defence of the soldier against the authorities and its emphasis on economic injustice – sought the Pantheon as a burial place because it would best show how the lowly would be raised up.[9]

What struck me in examining these parliamentary debates is that, whether it be too much to the left or too much to the right, the Pantheon sparked fear of division within the nation because it harked back to a history older than the present nation itself and, thus, implied its impermanence. It was not the 'left' or the 'right' per se that was to be avoided by rejecting the Pantheon, but rather the more destabilizing breakdown of neat divisions between left and right upon which present politics were built. The Paris Pantheon began as the church of Sainte Geneviève under Louis XV but was not completed until the French Revolution, when it became a mausoleum for its heroes. From its inception, before the revolution, and even more so after, arguments about the Pantheon's meaning were contentious, and political. In the post–First World War context, it might bring up unwanted debates about the meaning of the French Revolution, and whether France had lived up to it. In Italy, the Risorgimento was closer in time and present political structures more precarious. Thus, documents in the Ministry of the Interior reveal equal worry over possible disruptions by 'ex-legionnaires from Fiume,' '*fasci di combattimento,*' '*arditi,*' '*arditi del popolo,*' 'communists,' 'socialists,' and 'anarchists,' as well as 'monarchists,' but to a far lesser degree. Though they run the gamut from right to left, these categories were at the time unstable, and reports to the Ministry of the Interior reveal genuine confusion as to which among them might belong together. The arditi, in particular, as courageous Italian shock troops during the First World War were among the most vocal veterans, but also hardest to categorize. They mostly participated in d'Annunzio's takeover of Fiume; many entered Mussolini's *Fasci* afterward. However, many also joined the anti-Fascist Arditi del popolo. To complicate matters further, the monarchists were admirers of d'Annunzio for cultural reasons, whereas the anarchists were directly involved in his takeover of Fiume. In sum, moderate governments in Italy and in France found themselves confronted with opposition from extremes that might strangely merge into each other, and it was such extremism – and the danger of a revolution or civil war – that they sought to avoid.[10]

The French did not, in fact, make the burial of the Unknown Soldier under the Arc de Triomphe permanent until 28 January 1921, almost three months after the ceremony honouring him. Thus, possible controversy could be deferred and perhaps undercut by the success of the ceremony itself. Nonetheless, the Arc de Triomphe is striking for its association with Napoleon and, once again, with the French Revolution: it was commissioned in 1806 after Napoleon's victory at Austerlitz, and it lists revolutionary and Napoleonic military victories; its most famous sculptural group, 'Departure of the Volunteers of '92,' known as 'La Marseillaise,' depicts the allegorical figure of France calling forth her people. But here, unlike in the Pantheon, revolution is not about dissenting intellectuals, or about overthrowing the past; indeed, it is about following the *Patrie* rather than demanding its recognition. It is also about reconstruction, as is made visible by the insertion of the Arc de Triomphe in the 'Historical Axis' of Paris that runs from the Tuileries Gardens and Champs Elysées through the Louvre to the Place de la Concorde and, now, to the Grande Arche de La Défense.[11]

In Italy, the Vittoriano was proposed by the minister of war and approved by the government as a permanent location before the ceremony. The Vittoriano, designed by Sacconi in 1895, was meant to commemorate the unification of Italy under Victor Emmanuel II. As Catherine Brice has shown in elegant detail, it was controversial from the design competition onwards, not only because it destroyed a large part of the Capitoline Hill in order to create a new Roman geography, but because of confusion over the values it was to represent. Since the monument was not inaugurated until 1911, and remained incomplete until 1935, its construction presided over major changes in the nation. Later changes notwithstanding, placing the Tomb of the Unknown Soldier there in 1921 was seen by many as a way finally to complete the Vittoriano. Moreover, a survey by the newspaper *L'Epoca*, on 31 July and 3 August 1921, shows that a majority of people favoured the Vittoriano over the Pantheon because of the greater visibility and access, which were felt to be more democratic. Of particular relevance here are two ways in which, like the Arc de Triomphe, the Vittoriano embodies the new nation but shies away from revolution in order to emphasize reconstruction. The first is in how it eclipses the importance of Garibaldi, and of his memorial on the Janiculum, making his rogue vision of the Risorgimento a sideline to unification via the monarchy and existing laws. Garibaldi had been denied burial in the Pantheon, and burial of the Unknown Soldier in the same place would have linked them, making the new Italy more revolutionary and dangerously anticlerical. The cover on the 1 October 1919 *Il secolo illustrato* shows d'Annunzio entering Fiume and comments 'Garibaldi returns, remembers [...] and admonishes.' In contrast, for the state, burial in the Vittoriano implied a less inflammatory Risorgimento: democratic, as people wanted it to be, but not revolutionary. Thus, the law creating the Unknown Soldier Memorial in Italy states in the same sentence that it will allow the soldier to return to his widow or mother, and to be surrounded by the comforting 'emblems of the different phases of the national Risorgimento.' The second is how it replaces the historical monarchy, with its ties to the Church as seen in the Pantheon, with an allegorical representation of unity. Specifically, as the First World War ended, the very centre of the Vittoriano was still under construction because the design of the equestrian statue of Victor Emmanuel II and the sculptural groups below had not been finalized. When the Unknown Soldier Memorial was added it prompted a rethinking that ended up emphasizing the allegorical statue of Rome far more than the king, and put an end to debates about whether 'great men' (including Garibaldi, of course) should be represented around him.[12]

The Arc de Triomphe and the Vittoriano, then, incorporate the Unknown Soldier into a nation that is explicitly modern yet at the same time presented as permanent, via the allegorical statues that evoke an 'essential' Frenchness or Italianness that claims to be above political divisions.

Figure 10. Photo of Plaster Maquette of the Italian Unknown Soldier Memorial. © Archivio Centrale dello Stato, Rome

Another advantage of these monuments over the respective Pantheons is their clear dissociation from the Church which, as we saw in Chapter 3, for the most part did not dare to criticize the Unknown Soldier, but did in some cases wish to reassert the religious nature of sacrifice as distinct from the political. As we will see later, under Fascism, the Church was made more present at the Vittoriano, for it had agreed to become part of the developing religion of the nation. Finally, I believe two more factors made these monuments attractive. One was their massive scale, such that they became almost landscapes in their own right, which was fitting to the scale of the war. This was emphasized in photographs and newsreels that captured huge military formations dwarfed by the monuments (as seen also in Gance's film), and also by the 1919 flight of Charles Godefroy under the Arc de Triomphe in his Nieuport biplane. A maquette of the Unknown Soldier's Tomb at the Vittoriano depicts a person to be about half the size he would be in reality, thus exaggerating the immensity of the Monument. The other was that neither the Arc de Triomphe nor the Vittoriano were burial places, unlike, obviously, both Pantheons. As a result it could be made clear both that the Unknown Soldier's burial was different from all others and, perhaps more important, that his body bore a unique and new relationship to the nation.[13]

As we are about to see, however, this relationship was not necessarily identical to the spiritualization and permanence evoked by the allegorical figures of 'Patrie' or of 'Roma.' As we saw in Chapter 3, the definition of *corpus mysticum* evolved in the Middle Ages and into the modern era so that it no longer meant the body of Christ but instead the Church, and then, by extension and later replacement, the nation and its borders or, the land as a symbolic body. As a result, actual human bodies lost concreteness, as Kantorowicz shows. But this process of abstraction was undermined by the materiality of the First World War. First, because the symbiosis between the living and the dead prevented sacrifice from being disembodied and the agency of the state alone rather than of the soldier, as we saw at the end of Chapter 3. Second, as I will show here, the war's materiality countered not just the disembodiment of sacrifice, but the resulting devaluation of the body, its reduction to being the other or the outside of understanding. Hence, the association of the Unknown Soldier's body with the body of Christ had two aspects. First, as seen in Chapter 3, and again in the d'Annunzio quotation that begins this chapter, a Modernist insistence on the humanity of Jesus is used to question the spiritualization of suffering that a traditional rhetoric of sacrifice would insist on; there was sacrifice, in other words, but no salvation, or resurrection, or 'dawn' in d'Annunzio's image. Second, the association of the Unknown Soldier with Christ emphasized incarnation as an alternative to the resurrection that was rejected. Instead of the flesh becoming spirit, so that the soldier's body was replaced by the allegorical *patria*, the spirit became flesh and, most important, infused the *corpus mysticum* of the *patria.* The traumatic embodiment of war, in other words, could no longer be expelled from the corpus mysticum. Consequently the outside and the inside, the patria and the material body, flowed into each other. One of the central images for this, of biblical origin, is that just as through Jesus hearts of stone became hearts of flesh, through the suffering of the soldier a land of stone became a nation of flesh. Of course, this vulnerability of the body and of the nation no longer defended by abstraction caused great anxiety, which was open for intense political manipulation: the flesh was depicted as both liquid and horrific, as the opening d'Annunzio quotation shows, and this sparked an obsessive need to control it.

Discussions of the Unknown Soldier Memorial and, in particular, of its location show that traditional views of the integration of the body into the corpus mysticum of the nation are countered by a mixture of horror at and fascination with the fluid, more incarnate, and even abject body revealed by war. This is seen especially in the repeated proposals for monuments on the battlefield itself. As early as 1919, *Il secolo illustrato* published a project by Mario Palanti for a mausoleum at the heart of the battlefield, 'at the mouth of the Timavo,' whose scale was compared to that of the Vittoriano. This was part of a widespread feeling that the war-torn landscape should itself be part of any monument, and indeed was in its scarred surface symbolically equivalent to the mutilated bodies it held. Though the 1921 survey in *L'Epoca* came up with the Vittoriano as the front runner, it showed that a number of people wanted Trento or the Carso, instead. In 1920, the Trench of the Bayonets was inaugurated at Verdun. The trench itself was the place where a whole company of soldiers was buried by the collapsing ground in 1916; only their bayonets stuck up above the ground. Though, as Winter writes, it is most likely that the bayonets were placed there by Germans to indicate the presence of graves, popular myth depicted these soldiers fighting until they were buried alive. I already mentioned in Chapter 2 the minimalism of the Memorial created there, where a simple unadorned steel structure was built to protect the bayonets and the bodies beneath them. As Winter argues, this could hark back to 'a primitive age when man fought against the savage beast and the chaotic forces of nature' and thus bypass the association of Christian and political transcendence. What I would like to emphasize further, now, is that the Monument in fact engaged with the 'savage' nature that the dead bodies were associated with. Thus, the architect André Ventre, in discussing his design, stressed that 'the erect forms of nearly one hundred soldiers' and the broken earth above them gave the trench 'an impressiveness no monument could ever equal'; at the same time, he emphasized the 'durability' of his structure that he 'guarantee[d] [...] to last at least 500 years.' The image of erect armed soldiers recalled Italian Minister of War Rodinò's assertion that the 'dead soldier's gun' does not miss (Chapter 3), and suggests that these soldiers still fight the gallant fight against the animality evoked in the opening of this chapter: the image is that their bodies resist decay, mirroring the strength of the structure that protects them. Overall, what we see at the Trench of the Bayonets is the contradictory desire to engage with the fleshiness of soldiers' experiences, yet also to preserve the fleshiness from degradation: savage nature is acknowledged yet still contained.[14]

A more extreme proposal was Churchill's famous suggestion that Ypres be kept as a ruin. Though this posed too many practical problems to be feasible – mainly because things eventually grow back on war-torn land – it was from a symbolic standpoint very significant. As Laqueur notes, it was part of a general focus on the landscape that reflected the fear of erasure. But, even more, the contradiction inherent in the notion of a permanent ruin reflected a deeply ambivalent attitude towards the fleshy incarnation such a ruin evoked. That is, to be true to the horror of war, and to the bodies it mangled, the landscape should continue to fall apart: yet, aside from being impractical, this vision was too traumatic to sustain, and what we have instead is decay frozen into one particular moment, saved from erasure but also from degradation. D'Annunzio, too, proposed a ruin as the only appropriate commemoration of the war experience. In the July 1921 issue of *L'Ardente*, in 'La chiesa di Doberdò,' he evoked as an ideal memorial a ruined 'church without prayers' on the torn battlefield, in which soldiers would forever 'bleed' and 'seem like bodies formed of earth with at their summit a face of suffering flesh.' Similar to those in the Trench of the Bayonets in their pose, these bodies were nonetheless reanimated, subject to decay. Suffering was here embedded in the landscape according to the biblical model invoked by d'Annunzio and others who saw the Passion reflected in Golgotha, and saw all of creation suffering with Christ. This travail of creation had a Franciscan and Modernist element, as d'Annunzio's text began by evoking 'Saint Francis, lacerated and worn.' Later d'Annunzio would add 'I was a bloody trunk quenching the thirst of the Carso's stones,' implying now that it was only in the exchange between body and landscape that the land itself would be reanimated and incarnated.[15]

Such incarnation associated with the land is opposed to traditional monumentality, as d'Annunzio wrote already in 1919: 'today the soldiers, peasants and not, cry out in one voice from beneath the ground and above it: "We do not want any monuments."' Here he anticipates the Socialist *Avanti!*'s 1921 complaint: 'Maybe among the Alpine rocks those bones would have remained mute and free from eloquence: in Rome they speak. To us they are a warning: to the devotees of the altar of the "patria," a remorse.' The Unknown Soldier Memorial, it is implied here, betrays the truth of the battlefield with its eloquence; in contrast, the body among the rocks is mute, evoking horror without claiming to appropriate it with words. This silence was, however, made present during the the inauguration of the Unknown Soldier Memorial, as I discussed at the end of Chapter 2, not only via a minimalist inscription, but also via the two minutes of silence that were part of the ceremony, and via the almost total absence of speeches. It was further emphasized as bringing the battlefield into the City because it was followed only by cannons and bells, evoking the noise of battle, rather than by music. Even so, it is clear that official tributes were 'too eloquent' for many, and that evocations of the war experience functioned to undercut their appropriation of silence. Thus, in the 6 November 1921 issue of *L'illustrazione Italiana*, d'Annunzio's *Notturno* was advertised in parallel with a description of the Unknown Soldier as 'suffering flesh, [...] a reality of travails, of sweat, of bloodshed, a block of anguish, torture, and pain.'[16]

These are the very terms emphasized by d'Annunzio as he identifies with his dead comrade, Giuseppe Miraglia:

> Shuddering, in the vacillating flame, I glimpse a dark spot under the narrow bed [of Giuseppe Miraglia].
>
> The flag is removed from his feet; the bandage is removed from his unrecognizable face. [...] The corpse falls apart.
>
> A sort of petrified horror surrounds me. [...]
>
> I see again that dark stain under the bed.
>
> It is blood, blood and lymph that leaked through the mattress ... It feels like my heart is dropping, down under my heels. [...]
>
> Four sailors lift the coffin [...] I put my hands underneath it and bear its weight. [...]
>
> My chest is full of screams and I cannot hear my own voice.
>
> The bed oscillates, swerves, and then collapses. A stone desert, crumbling and pierced, precipitously assaults my eye that cannot close. [...] Here is the pale Carso.

D'Annunzio's 'petrified horror,' much like the frozen ruin, is at first a reaction to the unsustainable vision of the body's decomposition; yet it leads back to 'blood and lymph,' and now to the poet's own heart that mimics the very movement of his dead comrade's decay. He sustains this weight, both literally and metaphorically, but his cry is silent. Finally, back in his convalescent's bed, where he is writing and remembering, he is invaded by the war-torn landscape, which is made up of falling rocks similar to his heart and the blood of his comrade Miraglia. The 'pale' Carso recalls d'Annunzio's address to his companions, cited in the previous chapter, as 'men of sorrows' whose 'faces are pale,' for all the blood has gathered at the heart. In both cases the image is of an inner bleeding, fluidity, and decay, that is outwardly visible as petrification.[17]

Similar images are found in the poetry of Ungaretti. Thus, in 'Sono una creatura,' the poet begins with a petrification that seems to overwhelm both the landscape and all living things, yet then reveals that within it there is an invisible cry not unlike d'Annunzio's silent scream:

I am a creature
Valloncello of Peak 4, August 5, 1916

Like this stone
of San Michele
so cold
so hard
so drained
so unyielding
so totally
lifeless

So like this stone
are my tears
unseen

Through living
is death
atoned[18]

The horror of war seems to reduce everything to the inanimate, yet beneath it, in the body but also in the landscape itself, something wells up. In turn, in the poem 'Perché?' (Why?) Ungaretti presents his heart as an incarnation of the war-torn landscape, which invades it and is reanimated:

I hold my heart
that crashes within me
that bursts and rumbles
like a shell
on the plain
but leaves me
not a trace of flight

My poor heart

bewildered
not to know[19]

The heart becomes not only the landscape but also the projectiles that deface it, so that the poet's fear fuses with the violence that surrounds him. No distance or 'flight' is possible.

In both poets, the contrast between petrification and liquidity – of blood, of tears, of flesh – shows again how the vision of wartime decay of bodies is almost unsustainable, traumatic to the point that like the Medusa, to see it is to be frozen. At the same time, it is clear that for both, petrification, linked to monumentality, conceals a suffering less absolute, less glorious and more fleeting, for it is subject to time. This reminds us that an important part of trauma is losing the full memory of, and thus the control over, past events. A few more examples from d'Annunzio and Ungaretti show how this sort of loss was associated at once with decay and with animality, that is, with the loss of physical and mental integrity. As Alfredo Bonadeo shows in *Mark of the Beast: Death and Degradation in the Literature of the Great War*, animality or *imbestiamento* expressed for many the reduction of suffering to something no longer noble, no longer able to be spiritualized or to elevate the subject. Overwhelmingly considered a degradation (closer to 'bestialization'), such animality was nonetheless at times sought by soldiers or veterans as a truer expression of the human condition as war had revealed it to be. This questioned the boundary between animal and human, hence, between the outside and inside of both the body and society. Moreover, horrific as it was, animality was essential to the corpse's giving us 'access to death' (in Harrison's words, cited above).[20]

Thus, d'Annunzio writes in the *Notturno*:

> The soldier comes back with a bunch of furs spotted with brown, clumped with dried blood. [...] These are the remains of Alfredo Barbieri [...].
>
> The air has become like frozen crystal. It is like that spiritual mass of ice that encloses the head of a corpse in the first hour. My direct impetus cleaves it like diamond cutting glass. The screeching slices my brain in two. [...]
>
> The glorious head had two holes in it. From one, blood was poured onto comrades, from the other, it glistened in the morning wind ... [...]
>
> There is nothing that it does not pour over. The blood is coagulated now; yet it rushes over my head when I lean down between the cart's wheels, not daring to kneel in front of anonymous witnesses. The myriad droplets warm

> and awaken and redden, like the brown relic that suddenly flames up in the ampoule.[21]

The bloody furs, all that is left of Alfredo Barbieri, suggest animal rather than human life. The image of crystal recalls the frozen quality of reactions to trauma, but we see here that the poet rejects this, breaking the crystal but also his own head which becomes now like that of his companion, full of holes from which liquidity, and mortality, flow. Finally, identification between the poet and his companion brings the latter's blood back from stillness to movement, repeating the image of stone turning to tears, and associating it even more explicitly with incarnation. The last image, of the 'brown relic that suddenly flames up in the ampoule' refers to the blood of San Gennaro: kept dried in two ampoules in the Duomo of Naples, it liquefies three times a year. Such blood is, of course, miraculous like the blood of Christ, but d'Annunzio eschews more canonical religious references in favour of a more popular and transgressive religiosity. Incarnation is here less God becoming flesh than it is the flesh itself coming back to life. We see this also a little earlier in the *Notturno*, where decay is depicted as an alternative resurrection:

> The corpse is now separate from me, closed, alone, already in the grave. [...]
>
> I thought it still mine, before, though it was decaying, though it was deformed. Now it is in prison. [...]
>
> It could not rise up even if Christ called it. [...]
>
> It is there, I see it through the cover and the wood. Last night it was murky, smoky, swollen. One more night has gone by. It is the third day. The decay continues.
>
> I feel a terrible cold in my bones. To touch death, to mould death, with a living heart!
>
> And yet we are once again alone, the two of us, alone as in the cockpit in flight.[22]

Here '*sfacelo*' (decay) is what replaces being called out from the tomb and resurrected, and once again it demands identification from the poet, who associates his living heart with it. Anonymity and the Unknown Soldier come into play later, as d'Annunzio evokes an even more horrific vision of decay:

> The body of Roberto Prunas, swollen and pale, is stuck against a wall of torpedoes.
> It hangs and oscillates, macerating in its uniform like in an open sack.
> No one will write his name on that underwater grave.[23]

Finally, everything comes together in the image that gives the book its title, of the night in which death and life meet and interpenetrate:

> My comrade is dead, buried, dissolved.
> I am alive, but carefully placed in my darkness just as he is. [...]
> There is a place in the soul, where the black river and the light river meet.
> Blessed are the two comrade heroes whose unrecognizable bones are mixed in the stretcher like two smoking coals![24]

In Ungaretti also, the night is what erases boundaries, as it is both within and outside of the body, ultimately uniting the body, the land, and the violence of war:

> And I plummet into myself
> And go dark in my nest
>
> I witness the night brutalized [*violentata*]
> The air is riddled

In this last image of the night being raped (*violentata*), we see how animality is not just a reduction of the human to the carnal and to decay. Instead, it is also an assertion of the world as creature and creation, in symbiosis with the human condition rather that separate from it, as the night, too, is embodied. Access to death, to what humans call death, passes through this symbiosis, through an experience of the different yet related finitude of the world.[25]

In Italy, the effort to domesticate soldiers' suffering did not take place only through the choice of a location for the Unknown Soldier Memorial that would be as eternal and undivided as possible. In addition, unlike France, England, or the United States, Italy choreographed the journey of the Unknown Soldier's casket from Aquileia, where he was chosen from among the eleven, to Rome. Documents at the Ministry of the Interior reveal that much care was taken to time each stop of the train that transported it on an open wagon, and that of equal concern was the representative group of mourners that travelled with it, which included politicians and officers, but also mothers, mutilated veterans, and foot soldiers. Travel began at 8 a.m. on 29 October 1921, with arrival in Rome scheduled for 2 November at 9 a.m. The train was to stop for five minutes at every station along the entire way, with much longer stops scheduled in Venice, Bologna, and Florence. At night, the train was kept far from central stations, and access was limited. More important, instructions made it clear that '*all speeches are to be forbidden.* A religious silence must be observed.' Each town was to arrange for a 'welcoming committee' (*comitato d'onore*) that included 'local soldiers and mutilated veterans.' Newspaper accounts show that the state's emphasis on silence and order was countered by popular expressions of grief that went from wailing and throwing oneself at the coffin, to innumerable crowds wishing to touch it, to 'violins, mandolins, and guitars,' to a thriving memorabilia market, to the creation of 'light shows' (*luminarie*) reminiscent of country fairs. Labita describes this as an opposition between 'day (picturesque)' and 'night (mystical),' but the more significant contrast was between spontaneous, heterogeneous expressions of grief (picturesque, but also a mixture of pagan and Christian, of frenzied celebration and existential alienation), and minimalist, homogeneous ones (mystical in the sense of 'negative' rather than 'ecstatic,' sombre rather than expressive). The former were closer to the chaos of the trenches, and to the night of decay and loss of boundaries evoked by d'Annunzio, Ungaretti, and other veterans. The latter were important to the state because the journey was a symbolic enactment of Italy's unification through contested territories, paralleled in the soldier's return from bestiality and decay to humanity and civilization. Thus, as the coffin came closer to Rome, one commentator wrote, echoing state scripts, 'The opposition between a contemplative tendency and enthusiastic impulses is about to be overcome. While through the rural areas and during the pilgrimage from Aquileia to Rome expressions [of grief] were mostly sentimental, the imperial background of Rome exalts and induces the crowd to glorification.' However, the *Avanti!* claimed that austerity was marred by 'mad merriment: songs, music, clapping hands, hymns, "fancy uniforms," choreography.' The Ministry of the Interior also received reports of various unprogrammed honours, among them one from the Arditi del popolo, who sometime in the night of 4 November deposited a wreath on the Unknown Soldier's Tomb, with the following text: 'not false gilded pomp, not marketing of the glorious dead, but the flame of faith, and omens of high hope, are offered by simple proletarian souls to their unknown brother.' This was immediately removed. Nonetheless, dissent, whether goliardic or patriotic or both, shows that the Unknown Soldier's return home was not only a domestication, but perhaps, also a more unstable bringing of the outside, of heterogeneity, and of animality into the core of the City.[26]

The Darkness Within

To the degree that the Unknown Soldier Memorial destabilizes the relationship of land, body, and nation, it undoes traditional monumentality but does not offer a clear alternative, and thus scripted interpretations vie with heterogeneous personal relivings of the war experience. As we saw, for the state, much hinged on maintaining the distinction between a wartime experience of animality and suffering linked to the material body and peacetime ability to frame this with a humanity spiritualized in its association with the patria. However, the symbolic journey home of the Unknown Soldier, for participants along the way, but also for writers mourning their dead comrades, emphasized the breakdown of all distinctions, as images of fluidity and interpenetration melded wartime and peacetime, the battlefield and the city, and the inside and outside of bodies themselves, or also the material and the spiritual body. Though the Memorial may 'petrify' suffering, in reality stones are '*sgretolat[i] e forat[i]*' (crumbling and pierced), the land and the body '*s'incaverna / e schianta / e rintrona*' (that crashes within me / that bursts / and rumbles), and dried blood '*rifiammeggia nell'ampolla*' (suddenly flames up in the ampoule). The corpus mysticum of the nation is contaminated with the fleshiness of incarnation. To a great extent, authorities responded with denial, though inaugural ceremonies, mainly under popular pressure, sought to give the flesh a contained ritual expression. I will examine this ritual and reactions to it in detail in the next chapter, but before that three works of fiction – by Jean Giono, Filippo Tommaso Marinetti, and Louis-Ferdinand Céline – will help us to understand what is at stake in allowing the materiality of the First World War to penetrate the corpus mysticum of the nation. As we will see, the fear of contamination revealed in their texts resurfaces in continued debates about the location of the Unknown Soldier Memorial and protection from decay over time. Returning to Kirkwood's distinction between 'obstinacy of knowing' and 'quest for comprehension,' we will find that the latter is not only a long process, but one in which acceptance and denial of trauma, decay, and violence interact.[27]

I already discussed Giono's *Le Grand troupeau* in Chapter 1, in the context of the primal scene of improvised battlefield burial that was at the origin of the Unknown Soldier Memorial. During the 'watch over the absent body' of her husband, Félicie remarks, 'Oh, my poor Arthur, dead all alone against the earth like an animal!'[28] As we are about to see, in the general economy of the novel, this image does far more than decry the animalization of men on the battlefield. Indeed, Giono's novel is a complex meditation on animality and wartime that draws out the deeper implications of Gianikian and Ricci Lucchi's juxtaposition of Comerio's hunting and wartime footage. The novel's first part, 'War Spells Ruin for Your Rams, Your Ewe-Lambs, and Your Crops,' opens with a long slow sequence, very cinematic, in which we see with growing dismay flocks being pushed away from mountain pastures by war. They are first led by an old man, 'alone, old, and exhausted,' and by 'a completely black animal with blood under its belly'; another ram's 'teeth and chops were clotted with blood.' The flocks themselves are depicted as an overflowing river: 'outside the big herd was flowing,' 'the animals were ill now. The sight of the long herd was too much [...] the pain and so much life worn away on the road.' Suffering increases, as one of the rams begins to die: 'The ram was still there on the ground, legs spread out. Blood was flowing from him.'[29] Finally: 'The ram had just died. It jerked up its heavy branched head, once, as though obeying an order. It looked up at the sky from between the branches of its horns, a long, endless look. Neck stretched out, it moaned like a lamb. Then it spread open its thighs, extended its legs, and with the noise of a bursting balloon let loose a bundle of guts and black blood.' The scene is punctuated by onlookers in the village repeating, '*gâcher la vie!*' (such a waste of life!). It finally comes to an end when a shepherd asks '*le papé*,' the elder in the related families of the protagonists of the novel, to save his ram who is also bleeding by keeping him at his farm, while the shepherd returns to leading his flocks as best he can: 'when these evil days are over and if I'm still alive I'll come fetch him.'[30]

This opening sequence already establishes an implicit parallel between the waste of animal life and the waste of human life during the war. The materiality of both is emphasized through images of fluidity, of the flock as a whole and of the bleeding animals. In the context of Giono's pastoral world, animal and human suffering may not be identical, but they belong together in a single creation that war overwhelms like the vengeful God of the Old Testament. As the comparison of the dying ram to a lamb implies, a new incarnation will eventually have to save this world which, the apocalyptic tone of the chapter implies, is being punished for man's inability to care for it properly. We see already in this opening sequence that for Giono the pastoral and its popular religiosity is what provides an alternative articulation of land, body, and nation, whereby the latter is eclipsed in favour of the common destiny of peasants, who suffer a vengeance brought upon them by leaders so distant that they remain nameless and invisible throughout the novel. In other words, if there is a corpus mysticum here, whose materiality the war brings out, it is made up of this river of flocks which, later in the novel, flows into a river of men.[31]

The parallel is made clear by the recurrence of the phrase 'such a waste of life' in the central chapter, 'There Will Be No Pity.' Throughout this chapter, Joseph, whose departure to war I quoted in Chapter 1, and Olivier, the fiancé of Joseph's sister, make their way through a massacre, encountering numerous dead bodies and fleeing soldiers. 'He [Olivier] turned round. A man was stretched out on the other side of the hole. His face was completely black, his brains poured out of a large wound in the corner of his head. No, he wasn't looking at Oliver [...] the eye swarmed with rot and mud.'[32] Later, Joseph observes bodies during a short morning truce:

> The faces of the dead were buried in the mud, or they jutted partly out of the holes, peacefully, with their hands resting on the rim and their heads lying on their arms. The rats came to sniff them. They jumped from one corpse to another. They selected the young ones first, those without any hairs on their cheeks. They sniffed the cheek, then they crouched down into a ball and started eating the flesh between the nose and the mouth [...] when they came to the eyes, they scratched them out slowly and licked the eyelids [...] the crows arrived on their strong, calm wings. They sought out along the tracks and paths the horses which had been knocked over. Close to these horses, with their stomachs burst open like the flowers of the caper-plant, there were wagons and overthrown cannons mixed up with scrap iron, bread, meat provisions still wrapped up in gauze cloths, and yellow sticks of cannon powder. [...] The remains of all the men from the night fatigue duties were lying here. They were stretched out, soup pans overturned on their legs, in a mush of blood and wine. Even the bread which they'd been holding was full of shrapnel and bullets. [...] the area of the neck between beard and chest hair was exposed. The meat was tender there and quite fresh, the red blood still formed into a little ball. The crows started pecking at once, tearing off the skin [...] The dead stirred. Nerves twitched in the stiffness of rotten flesh, an arm rose up slowly in the dawn. A

> black, splayed-out hand pointed to the sky. Overswollen bellies burst open. A man writhed in the earth, trembling, all the strings in his guts let loose. An iota of life stirred again in him.[33]

I quote this passage at length because it tends to produce strong reactions, from Redfern's remarking on its 'horrible beauty' to Bonadeo's rebuttal that 'it is certainly unique in the literature of the Great War in representing the desecration of death through a kind of second death – the destruction of human remains by beasts.' For Bonadeo, this page is a 'warning' against 'a protracted belief in, and practice of, violence' that came from the war. Both, however, emphasize in this passage a violence that is 'against nature,' by which, crucially, they mean *inhuman*, against the proper hierarchy of men and beasts. To be sure, this is the shock value of this page, in which Giono (writing more than ten years later) dares to bring out yet another horrific and hidden aspect of the dead bodies of soldiers, and thus also of the Unknown Soldier's body, beyond dismemberment or mutilation. His imagery again destabilizes all distinctions between outside and inside, but also between beasts and men, between the body and spirit: brains become liquid, blood gushes again from the dead (as in d'Annunzio); the animal flesh the soldiers were supposed to eat is humanized as it has gauze over it, while human flesh is animalized since rats eat it; dead men come back to life as their arms seem to imprecate against the sky; finally, the mixture of blood, bread, and wine evokes the Eucharist, but it is riddled with holes and pierced by shrapnel, evoking once again a suffered incarnation without any resurrection. Thus, even as I agree that Giono's pastoral tends in the end to uphold a certain order in nature, which war disrupts, I think heeding Kirkwood's warning against understanding too quickly allows us to sense that this page is perhaps most disquieting when the pastoral and nature do not meet, when the very possibility of order is questioned. In other words, Bonadeo and Redfern (and many other readers of First World War imagery) would see this page as a depiction of a disfigured nature, emphasizing the horror of rats biting into human cheeks. But even more disturbing is the way the 'peaceful[ness]' of the dead men mirrors the 'calm' of the crows; similarly, both the open bellies of horses and the bloody hands of men are compared to flowers, implying once again that they are of the same, not corrupt, though perhaps beyond good and evil nature. Ultimately, if humans and animals are part of the same corpus mysticum, it becomes difficult to decry violence as beastly: it seems to become part of the condition of life itself.[34]

As Gumbrecht has pointed out in his investigation of latency in post–Second World War Europe, confronting a degraded, less than idealizing notion of the human is akin to confronting trauma, in that it is rare and perhaps impossible to gaze at directly. This was true already during the First World War, when, I believe, the contrast was even stronger between the reality of war and the ideals many still upheld, at least when the war broke out. Winter shows very well that solders and civilians believed in the justice of the war at its start and only with difficulty could revise that view. In this light, the ending of Giono's novel, which brings us back to the pastoral order, is an attempt to preserve some sense of justice, or some ideal vision of the human, separate from bellicist nationalism, but also from a more savage nature. At the same time, as Michel Serres writes, these years witnessed the radical reduction of agriculture in Europe, to the point that the pastoral world evoked by Giono had almost disappeared by the time he was writing his novel. Thus, a careful reading shows that elements of wartime trauma lurk under the nostalgic, comforting ending.[35]

Shortly after the passage I cited above, Joseph is wounded and will eventually lose his right arm, so that by the end, he has just returned home, to his wife Julia, who has been betraying him but now sends away her soldier lover, relieved to return to Joseph and to adapt her body to his new shape. Soon after, Olivier returns, as well, to his fiancée Madeleine who, we learn, has given birth to another man's child, which she tried to abort, and now as a result has 'two dead legs.' At the very end of the novel Madeleine gives birth again, now to Olivier's child, who is a healthy boy. The shepherd who left his ram at the farm at the beginning of the novel returns to collect it, having survived the war, and he and the animal bless the child:

> It's a good sign for births when animals are there. One more animal on the earth. [...] He blew on the little one's mouth. 'The green of the grass,' he said. He breathed on the little one's right ear. 'The noises of the world,' he said. He breathed on the little one's eyes. 'The sun,' he said, then he turned to the ram. 'Come here,' he said. 'Breathe on this little man so that he'll become like you, one who leads, one who goes in front, not one who follows.' [...]
>
> 'If God listens to me, it'll be given to you to love slowly, slowly in all your loves, like one who holds the arms of the plough and goes a little deeper each day.'[36]

This birth of this child, of course, recalls that of Christ; references to the grass and the sun further recall Saint Francis' praise of the elements and thus a popular yet also Modernist religiosity that emphasizes the poverty and humanity of Jesus, consistent with Giono's pastoral world. It would seem that this new incarnation can finally redeem wartime suffering, not transcending it as either national or Church unity might, but through the more horizontal, natural mystical body that according to the pastoral holds humans and nature in fruitful symbiosis. Yet the prevalence of mutilation at the end cannot quite be ignored nor, more important, can the suspicion that it points to a nature – human and animal – less benevolent than it seems. Olivier accuses Madeleine of being responsible for her illegitimate girl's inanimate legs, as the latter are said to 'bear the mark' of her mother's behaviour; by implication, Julia's infidelity is perhaps also in some way responsible for Joseph's mutilated right arm (clearly an emasculation); Olivier himself, we also learn at the very end, has come home with a mutilated right hand because he had a friend shoot at it so he could return home, having heard of Madeleine's pregnancy. Though Giono does not pass moral judgments on wartime infidelity, since he depicts it as a force of nature (albeit one close to the abject: 'I've got too much blood,' notes Julia), all these mutilations end up being the responsibility of nature itself, which may bless, but may also maim. Perhaps most telling, the little girl's legs which hang unnaturally from her crib at the end of the novel recall the unnaturally spread legs of the dying ram in the opening pages. As he was compared to 'the lamb,' we are now led to wonder whether beneath the reassuring image of incarnation provided by the healthy baby boy, there lies another, repressed image, where the dying ram and the mutilated girl come together in a suffering that points to a violence inherent in all forms of life.[37]

If in Giono nostalgia serves to veil a terrible glimpse into pervasive violence, in Marinetti this same vision is undercut by aestheticization. Rather than pastoral order, it is the beauty of wartime violence that constitutes a disavowal – denial, yet oblique acknowledgment – of the contaminating materiality of the body. I have written elsewhere about Marinetti's 1909 *Mafarka le futuriste* and its manuscripts, where we find that the mechanical man of Futurism can never vanquish either the mortal and decaying body, or the bleeding heart that longs for metaphysical stars. Here, however, I will focus on a short story by Marinetti, 'La Carne congelata' (Frozen flesh), first published in 1922 in the collection *Amori futuristi.* In it, latent violence emerges again in images of liquidity and mutilation, linked to the feminine body; aestheticization seeks to keep this violence separate from the more efficient courage of soldiers, who are 'untamable but most generous beasts.' Aside from the theme of *imbestiamento*, and its publication one year after inauguration, this story connects to the Unknown Soldier Memorial because its protagonists are arditi, the shock troops often associated with the Unknown Soldier's supposed heroism, and further, they are involved in fording a river, like d'Annunzio's peasant to whom he compares the Unknown Soldier (as cited in Chapter 1).[38]

Marinetti, too, was a veteran, and a writer who, like d'Annunzio, had been part of the interventionist campaign; both of them took part in real fighting, but also in symbolic actions and speeches. While their long-standing rivalry was purportedly set aside during the war in a mutual acknowledgment of courage, by 1922 Marinetti was jockeying for the mantle of Italy's postwar poet. Futurism, founded in 1909, and its attack on things decadent, nostalgic, or sentimental, as well as its celebration of speed, technology, and abstraction needed to be reasserted. More specifically, *Amori futuristi*, whose subtitle is *Programmi di vita con varianti a scelta*, along with other Marinetti publications from this period, seeks to counter the mournful tone of so much postwar literature by emphasizing a kind of vitalistic *polemos*, which sees war as part of a cycle of destruction and creation. He wants soldiers to return from carnage thirsty for life and empowered by variety (hence, '*varianti*'), and his short stories pit a healthy, and violent, heterosexuality against the obsession with mourning and symbiosis with comrades found in d'Annunzio's 1921 *Notturno*. As much recent work on Marinetti shows, however, in reality Futurist vitalism cannot be separated from its decadent roots which, after the First World War, become even more haunting. This is expressed most disturbingly in Marinetti's recurrent need to go back to war, in a desperate attempt to finally experience in it a vitalism whose phantasmagorical nature becomes ever more evident. Marinetti himself volunteered for the Russian front in 1942, but already in the 1922 'La Carne congelata,' his arditi are depicted not fighting in the First World War, but in another, new war.[39]

The short story opens with a typically Marinettian contrast between the soldier, whose will he encourages, and his lover, a 'fragile camellia' (*camelia fragile*) who cannot live without him: 'don't exaggerate your anguish [...] our will must harden our nerves against any vile sentimentality,' 'here we are, thrown into a new battle.' Throughout the story, the reader is addressed as a 'you' (*tu*) who is given instructions on how to be an 'ardito leader' (*capo ardito*) who can manage 'ugly wild defiant merry characters whose ferocity surpasses that of all Marats and whose heroism is greater than that of all the grey-green arditi of the last war,' who need to be 'punched' regularly, as well as 'rewarded,' and most of all 'unleash[ed] against the enemy' as often as possible. The Terror is invoked along with a kind of animal savagery that makes violence a fact of nature rather than a product of war. To be equal to it, Marinetti instructs us, we must learn from one ardito in particular, Guzzo, 'lean, agile, and fast, a true Saracen with the sweetest black eyes, but a mean mouth, deformed by a razor slash across the left cheek.' The 'razor slash' invokes not a war wound, but a more sinister fight; and as earlier in the collection, in the story 'Il rasoio voluttuoso' (The voluptuous razor) Marinetti proposes making both sex and shaving more spicy by combining them, it also invokes sadomasochism. This is further reinforced by Guzzo's split into sweetness and evil which in its psychological doubleness points more to the aestheticized, formalized, disavowal of violence of Sacher-Masoch than to the apotheosis of negation we find in Sade. As Gilles Deleuze has argued in 'Coldness and Cruelty,' this distinction is important because it questions the notion of 'sadomasochism' as a dialectic of opposites, allowing us to understand them as two separate worlds, one where violence is absolute negation (Sade), the other where it is represented disavowal (Sacher-Masoch). When it comes to both the First World War and Futurism, the first sort of violence is also what Deleuze calls 'molecular rage,' an impersonal force that wipes out the human (whether you decry this or relish it). In contrast, the second sort of violence is inseparable from the human imagination's fascinated anticipation and denial of it, as represented, for Deleuze, in the suspended gestures of Sacher-Masoch's heroines. We find, then, that already at the outset of Marinetti's story the attempt to associate the arditi with impersonal or 'molecular' rage is threatened by the aestheticized cruelty of Guzzo.[40]

As the story unfolds, a fight breaks out because Guzzo purportedly has a large piece of cured meat (*carne salata*) in his backpack. Arguing that he has not stolen it and that it is his property, Guzzo refuses to get it out, because, he says, 'it will spoil if I show it.' The commander has to 'tame' the men who 'growl' like wild animals sniffing a prey. After he succeeds, we are advised to follow Guzzo as the group wades across a river, near a bridge that has been bombarded. The scenery suddenly shifts to nocturnal moonlight, as we are bathed in a dreamlike atmosphere: 'The moon rises with jewels stolen from the water that bubbles with laughter. Sly reflections everywhere.' Guzzo, kneeling in the mud, is opening his backpack, as we are urged to watch and come closer, unseen by him. We, but not he, watch the landscape suddenly transformed by two searchlights: 'they cross, miraculously awakening fantastical light silver leaves and stems, emerald houses, mother-of-pearl beaches, spectral forms, islands of ice, landscapes of liquid gold. Guzzo is still rummaging in his backpack, while the sharp scissors formed by the two projectors open with slow mystery. They are two long brooms of dusty silver. The water catches fire in a thousand ecstasies. [...] fleshy velvets with human pallor vibrate, implore.' The war-torn landscape is not just aestheticized here, but 'suspended' according to Deleuze's use of the term: thus, mud becomes all iridescent reflection, and nature evokes velvet flesh, yet they are frozen in the moment the searchlights catch them; and the comparison of these to scissors indicates imminent violence.[41]

This scene is recurrent in Marinetti's works, as it appears also in Marinetti's 1922 novel, *Gli Indomabili*. The 'untamables' of the title are by day savages bound up in spiky chains with which they can wound, but not kill, each other, whereas by night they are unleashed and ravished in a 'lake of unctuous dissolving moonlight,' which infuses them with sensual freedom and sentimental nostalgia, a 'gentle music of affections,' which they reject when they return to the desert that they inhabit by day. In their doubleness, the untamables aid the revolution in a future megalopolis by helping the oppressed underground workers against the idealistic but oppressive '*cartacei*' (paper people). Cinzia Blum rightly reads this as an expression of Marinetti's postwar disappointment with social change, as the 'paper people' are called to speak true Futurists' words-in-freedom and not banal abstraction; she adds, however, that Marinetti's critique is regressive insofar as the only alternative he proposes is 'the past of reverie [...] poetic oasis.' At the end of the novel, Marinetti notes that even as the untamables must return to the animal violence of the day, the night remains within them, in the form of the 'superhuman flightfresh distraction of Art that gradually metamorphose[s]' them. At the same time, as Blum points out, they cannot be 'harnessed by "goodness" because they are the vital forces upon which progress is predicated.' For her, 'such a cyclical conception of history' means that 'the sole possibility for redemption [...] is escape into the consoling realm of aesthetic catharsis.'[42] Yet I would submit that catharsis, too, is presented as temporary, so that what remains is an unsuccessfully split subjectivity, in which daytime or impersonal violence can no longer be separated from night-time refreshment: that is, the night-time oasis suspends violence, again in the Deleuzian sense, rendering it latent, imminent, yet also fully human, so that it can no longer be banished to the 'bestial' outside of daytime.

Marinetti's unpublished diaries from the First World War reveal that *Gli Indomabili* began as a dream, from which the megalopolis and the revolution were absent; more palpable was the anxiety of the passage from day to night and the return to day. Similar imagery is present in another moment of anxiety from Marinetti's unpublished diaries: 'Always, always in my life these two tormenting states of mind have alternated with one another. In the happiest and most voluptuous bed I have suddenly torn myself away from the arms of a seductive and delightful lover, and trembling seated at the edge of the bed I have strained my ear to catch a distant, imaginary shelling. Nostalgia, desire for heroism and violence. In the muddy broth of the trench dug by the shelling, a monstrous ivy of women's naked bodies climbing from my temples up to the moon. Torrid nostalgia of lust. Why? Why?' On the one hand, night-time voluptuousness is haunted by a distant shelling; on the other, the trenches fill up with moonlit bodies that recall the 'fleshy velvets with human pallor that vibrate, implore' described in 'La Carne congelata.' For Marinetti, there seems to be no escape from this tormenting alternation, which is a forever unstable split within the self. Significantly, the editing of Marinetti's war diaries in their published version has done much to obscure this instability, for what has been left out are all the passages considered too literary, or also too imaginary, in favour of ones that concentrate on descriptions of the fighting or analysis of strategy. What we miss, then, is what Scarry calls a 'visionary solution' or also what Harrison refers to when he says a dead body is not merely a 'thing': the work of the imagination to make sense of, yet also to deny or at least make more palatable, a traumatic reality. When we bring passages such as the one above back into Marinetti's description of the war, we grasp that Marinetti cannot keep 'aesthetic catharsis' or also decadent deliquescence separate from vitalistic violence.[43]

The second part of 'La Carne congelata' reveals even more Marinetti's failure to render violence impersonal, to separate it from human desire and guilt, as wartime materiality awakens, comes back to life, and is incarnated under the caress of the searchlight:

> Kneeling, lit with splendour, Guzzo extracts from his backpack something that is even more resplendent. Come closer! He doesn't hear you, he is absorbed. [...] He is really holding a piece of meat. A piece of woman, decapitated, without arms and legs! The gracious stump of a small woman! It seems chiselled and encrusted with scintillating precious salt. Her small round breasts tremble, perhaps are about to speak since the lost faraway

> head no longer does. Her humble, shy, naïve, meek belly curves towards the delicate, dreaming garden, between the sensual thighs that are cut in half. The two cut thighs are covered by tight black cowls. So is the neck.
>
> 'My dear, I am here with you, with you. I have fallen down, down, from the heights of my heroism! Death gives us this last hour of joy, see? Everything is beautiful around us. [...] Now you open the arms you no longer have and you proffer the lips you no longer have! [...] I have fallen from the black desperate butchered twisting sky that is my dead life, down down onto you, in you, in your hot fiery humid sucking heat! Paradise hell god of mine, mine, mine. I drink infinity in you, tiny, immense divine one [...]'
>
> He cries, and in a feeble female voice repeats part of a letter he has memorized: 'Take me with you to war. Take my flesh with you and kiss me day and night, kiss me, eat my face with kisses, devour me, I am yours! Yours!'
>
> Then Guzzo continues, changing his voice again: 'I did what you wanted! [...] I ate it all with millions of kisses! And now you are my self! In my veins my blood cradles you, little one, with all the tenderness that dead mothers have in Paradise!'[44]

As Blum writes, at first read it is seems apparent that Guzzo devours his lover in a 'melancholic attempt to deny the loss, or impossibility, of an absolutely faithful, ideal love' and that 'she is made to assume man's losses as her own, thus making it possible for the male subject to reclaim his mastery.' Molecular violence, then, is separated from feminine instincts and desires that are cut to pieces, packaged, and sealed up for easy consumption, and ejection. Yet, as Blum also points out, at the end, 'unresolved conflict and a fundamental affective ambivalence speak through the body of the "maternal" soldier who loses mastery and identity to the object he needs to sustain his fantasy of absolute love. Ultimately, he becomes the woman he has ingested,' according to the master-slave dialectic. But, I would add, there is even more to this scene, beyond the fantasy of love and anxiety about autonomy noted by Blum, in which 'love of death takes the place of primal continuity with others.'[45]

Specifically, the aestheticization of the woman's mutilated body reveals her to be frozen or suspended like Sacher-Masoch's heroines, who do not merely take on but spectacularly dramatize anxiety that is about autonomy and love, but also about violence and sexuality. Particularly noticeable are the 'cowls' (*cuffie*) – a garment normally reserved for the head – that cover yet, of course, also bring out in dramatic black against white crystalline skin the traces of mutilation. They recall the 'black desperate butchered twisting sky' of Guzzo's life, and thus, we begin to intuit that mutilation is already within him before it is externalized. Moreover, as her 'meek belly' becomes 'hot fiery humid sucking heat,' we see that her sexuality may be frozen yet its dangerous liquidity is still latent (as was suggested earlier by the claim, 'it will spoil if I show it,' as though she might melt, or worse yet, decompose). Guzzo's fantasy is to counter the horror of war within his body by possessing an intact and incorruptible one (intact, because free from imagination and longing). Instead, his 'corruption' contaminates and reawakens the 'frozen flesh.' Significantly, Marinetti in his manuscripts describes his own heart in very similar terms:

> I carry within myself, in my chest, a black, viscous, beast, sticky and with little elastic limbs armed with suckers, [...] my little lustful soul [...] Dear God, what male will she demand? what male could mate with such a disgusting, sticky, and monstrous beast? For there are no animals that resemble her. And in any case, all is enclosed in my sealed chest.[46]

As we see here that the 'viscous' soul, like the hidden heat of the frozen woman, is monstrous in its desire for the infinite: it is this animality, and not only continuity with others, that Guzzo would like to ingest and eliminate, yet cannot. In other words, what Guzzo now carries within his own veins is not only loss of mastery over desire, but specifically over the nexus of sexual desire, metaphysical resentment, and violence. What is traumatic here is not just the loss of autonomy implied by love, but even more the way war connects sexuality and violence (as seen in Marinetti's diary entry cited above), haunting any possible recognition of our continuity with others, or any possible redemption.

The very end of 'La Carne congelata' confirms this, as we find that Guzzo passes on to the reader not only the death wish that leads him to court a stray bullet, but the cruelty he exhibited towards his lover:

> The arditi are piling up at the front in the ford that dreaming tramples sucks chews the starlight. [...] Guzzo with a leap will put himself in front of a bullet that jokingly was looking for him trying to miss him. Hit, he will fall in the water. Do not hesitate, take his backpack and put it on your shoulders, then untangle his corpse from the weeds and push it out into the current that will take it to sea. Then advance! As you walk through the shameless scandalous splendours of the projectors that seek you, you will hear a cry

near you, over you, in you! The body of Guzzo's lover will cry, already adulterous, trembling with love for you! And her tears will trickle down your back in slow, voluptuous drops.[47]

Pushing Guzzo's cadaver away to sea recalls the decomposing, faceless cadaver of Roberto Prunas imagined by d'Annunzio, cited above, and connects Guzzo one more time, albeit indirectly, to the Unknown Soldier. More important, it shows that the story is not just about controlling feminine continuity with others and replacing it with love of death or masculine courage, though it is certainly about that. More deeply, it is also about disavowal and the aestheticized, frozen representation of a violence that cannot be separated from our continuity with others: to carry Guzzo's backpack is to carry mortality and decay and the impossibility of 'infinito,' and the tears that infuse the carrier are like those within Ungaretti's stone from the Carso; yet these tears are now also associated with sexual pleasure and with the violence of taking Guzzo's place.

Louis-Ferdinand Céline's 1932 novel *Voyage au bout de la nuit* offers us in the very first pages a typical 'gem' by this author: 'love is the infinite placed within the reach of poodles' (*l'amour c'est l'infini mis à la portée des caniches*). The novel is largely autobiographical, recounting the story of Bardamu, who haplessly volunteers for the First World War, seduced by a passing marching band, even as he is, in the opening pages of the novel, ironic not only about love, but also about the 'French race,' and about what he calls '*Patrie* #1' and '*Patrie* #2.' Like his main character, Céline was wounded in 1914, declared unfit for duty in 1915, and dismissed in 1916. Also like his main character, Céline is known for his cynicism and misanthropy. *Voyage* was his first novel, and though it was a great public success, it became controversial when it was denied the prestigious Prix Goncourt. Céline continued to write fiction that was as provocative stylistically as it was uncomfortable thematically until his death in 1961. Though he remains infamous for the anti-Semitic pamphlets he published in the late 1930s and early 1940s, Céline is also one of the most innovative and underread twentieth-century French authors – in many ways the ultimate outsider, whose bleakness is too blunt to become mainstream. Thus, in *Voyage* desire for 'the infinite' – not just love, but any value or redemption – pervades the novel in that it makes the body move, and awakens enthusiasm or at least a vitalist will to live. Yet it does so always under erasure, negated the minute it appears. Céline's language itself presents this same tension, as his disgust not just at war but at humankind as a whole is expressed in a prose that mixes popular expressions and argot with acute powers of analysis and, most of all, draws us into its chaotic rhythm, dragging us along as we are horrified yet delighted by its rawness and rage. In other words, neither Bardamu nor Céline seem to be able to keep to the spare nihilism suggested by the aphoristic condemnation of love I cited above: instead, fascinated, they watch the animality of love unfold, participate in it, and draw us in, ultimately implying that no one, no matter how cynical, is safe from the human propensity to self-delusion, covert violence, and idealism. In Céline, we find an extreme attempt to unveil the trauma of the First World War or more accurately to ridicule any attempt at understanding it so that the word 'trauma' itself is already too portentous, too meaningful. Instead of the nostalgia we saw in Giono or the aestheticization we found in Marinetti, we have a nihilism that would expose the violence of war, and of humans, as utterly meaningless. Just as nostalgia and aestheticization cannot contain anxieties about embodied mortality as the place where violence and love meet, neither can nihilism. Specifically, in Céline, nihilism fails to cut off the imagination and we find, even more than in Marinetti or Giono, that the body refuses to be a mute 'thing.'[48]

The body's revulsion at, and yet participation in violence, along with the self-disgust and misanthropy this occasions, is a recurrent motif in the novel, which evokes, in turn, Bardamu's war experiences, on the battlefield and then in Paris, his stint in the African colonies, his attempt to make a life in America, and finally his return to a poor Paris suburb where he practises medicine, having finished his studies that the war interrupted. Céline presents colonialism, American capitalism (Bardamu works for a Ford assembly plant in Detroit), and French bourgeois capitalism as reinventions of the war itself, since in all of them 'cruelty' dominates. Significantly, though violence is no longer as visible because 'the rich don't have to deal personally with killing to eat,' the stench of war remains pervasive, like death itself: 'People, countries, and objects all end up as smells.' The stench begins early on, as in his first month as a soldier Bardamu spends most of his time wandering in the night, in which 'unknown soldiers' surround him with 'a thousand deaths.' Immediately after he experiences one of his first real horrific battles: 'I'd turned into noise and flame myself. After a while the flame went away, the noise stayed in my head, and my arms and legs trembled as if somebody were shaking me from behind. My limbs seemed to be leaving me, but then in the end they stayed on.' Fear is such that his body almost falls apart. A few pages later, Bardamu is sent to get food, and confronts another horrific scene: 'On sacks and tent cloths spread out on the grass there were pounds and pounds of guts, chunks of white and yellow fat, disembowelled sheep with their organs every which way, oozing intricate little rivulets into the grass round about [...] Blood and more blood, everywhere, all over the grass, in sluggish confluent puddles, looking for a congenial slope [...] [I was] overcome by an enormous urge to vomit, which I did so hard that I passed out.'[49] Unlike Giono, Céline does not poeticize the continuity of animal and human life in order to decry the barbarity of war; Bardamu's reaction is a visceral revulsion, another falling apart of the body, when confronted with bleeding animal flesh that is indistinguishable from human flesh.

Yet the body is also what demands to live, no matter what, when Bardamu returns to Paris, after being wounded. He comments ironically, 'The one thing that's really indecent is bravery. You expect a body to be brave? Try asking a worm to be brave, he's pink and pale and soft, just like us.' A few pages later, he adds: 'The soul is the body's vanity and pleasure as long as the body's in good health, but it's also the urge to escape from the body as soon as the body is sick or things are going badly. Of the two poses, you take the one that suits you best at the moment, and that's all there is to it! As long as you can choose between the two, you're all right. But I couldn't choose anymore, my die was cast!' For Bardamu, war has made him a 'monster' because he can no longer keep animality or the body separate from the soul and bravery: courage becomes impossible because inseparable from both visceral fear and violent rage. Another way of putting this is that the 'image' or 'soul' cannot transcend the body or animal flesh, because imagination itself has become too embodied or, as Leed writes so suggestively, 'symbols become experiences': 'courage' becomes 'slaughter.' A particularly salient example of this traumatically embodied imagination is what finally seems to lead Bardamu to become mentally ill. He had already noted that 'in this business of getting killed, it's no use being picky and choosy ... You've got to act as if life were going on, and that lie is the hardest part of it.' This is what Bardamu cannot do, because he has too much imagination, as we see shortly after, when he is visiting a fair with Lola, and they come to the 'Stand of the Nations,' where you can shoot at tin soldiers with play guns. As Lola incites him, saying he must be good at shooting, Bardamu has visions of his own regiment, and finally yells 'They're shooting at me, too, Lola!' She sees this as 'nonsense,' but Bardamu reflects, 'I think it was then that my head became so agitated, with all the ideas going around in it.' Later he adds that to see men for what they are, 'will give you another self. There will be two of you.' In theory, such duality could save Bardamu allowing him to deal with war yet also to question the 'obscene mystical fascination' of heroism that the war machine would like him to accept; and yet as the novel progresses, and he is dismissed from active service yet never truly cured, we find that this neat split falls apart.[50]

On his way to Africa, Bardamu notes that the heat excites the white passengers' cruelty, because 'the North at least preserves your flesh,' whereas Africa 'sucks' 'devours' 'cleaves' 'sticks' 'burns' to the point that 'rot sets in quickly, out in there the greenery'; 'the White man, encased between his acid sweat and his tropical shirt, poisons himself. It's not safe to go near him.' These images of humans as 'meat' that falls apart, of course, recalls the butchery of the war, and the literal butchery described above. In the end Bardamu sets the rubber-collecting plant he is supposed to be in charge of on fire and disappears. In America, Bardamu's disgust becomes more specific:

> An Elevated Railway train was passing. It bounded between two streets like a cannonball filled with quivering flesh, jolting from section to section of this lunatic city. [...] You can see they don't give a damn whether they're getting what they want out of life or not, you can see they don't even try to understand what we're here for. [...] [Americans have] no trouble with their conscience. I'd seen too many puzzling things to be easy in my mind. I knew too much and not enough. [...] Not much music left inside us to make life dance. And where, I ask you, can a man escape to, when he hasn't enough madness left inside him? The truth is an endless death agony. The truth of the world is death. You have to choose: death or lies. I've never been able to kill myself.

Humans are 'meat' once again, but they can only live by pretending they are not. Bardamu cannot pretend, yet at the same time cannot confront the truth: caught between too much and not enough, he is also caught between a suicide he can't commit and a dance or a delirium he can't quite produce. Once again, he has too much imagination, and in the end loses his job at Ford because 'we don't need imaginative types in our factory.' Significantly, this imagination is once again linked to the body for, feeling himself slowly become 'a different man [...] a new Ferdinand' at the prospect of spending his entire life making the same gesture in the factory, and finding the noise of the machines has come to inhabit his head, Bardamu adds, 'I wanted to touch a real body, a pink body made of soft, quiet life' in order to 'restore my soul.' Yet, in the end, though he is loved by the generous Molly who has 'an infinite heart,' Bardamu insists that 'it was too late to start being young again' and gives in to 'my mania for running away from everywhere in search of God knows what.'[51] We thus find him even more unable to split body and soul in order to make life livable, and caught in the in-between, the no man's land, where the body regenerates a soul in search of love and yet finds itself confronting death and abjection again and again.

In the last section of the novel, Bardamu's vocation as a doctor, when he returns to Paris, seems to come directly from his fascination with this 'endless death agony,' in which the meaning of flesh or incarnation is never given, and yet cannot be denied, in which love and abjection cannot be separated. For in Paris, he works in a poor quarter, and almost seeks out not just illness, but abortion, child abuse, attempted murder, and the death of innocent children, as though to prove that humans are just meat, and an expendable, violent, hopeless one at that. At the same time, his imagination cannot be silenced, or, his embodied experiences keep generating new souls, new experiences of that visceral revulsion the novel opens with, and cannot seem to exorcise. Thus, even as he caricatures Montaigne's famous letter of consolation for the death of a child, Bardamu also admits, 'since Robinson's return I'd been feeling very funny in body and mind, and the little innocent's screams made an abominable impression on me. What screams! Heavens above, what screams! I was at the end of my rope.'[52] This episode recalls the one at the 'Stand of the Nations' cited above, as Bardamu cannot separate himself from the reality of innocent suffering, feeling it in his own body. The return of Robinson is also important, for he is like Bardamu's double throughout the novel: Bardamu meets him during the war, when he is trying to find a way to turn himself in to the Germans to be a prisoner (a thing he fails to do, in the end); rather surrealistically, Robinson is his predecessor at the rubber-collecting plant in Africa, though Bardamu barely recognizes him; he reappears in America, where he is trying to become rich by selling liquor during Prohibition; finally, back in Paris, Robinson takes part in a family's scheme to murder their elderly grandmother so they can have her property. Robinson's cynicism, then, seems to succeed where Bardamu's fails: he is all of one piece, choosing cowardice, oppression, illegality, and finally murder without apparent qualms, a true 'beast' instead of the dual 'monster' that is Bardamu with his overactive imagination. Robinson seems a prime example of why it is better to stop asking the why of things. But every time he reappears Bardamu is by contrast confronted with his own inability to stop asking questions.

The novel ends with Robinson's death which, remarkably, turns into a moment of – albeit minimal – authenticity, not quite an answer to Bardamu's questions but an acknowledgment of their truth. Like Bardamu with Molly, proving that he is not made for happiness, Robinson abandons his fiancée Madelon. Enraged, she pursues him for months, until in a final showdown, he exclaims, 'Wrong! [...] I've got plenty of courage, as much as you! ... Only, if you want the whole truth ... everything, absolutely everything! disgusts me and turns my stomach! Not just you! ... Everything! [...] You wallow in sentiment when you're a worse brute than anybody ... You want to eat rotten meat? ... With love sauce? ... Does that help it down? ... Not with me!' Madelon shoots him in the belly in response. But Robinson is suddenly filled with 'a sort of calm,' and Bardamu concludes, 'He wasn't like the usual sick man, we didn't know how to act in front of him. It looked as if he were trying to help us live. As if he'd been trying to find us pleasures to go on living for. He held us by the hands. One hand each.'[53] This ending makes Robinson the mirror image of Marinetti's Guzzo, and in doing so brings latent violence to light as much as may be possible. First, of course, Robinson is killed by his lover instead of killing her, and his provocation of her rage is so extreme that it has to be read as a death wish, the same one Guzzo cannot quite acknowledge even as he courts a stray bullet. Instead of projecting fear of and fascination with decay outward, Robinson thus, at the very end, recognizes it as an undeniable component of the self. Second, then, whereas to witness Guzzo's death forces us to repeat his violence and denial, Robinson's death is presented as a moment of acceptance that undercuts nihilism, since he reaches out to Bardamu and his friend, with the calm of having had the courage to spit out his disgust to the very end. Ultimately, the physical continuity between Robinson and Bardamu, combined with the former's acknowledgment of self-disgust and the death wish that has been driving him, is as close to a revelation as we can have of how the death instinct unleashed by war and continuity with others are not separable. This is adamantly not any form of redemption but rather a moment of respite from the desire for redemption that, as so much of the novel shows, causes so much hypocrisy and violence.

To conclude, we find in Giono, Marinetti, and Céline a similar anxiety about the violent animality of the human flesh, which is also a glimpse into the traumatic reality of the death instinct as war awakens it. As in Freud, the death instinct can be felt but remains mysterious, subject to various forms of denial, even as it re-emerges in the embodied imagination and, most tellingly, invests sexuality, love, and the desire for the infinite. The First World War appears to destabilize the symbolic order that kept animality at bay, expressing it yet also successfully containing it: thus nostalgia, aestheticization, and nihilism all fail to prevent the dead bodies of wartime from coming back to life. Asserting their continuity with live bodies, they contaminate the living with a violence that prevents the reassuring separation between the inside and the outside of the human and the social body. The corpus mysticum of the nation, as it becomes more fleshy, also becomes more animalistic, more violent, more subject to decay.

That such contamination by animality – as an expression of physical and social disintegration – haunts the Unknown Soldier Memorial is made clear by one last element of his tomb, once again unique to Italy. The casket containing the remains of the Unknown Soldier, initially placed in the stone sarcophagus we see from the outside of the Vittoriano, beneath the statue of Rome, was moved into a crypt located somewhat further up and behind the statue. Most revealing are the three reasons given for this move. Contrary to what historians have claimed, 'introduc[ing] a Catholic presence' after the Lateran Accords in 1928 was only a third one, though it was also the last push towards completion of the move, which began in 1923 but was not final until 1935. The final reburial on 24 May 1935 was, as Atkinson and Cosgrove argue, an occasion for Fascist unity to be consolidated even further in the 'real presence' that a single body could conjure. Nonetheless, before the crypt was decorated with a crucifixion and other decidedly Christian motifs, it had already existed as part of Sacconi's original design for the Vittoriano, which was far more secular.[54] Already in 1924, nineteen years after Sacconi's death, but well before the Lateran Pacts, Sacconi's friend Primo Acciaresi, author of a monograph on the Monument, was demanding that the Unknown Soldier be moved to the crypt, which he considered to be more appropriate because 'in all eras the divinities or heroes to whom cults were dedicated, or the illustrious men for whom mausoleums were erected, had their propitiatory altar or their funerary locus in the most hidden and noblest part of the temple.' Returning to the 1921 tension between heterogeneous, spontaneous expressions of grief, and state desire for a more sombre atmosphere, Acciaresi argued that Sacconi's crypt, which he 'prepared' for the Unknown Soldier as though intuiting the future, would allow for quiet contemplation in a way the outside location could not.[55] The Direzione Artistica or Artistic Committee of the Vittoriano appeared to respond positively to the idea, in a letter of 30 March 1924, which explained how the crypt's Greek cross plan would be adapted to create access. Along with an 'estimate' and detailed 'plans,' the letter insisted on the spare quality of the decoration, which would be mostly white marble, stone from Mount Grappa (a major battlefield), and 'civil and patriotic symbols.' This second reason for the move to the crypt was to intensify unity, then, but not under the aegis of the Church. Even the 20 October 1925 printed leaflet 'For the Sacred Crypt and Perpetual Honours for the Unknown Soldier,' which made official the decision to move the coffin, emphasized patriotism far more than Christian religiosity.[56]

More important, the pamphlet claimed to rectify the fact that the 'tombstone of the Unknown Soldier' was 'deplorably neglected.' This most likely referred to the primary reason for the move, mentioned in a letter of 14 March 1924 from Mussolini to the Minister of Public Works Carnazza: 'In other words we need to find a location that will ensure the perfect preservation of the corpse (which, near the wall where it was immured, appears to have been exposed to leaks that in the long run would have severely damaged it).' The letter mentioned the crypt as a good solution, and concluded by emphasizing the 'demands of piety, patriotism, proper preservation, and aesthetics.' On the one hand, such leaks may have been a real problem, in particular because the statue of Rome and the friezes around it were initially only plaster, created in a great hurry for the inauguration of the Memorial on 4 November 1921. At the time, this fact was remarkably well concealed by the government and the media. Thus, the first mention of moving the Unknown Soldier's casket that I have found dates to 12 January 1923, when plans were being made to replace the plaster statues with permanent stone versions: as this required a temporary removal of the casket (and it was indeed placed inside the Vittoriano, in a gallery, for a number of years), it also made possible a revision of its permanent location.[57] On the other hand, the leaks mentioned in Mussolini's letter reveal a phobia for physical decay, a desire for 'perfect preservation' that is doubly unrealistic: the body had presumably decayed in its first stay in the earth of the battlefield (as mentioned in Chapter 2, it was often referred to as 'bones') and, as it was not embalmed, could only be subject to further decay (though, as we saw also in Chapter 2, people imagined it to be impervious to decay, like the body of saints). Fear of leaks, then, is also imaginary fear of the dissolution of the body politic, and of the outbreak of the sort of animalistic violence that we found in Giono, Marinetti, and Céline. From this perspective, removal to the crypt and arguments against it that took place in 1924 can be seen in a new light, that is, not about the Catholicity of the Memorial, but rather about Mussolini's wish to assert the intactness of the body within. The Artistic Committee of the Vittoriano, though it obediently came up with a plan, also objected in a letter dated March 1924, noting the cost of changing work that was well under way, but most of all insisting on the suffering involved in the original decision. More overtly, editorials in *La Tribuna* of 3 April 1924 announced with a front-page headline 'The Unknown Soldier Must Return to His Tomb.' A dramatic photograph showed 'the darkness of the crypt where they want to bury the Unknown Soldier,' and various articles decried the removal of the tomb from the people, who had devoted so much humble attention to it since it had been inaugurated. We may recall, by contrast, the insistence in 1921 that he return to 'the sun of Italy.'[58] We find, once again, the resurgence of heterogeneous expressions of grief, in opposition to state homogeneity, but with an additional emphasis on the different sort of history created by the first in contrast with the second. The tomb outside allows for multiplicity, accretion, layering, and decay, whereas the symbolic value of the crypt's intactness is to create a history that is outside of time, a history that, as Claudio Fogu has written, 'make[s] [...] present' the past in order to appropriate its power. We are about to see in the next chapter that a more layered history, and the real cost of violence, were part of the Memorial's inauguration in 1921, and remained latent for years to come, even as much was done at the inauguration itself and even more under Fascism to deny them. This latent violence was associated with embodied feminine flesh – the very flesh that the texts of Giono, Marinetti, and Céline, cannot quite come to terms with. As Leone Gessi remarked in the *Corriere d'Italia*, 2 November 1921, 'the idea of honouring an unknown soldier [...] seemed, initially, a rather bad and very dangerous one.' It is to this danger inherent in the nexus of body and spectacle that I now turn.[59]

5 Mutilation and Spectacle

Phantom Pain, Mutilation, and Repetition

> Mutilated soldiers and handicapped veterans [...] [will be placed] not in the area reserved for veterans (Piazza Cinquecento) but instead on the platform in the Station on the arrivals side, near the Saletta Reale, [...]. This is a concession the Military Authorities had to make in order to calm the loud protests by mutilated soldiers, who were not satisfied with the place previously assigned to them.[1]

> People have been putting up posters in the area in the name of the National Soldiers Association, signed Temporary Committee, that incite the crowd to break through the military lines and take possession of the body of the Unknown Soldier, as a protest against the rigidly official quality of the celebrations that exalt him.[2]

The materiality of soldiers' bodies, repressed in the official story of the Unknown Soldier Memorial because it undercut the abstraction of the mystical body of the nation, returns nowhere more saliently than in the role of mutilated soldiers in the Memorial' inauguration. The first telegram above, from the files of the Italian Ministry of the Interior, refers to a last-minute decision to give in to agitation by mutilated veterans who felt they were being positioned too far away from the Unknown Soldier's coffin. A day before the entombment ceremony, when the Unknown Soldier was already in Rome, at Santa Maria degli Angeli, another telegram warns about a pamphlet put out by the Associazione Nazionale Combattenti (ANC, National Soldiers Association) urging people to protest excessive government control over the ceremonies by taking possession of the body. Mutilated soldiers were only part of the ANC, which was founded in 1918; however, this association was an offshoot of the Associazione Nazionale Mutilati e Invalidi di Guerra (National Association of Mutilated and Handicapped Soldiers), founded as the war began, and they are thus closely associated. Overall, it is clear that veterans in general felt they were being kept under control and that mutilated soldiers became the centre of this debate because their bodies' material link to the Unknown Soldier's body via a common suffering was so visible.[3]

Government control was in fact quite stringent since, as the Ministry of the Interior files show, the state was worried about dissent from veterans and their families, regardless of their political affiliation. The state was pushing a far more abstract version of sacrifice, as we have seen in previous chapters, than participants in the war remembered, and it was well aware that it needed to frame any possible eruption of materiality and suffering. Remarkable in this respect are the five large maps – most of them folio-sized, all of them printed, in multiple colours – that block out the space to be occupied by *mutilati* (mutilated veterans; always mentioned as a special category, almost always first), as well as *ex-combattenti* (veterans), *madri e vedove* (mothers and widows), soldiers on active duty from the different parts of the armed forces, various government representatives, and the population at large. The archives of the Italian Ministry of the Interior reveal that special barriers were built in order to enforce these maps, for the government feared 'the public would enter the ceremonies.' As Donna Haraway shows, maps can serve as fetishes when they swerve away from actual life to 'make things *seem* clear and under control'; for her, maps of the 'new world' or of territories to be colonized are prime examples of such fetishism because they replace unknown disorder with a semblance of order. Marie Louise Pratt, in turn, argues that biological maps, such as Linnaeus' plant taxonomy, function in a similar fashion to regulate the disorder of nature via categorization that appears objective but is in reality more arbitrary than it seems. Both of these precedents are relevant for these 1921 Italian government maps, since what needed to be fit into an orderly framework was both a territory or patria that was seen as mutilated and a biological body whose integrity had been undone.[4]

As I am about to demonstrate in the first part of this chapter, maps and mutilation were intimately related in the inauguration and in the symbolic interpretations of the Unknown Soldier Memorial. Already, we can see that they complement each other in an uncanny fashion that, once more, recalls the ambiguity of the corpse as described by Harrison.

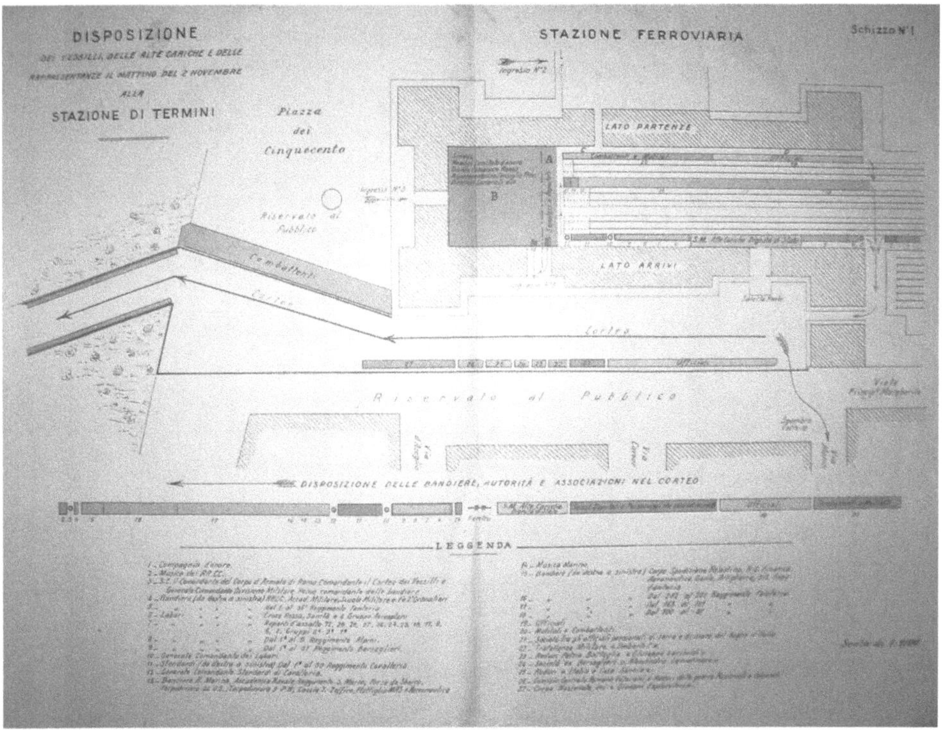

Figure 11. Map of train station for arrival of the Unknown Soldier in Rome. © Archivio Centrale dello Stato, Rome

In the case of mutilation, the absence of a part of the body renders what remains more of a 'thing,' not unlike the corpse in that it no longer has the wholeness or integrity that makes it fully human. In the case of the map, an actual presence is given, but it is a simulacrum, an 'image' not unlike the imaginary afterlife of the dead, which re-establishes order and separation. As mentioned previously, separation is necessary for the completion of mourning; but mourning fails if we impose it too soon, if we refuse to dwell with the corpse, or in this case, if we fetishistically expose the 'thing' yet fail to engage with it. Hence, though government maps might seem merely a question of crowd control, and the presence of mutilated veterans merely a concession to agitators, I submit that *both* were an essential part of the ceremonial inauguration of the Unknown Soldier Memorial, as they worked together to suggest a vision of

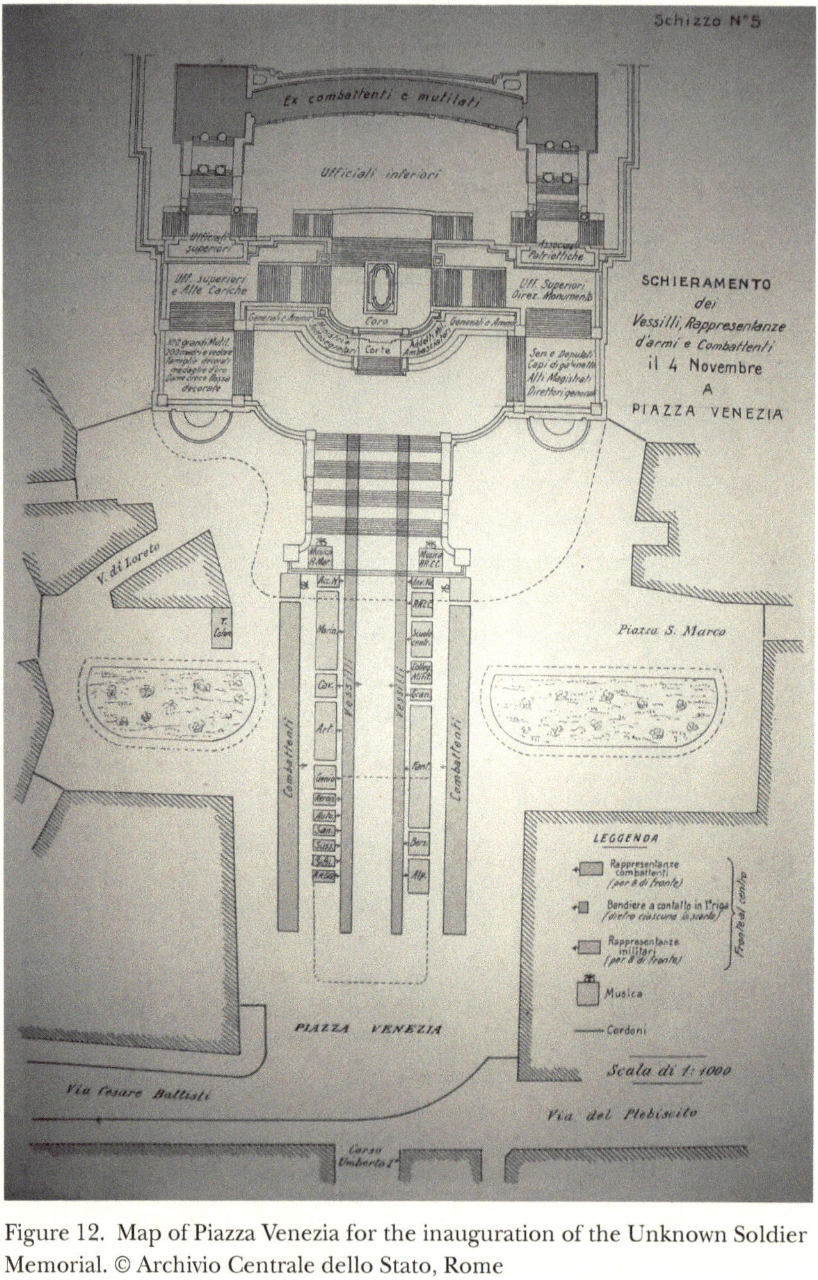

Figure 12. Map of Piazza Venezia for the inauguration of the Unknown Soldier Memorial. © Archivio Centrale dello Stato, Rome

Figure 13. Crowds at Piazza Venezia at inauguration. © Courtesy of Department of Special Collections and University Archives, Stanford University Libraries

horror and at the same time to reassure spectators. Thus, though there was some argument about how close mutilated soldiers were to be placed, it is remarkable that planning for the ceremonies always included them, to the point that one of the many laws passed regarding the creation of the Unknown Soldier Memorial in Italy, proposed by General Diaz, was specifically concerned with the 'placement of mutilated soldiers and wounded veterans.' Similarly remarkable was the organization of the choosing ceremony in Aquileia as, before Maria Bergamas was selected to pick the one to be the Unknown Soldier, it was already decreed that her making the choice would be witnessed by 'the Minister of War, 2 Senators, 2 veterans from the Chamber of Deputies, the Mayor of Rome, 10 mutilated soldiers, 10 veterans, and 10 mothers and widows of missing soldiers (*caduti non identificati*).' Travelling with the Unknown Soldier on his train car on the way to Rome, we find again that along with a handful of politicians and a dozen or so policemen, there are 'two mothers, two widows of fallen soldiers, two soldiers, two mutilated soldiers.' Upon arrival in Rome, as we saw above, mutilated soldiers ended up being right next to the train itself, as well as outside, as originally proposed. At Santa

Figure 14. Soldiers in wheelchairs at the Unknown Soldier inaugural ceremonies in Florence. © Courtesy of Department of Special Collections and University Archives, Stanford University Libraries

Maria degli Angeli, the Unknown Soldier's honour guard included 'mutilated soldiers' again, and '10 mothers and 10 widows'; 'mutilated soldiers and veterans with flags' also accompanied him to the Vittoriano. Finally, for the entombment itself, not only did 'veterans and mutilated soldiers' occupy the entire background, under the colonnade behind the equestrian statue, but 'mothers (100), widows (100), greatly mutilated soldiers [*grandi mutilati*] (100),' as well as 'the families of holders of war decorations' and 'Red Cross women holders of war decorations,' were also placed on the front left side of the tomb (symbolically, they were seated at the Unknown Soldier's, and Rome's, right hand, and thus also associated with Christ).[5]

This veritable obsession with mutilation was, as the list above shows, accompanied by a highly unusual presence of women. Historians have noted the latter as a particularity of the Italian case, not seen in other countries, but they have largely interpreted it as an expression of Italian culture's attachment to mothers. For Claudio Canal, it is part of a dominant rhetoric in Italian monuments that casts the dead as '"sons" to the *patria* who is "Mother"'; this tends to 'dissolve the presence of the concrete woman' who mourns in favour of a '*donna angelicata*' (a spiritualized woman in the Medieval tradition) who like the Virgin sustains her son in death and helps him to be reborn spiritually in the afterlife. As Canal argues, this re-establishes 'the patriarchal order' (the woman is an instrument) and, I would add, associates mourning with abstraction (death is immediately reconfigured as spiritual rebirth: the patria does not actually mourn, for her sons are always already reborn in her). However, the association of women with mutilated soldiers counters the abstraction that Canal mentions. We already saw in the last chapter that we find in authors as different as Giono, Marinetti, and Céline images that conflate the darkness within the soldier's body – an animalistic darkness of continuity with others, fluidity, violence, and mortality – with the female body. What we will see in the coming pages is how this female body, in all of its materiality and transgressive sexuality, returns to haunt the proper mapping of mutilated soldiers. In particular, records show that women did not behave as *donne angelicate*, but rather as 'furies' or, at best, as the *mater dolorosa* who acts out suffering. Thus, Susan Grayzel shows that in England and France (but this was true also in Italy), the more controlled presence of mourning women was undercut by mothers who defied the state to practise 'clandestine exhumations,' which she associates with the exceptional memorials that depict a mother 'not sobbing with grief or love of France but [...] vividly demonstrating her animosity to those who have killed her son'; one of her examples, the Péronne War Memorial, shows a mother crouched over her son's dead body, head held up in anger, shaking her fist towards the viewer. This female body is not abstract but expressionistic, and we will find it in literary texts as well. It is still associated with the Virgin: not the angelic one, however, but the very pagan goddess of generation, sexuality, and death that was so alive in Italian popular religiosity; to the degree that the Unknown Soldier is associated with an incarnation that eschews dogmatic definitions of transcendence, she presides over it.[6]

Though the situation was similar in France, England, the Unites States, and Italy, the actual presence of women and mutilated veterans was most pronounced in Italy, while in the three other cases they seem to have been kept away more effectively. Thus, in France, reports decrying the dreary parades of 'handicapped veterans' on 14 July 1920 were brought up to insist that a strong showing of active duty soldiers enliven the Unknown Soldier's inauguration a few months later. Le Naour discusses a polemic that erupted over the mutilated soldiers' presence, as depicted and made popular in a painting by Galtier-Boissière, and he also adds

Figure 15. Woman weeping near the Unknown Soldier's casket (1921 photo). © Courtesy of Department of Special Collections and University Archives, Stanford University Libraries

that it was the *Journal des Mutilés* that launched the campaign to add an eternal flame to the French Unknown Soldier Memorial after its inauguration, thus pointing to the mutilated veterans' feeling that they had been unfairly given a back row spot. Overall, aside from feminist works, the majority of those who write about the inauguration of France's Unknown Soldier do not even mention mutilation or women, stressing instead its military order. In England, an official order of precedence for mourners at the inauguration was drafted, but it did not mention mutilated veterans specifically, limiting itself to creating a spot for 'ex-soldiers and sailors.' Hanson is one of the few to mention that this happened only after an initial exclusion of ex-servicemen, undone because of widespread discontent. The main problem at the British inauguration was the limited size of Westminster Abbey, where coveted seats not reserved for royalty, government officials, or the military were distributed through a lottery to soldiers, veterans, and their families. As Hanson points out, the result was a disquieting closeness not so much among the mutilated and the healthy, as between different social classes that had never mourned together in this way. Hanson does, however, stress that along the route of the train that brought the Unknown Soldier from Dover after he had been repatriated from France, women were numerous and noticeable. In the United States, crowds came to pay tribute to the Unknown Soldier at the Lincoln Memorial and accompanied him to Arlington Cemetery, but the official parade description mentions only, among veterans, the importance given to soldiers 'who had received the medal of honor,' and notes that each state and territory was asked to send only three men to represent those who fought in the war; this was supplemented by representatives of 'forty-four patriotic, fraternal, and welfare organizations.' Medals, of course, recognize valour and were often awarded to wounded soldiers, but the documents never discuss mutilation specifically.[7]

In all four cases, what was at stake, ultimately, was the widespread perception that the war had made suffering spectacular or even 'obscene' in ways that transgressed acceptable social norms. Could such transgression be ritually healed, and if so how? The association of the Unknown Soldier with mutilation was an attempt to repeat battlefield trauma, and the question therefore regarded the impulse behind, and the result of, such repetition. There is, on the one hand, the fetishism of maps that gives an illusion of control. Here, from a psychoanalytical perspective, the fetish replaces or covers over an unbearable absence. The Unknown Soldier, from this point of view, replaces not just all the missing bodies, but all the missing limbs the war took away. Wartime trauma is repeated, but in a situation where control can be re-established. On the other hand, the expressionistic and transgressive female body reminds us that from an anthropological perspective the fetish is an object temporarily animated by the divine. Condemned by Western religions as a 'fictitious' (which has the same root as 'fetish') incarnation, it, too, is seen as an illusion, and thus its ritual function, of rendering the spirit, for a time, material, is misunderstood. From this point of view, the way the Unknown Soldier reanimates dead bodies and makes new limbs grow is not a regressive refusal to mourn. Trauma is repeated, its monstrosity is magnified or made spectacular, so that we can finally touch it, embody it, and dwell with it – so that it, too, can descend into the human, becoming itself humanized, instead of haunting us from an impalpable beyond. The crucial difference between these two points of view has to do with the distancing quality of vision in contrast with the animating quality of touch. In the first case, the Unknown Soldier's ritual return amounts to voyeurism. Suffering is on display within an orderly frame so that we can watch with detached fascination. This is very much how Fascism capitalized on wartime trauma: in 1922 and later ceremonies, women and civilians were removed, and eventually the Unknown Soldier himself was displaced by less fleshy memorials. In the second case, suffering breaks out of its frame and seems to become contagious. Not unlike the 'melancholy' seen by Freud as a failure to mourn properly, this contagion makes manifest a mortality common to the dead and the living. As Amy Hollywood has argued, when such melancholy remains embodied in experience, when it is not abstracted or reified into a permanently lacking subjectivity, it can also become the 'spectacle' and the increased awareness of the 'historically marked transience of the subject.' In other words, our push to deny mortality, and in particular the violent mortality brought out by warfare, may be temporarily relieved by the cathartic experience of embodying, of physically acting out the death of an other. In this case, the Unknown Soldier's return is not something for us to witness, but something that animates us.[8]

As we have seen, d'Annunzio and Cendrars identified with the Unknown Soldier in their literary texts to which I shall return later in this chapter. As regards mutilation, life and literature converged, in that both writers were wounded in the First World War. D'Annunzio lost an eye in a flying accident in 1916, and spent many months in virtual blindness to heal the eye that remained, whereas Cendrars' right arm was amputated after he was wounded in 1915. Both writers considered themselves to be mutilated (and were officially recognized as such by the state), and crucially, both derived a powerful symbolic authority from this mutilation, which, they claimed, rendered them similar to the Unknown Soldier. Thus, d'Annunzio dubbed himself '*l'arcangelo mutilato*' (the mutilated archangel), and insisted on the imaginative continuity between his broken body, that of 'mutilated victory,' mutilated Italy (without Dalmatian territories), the Unknown Soldier's body, and the bodies of soldiers in general (fissured, falling apart, like d'Annunzio's dead comrades in the *Notturno*). This is confirmed by reviews of the *Notturno*, including one that dubs it 'the great healing book for those mutilated in the body and those mutilated in the soul.' Because vision is part of writing, and at the same time blindness has been associated with poetic inspiration (most famously, with Homer), d'Annunzio could interpret his mutilation as a simultaneous representation of the postwar desire to silence veterans and as an expression of their access to a higher truth that the war experience had revealed. Thus, when d'Annunzio wrote the *Notturno* on strips of paper while blind, this was also a triumph of wartime truths (higher vision) over postwar lies (actual blindness). Finally, as the visionary experience of war is associated with the 'archangel,' we see that for d'Annunzio it constituted a sort of Luciferine annunciation (already present, of course, in d'Annunzio's own name), a new incarnation that was at the same time a revolt against traditional transcendence. Its pagan and feminine undertones are shown in d'Annunzio's reference to his strips of paper as 'sibylline.'[9]

With a very different rhetoric, yet some surprising parallels, Cendrars called himself 'Prométhée mutilé, avec [sa] main de voleur de feu' (mutilated Prometheus, with [his] hand that steals fire). He associated the Unknown Soldier with Prometheus in his *La vie et la mort du soldat inconnu* (as we will see in detail shortly), but also, as a mutilated veteran marching above the Arc de Triomphe in Gance's *J'Accuse*, presented his own mutilation as parallel to that of the soldier below. The right hand, of course, is also essential to writing for most people, and right-handed Cendrars, who was wounded and amputated in 1915, did not return to writing until 1917 when, however, he reported an experience of death and rebirth linked to the sacrifice of his hand, and also to his abandonment of poetry for other forms of writing. His own pseudonym prefigured this, as it indicated a desire for rebirth from the ashes (*cendres* means ashes or cinders in French). He began an initial version of his novel *La main coupée* in 1918 but, significantly, did not complete the work until 1946; moreover, as became typical of Cendrars, it remains uncertain whether the hand in this work is his own, even though he wonders 'whether it has been buried, or instead thrown into the garbage in a mass grave with other remains, bodies, or bits of flesh,' and describes it as 'a great blossoming flower, a red lily, a human arm gushing with blood, a right arm cut off at the elbow, whose hand, still alive, dug into the soil as though to take root.' The two images show, first, that Cendrars' hand could have been buried beneath the Arc de Triomphe (with some hefty coincidence at work), and second, that it lived anew in symbiosis with the land, enacting the kind of incarnation associated with the Unknown Soldier and his materiality. Cendrars' mutilation, then, is also a new incarnation that refuses the traditional spiritualization of suffering: associated with Prometheus and the phoenix, it steals the experience of death and rebirth from the gods and brings it to a human dimension.[10]

The second and third parts of this chapter analyse in detail how emphasis on mutilation proposes a new understanding of incarnation and mortality, in d'Annunzio and Cendrars, as well as other writers, and also in contemporary reactions to the Unknown Soldier Memorial. Before I get to this, however, a reflection on how both writers mythologize their mutilation is necessary. We tend to think of trauma as something too violent to be confronted in the moment, which we slowly inch towards via displaced repetitions, until its truth finally comes to light. Yet, as I showed with reference to the films of Gianikian and Ricci-Lucchi, the delay inherent in trauma, and its excess with respect to available frameworks of understanding, means that we cannot return to a putative unmediated origin of trauma. This is not to say trauma has no 'real' cause, be it physical or emotional or both, but rather that we only have access to trauma retrospectively, through cultural mediation. This is bound to be an uncomfortable realization, for it brings up the spectre of simulation, and the possibility that genuine suffering will be denied. Yet recent work on trauma suggests that in order to heal it is better to relinquish the pursuit of an origin, especially a single origin, and to focus instead on its different interpretations and cultural mediations. As cited early in Chapter 4, Kirkwood thus warns against the 'obstinacy of knowing'; later in the same book, Daryl Paulson and Stanley Krippner show how the trauma of Vietnam veterans can be alleviated by transforming a narrative of humiliation (in which veterans represent a 'shameful episode' in American history) into one that sees their suffering as an initiation ritual (veterans represent a chance America had to 'grow up'); they echo Tick, who in *War and the Soul* describes using the 'Warrior archetype' to help veterans complete their 'rite of passage.' From this perspective, it is quite remarkable that both d'Annunzio and Cendrars seem far less concerned with the literal truth of losing a body part than they are with immediately transforming that loss into a myth that is adamantly their own creation.[11]

In her introduction to *Unclaimed Experience: Trauma, Narrative, and History*, Cathy Caruth explores the 'complex relation between knowing and not knowing' that is central to trauma. To the extent that trauma is an experience not fully assimilated as it occurs, which thus engenders repetition, for Caruth our return to trauma is 'in a language that is always somehow literary' in the sense that 'it defies, even as it claims, our understanding.' This foregrounds once again the inevitability of trauma's cultural mediation, but in evoking the 'literary' implies a connection between healing and creative invention: Paulson and Krippner helped their veteran patients by providing them with anthropological descriptions of initiations that they could use as models for their own narratives. To describe this type of modelling, Caruth comments on Freud's citation of Torquato Tasso's *Gerusalemme liberata* in his *Beyond the Pleasure Principle* (his seminal reflection on traumatic repetition, which I shall turn to shortly). In the passage in question, the character Tancredi unwittingly kills his beloved Clorinda, who is imprisoned in a tree, and thus in Freud's analysis reveals 'the possession of some people by a sort of fate, a series of painful events to which they are subjected, and which seem to be beyond their wish or control.' For Caruth, what is crucial here is the 'voice that is paradoxically released *through the wound*': Clorinda's voice, which makes Tancredi aware of what he has done, and becomes also the expression of Tancredi's own trauma, of the violence to which he, too, is subject. Caruth's book emphasizes how in trauma it is the voice of an 'other,' figured by the wound which undoes the closure and integrity of the self, who 'is asking to be seen and heard.' Woundedness, we might say, needs to speak here, but, as Caruth argues, we need an other to speak our woundedness, to help us even to recognize it as part of the self. We need someone else's narrative to make sense of our pain.[12]

However, in *Trauma: A Genealogy*, Ruth Leys takes issue with Caruth's depiction of the 'contagious effects of trauma,' arguing that her reading of Tasso is tantamount to making Tancredi a victim just as much as Clorinda. In Leys' view, Caruth ends up making trauma 'unlocatable in any particular individual,' thus contributing to a 'collapse of distinctions.' Leys here points, once again, to the spectre of simulation, the danger of denial inherent in saying we have no unmediated access to the truth of trauma. More important, though, is Leys' response to this problem: instead of proposing a 'more correct' interpretation of trauma, or a 'truer origin' for it, she seeks to historicize our understanding. She thus begins with Allan Young's claim that both Post-Traumatic Stress Disorder (PTSD) and trauma are a construct, 'glued together by the practices, technologies, and narratives with which it is diagnosed, studied, treated, and represented by the various interests, institutions, and moral arguments that mobilized these efforts and resources.' Two consequences obtain from this. First, 'mimetic' understandings of trauma (as constructed, or even induced, at least in its specific expressions, by the suggestions of the psychoanalyst, by hypnosis, or by any other type of suggestion), and 'anti-mimetic' ones (which see trauma as an external, objective event) are *both* historical constructs, which arise in reaction to each other. Caruth's emphasis on listening to the other and on the 'literary' is thus an attempt to counter the notion that trauma has to be proven by neurobiological effects. Second, then, for Leys what is more deeply at stake in our interpretations of trauma is not empirical truth but ethical choice. That is, we must look with a critical eye at how our definitions serve given 'interests, institutions, and moral arguments' and ask ourselves what understanding of trauma will be most healing, and to whom.[13] We see here that the initiation narrative offered by Paulson and Krippner to their Vietnam veteran patients contains a complex, if implicit, ethical thrust: the senseless violence of an unpopular war can make sense, but only insofar as it transforms the individual, who becomes an adult capable of making his or her own moral judgments, no longer dependent on those of the state. The ethics of Paulson and Krippner reflect the ideals of representative government, then, 'by the people,' even as they also further a therapeutic self-sufficiency for the subject. When we look back to the First World War, the Unknown Soldier Memorial, and the interpretations of trauma and mutilation it provoked, we must thus, on the one hand, appreciate the healing power of mythologizing suffering and, on the other, ask ourselves which interests are being served and how.

A few remarks on early discussions of trauma, in Freud's *Beyond the Pleasure Principle* and in Jung's response to Freud in 'Confrontation with the Unconscious' are pertinent. Their debate is tied to mutilation and thus to the Unknown Soldier Memorial via the problem of phantom pain. Specifically, Freud's description of trauma that Caruth begins with, cited above, mentions 'possession' and 'being subjected' by something that is not understood. Trauma here behaves like a phantom limb: physical and emotional symptoms remain even as their cause has disappeared. More important, that cause can never be fully known, as is evinced by the notion that phantom pain is an 'illusion.' Indeed, the First World War caused a major shift in the understanding of trauma from physical lesion to wounding of the mind or hysterical shattering of the personality, so that the physical loss of a limb would not, by itself, without the psychological scars related to it, be traumatic.[14] A major image for this idea is that what is lost in mutilation is far more than the actual limb or part that is missing: it is the entire integrity of the body, or indeed the person, who can no longer be whole. The same thing is revealed in d'Annunzio's and Cendrars' mythologizing of their mutilation, for in both cases what is required for some sort of integrity to be re-established is far more than a replacement or prosthetic part, a new eye or a new arm: nothing less than Lucifer's wings or Promethean fire will do. Phantom pain, then, expresses the crucial difference between lack and loss, and suggests that trauma is not about a missing object in our 'memory stores' but about how an experience not fully assimilated, perhaps unassimilatable, causes a breakdown of the self, undoing its boundaries. Another way of putting this is that while lack is still known as part of the self, in experiencing loss the self becomes a part of something that it cannot fully control or understand.[15]

This distinction is also central to Freud's famous discussion of trauma in relation to the child's game of '*fort/da*' (gone/there) in *Beyond the Pleasure Principle*, published in 1920, just as the Unknown Soldier Memorial was being created. In this game, a little boy throws his toy away so that it disappears, but then pulls it back again, only to start over, pushing it away anew. For Freud this repeats the traumatic severing of the child-mother bond. He then gives three hypotheses as to why we have the impulse to re-experience something that was painful. First, he suggests, we repeat in order to modify the final outcome: though we cannot make what was lost in the past come back we can re-enact it, and this time, during re-enactment, what is missing will return, at least symbolically. In terms of the Unknown Soldier Memorial this would mean that recomposing a single body and reburying it at home could successfully make the soldier's death more like a normal pre-war one, undoing its animality and violence, as we saw in Chapter 2. Freud's second hypothesis is that even if we cannot change the outcome, that is, what was lost does not come back, nonetheless by repeating we put ourselves in charge, whereas in the past we were passive victims. This, too, could be the case with the Unknown Soldier Memorial, especially as regards anonymity: for during the war, anonymous death was imposed on passive soldiers, whereas after the war, in the case of the Unknown Soldier, anonymous death is presented as a sign of sacrifice and heroism, a fate he chose because he was this very special type of victim-hero, as we saw in Chapter 3.[16]

What is important about these first two hypotheses regarding traumatic repetition is that in both instances we remain within the pleasure principle, and within lack rather than true loss: repetition undoes, covers over, compensates for, a loss and a passivity that we find utterly unacceptable, and indeed, will never be able to accept, driven as we are by the need for pleasure, defined as control and equilibrium of libidinal forces. Thus, loss is never confronted as greater than we are. But there is a third hypothesis. As Freud writes: 'We are therefore left in doubt as to whether the impulse to work over in the mind some overpowering experience [...] can find expression as a primary event, and independently of the pleasure principle. For, in the case we have been discussing, the child may, after all, only have been able to repeat his unpleasant experience in play because the repetition carried along with it a yield of pleasure of another sort.'[17] This 'other sort' of pleasure remains mysterious to Freud himself, and has, of course led to much discussion throughout the twentieth century about the stakes of traumatic repetition. First, then, Freud's hesitation among different explanations reveals a burgeoning sense of trauma as a historical construct, and one that, as both Young and Leys insist, depends on political and ethical necessities. As I have argued throughout this book, in the Unknown Soldier Memorial we find an unrelenting tension between older forms of mourning (which would claim to be eternal) and the need for new forms (which, constructed, are also unstable and frightening, and too obviously at the mercy of politics as well). To understand trauma, or First World War deaths, as constructed, already brings us closer to loss than lack, for it suggests that no discourse will contain it.

Second, and perhaps more important, Freud's analysis is itself an instance of traumatic repetition: as Luce Irigaray and Jacques Derrida have pointed out, he fails to mention that the child in his example is his grandson, and, more telling, omits that this child's mother, Freud's own daughter, is now dead. You could say, then, that Freud's entire text, his search for a mysterious 'beyond' to the pleasure principle is a case of him repeating, but not entirely or authentically, the loss of his daughter. What is most significant about his analysis, then, is not the positive or rational understandings it offers (the first two hypotheses), but much more so the recurrent instance of a third hypothesis, which is not a hypothesis at all, so much as an expression of obscurity or perhaps even denial, a feeling that something of 'another sort' is going on, such that we are 'left in doubt.' Traumatic repetition points beyond the pleasure principle and, in fact, beyond the notion that the libido always seeks equilibrium; but what it points *to* remains mysteriously absent, or in some way beckons without coming into full visibility. Significantly, we can begin to glimpse it via Freud' own repetitions, and specifically, I propose, in the similar terms he uses to express his doubt about trauma in 1920, seeing it as a shrouded mystery, and to discuss another topic, in his 1930 work *Civilization and Its Discontents.* In this work, Freud reverses his earlier views as he faces rising violence in Europe, and argues that repression is necessary to the preservation of culture and peace. The book opens, however, with a discussion of the 'oceanic feeing,' the feeling of dissolution into the world, of a sort of deeper oneness, which religion, and mysticism especially, often invoke. In Freud's view this is not only a narcissistic illusion, but a dangerous fascination with self-dissolution, with death and violence: under the guise of oneness or symbiosis, then, what the oceanic feeling really expresses is the danger zone where we are no longer motivated by pleasure, and where, by implication, normal mechanisms of repression and containment no longer function.[18]

Freud's description of the oceanic is strikingly similar to his evocation of the mystery that lies beyond the pleasure principle:

> It is always possible that even in the mind some of what is old is effaced or absorbed [...] to such an extent that it cannot be restored or revivified by any means [...] It is possible, but we know nothing about it. [...]
>
> [...] the part played by the oceanic feeling, which might seek something like the restoration of limitless narcissism, is ousted from a place in the foreground. The origin of the religious attitude can be traced back in clear outlines as far as the feeling of infantile helplessness. There may be something further behind that, but for the present it is wrapped in obscurity. [...]
>
> Let me admit once more that it is very difficult to work with these almost intangible quantities.[19]

Notice how the oceanic always lies 'behind' what can be known, beneath 'clear outlines,' and most important, notice how it is 'effaced' from 'the mind' so that if it exists at all – as Freud has to admit it does – it must lie elsewhere, in some more obscure recess of the body. If loss lies at the origin (for Freud, the loss of a father's protection), there is something now beneath or perhaps better *within* that origin, something 'wrapped' inside the 'obscurity' of loss, that is not merely loss itself (linked perhaps to the loss of the mother, left unmentioned in *Beyond the Pleasure Principle,* as Irigaray suggests in her reading). The mind knows loss, or more accurately lack, according to the terms I am using, but within that, the body experiences something other, and 'almost intangible.' Freud considers this *something* to be contrary to pleasure, and to self-preservation, and thereby dangerous; he imagines it to be something like molecules wishing to return to their molecular state, undermining the identity of the organism and the self.

Jung's response to Freud is in this context illuminating. In his 1957 *Memories, Dreams, Reflections,* Jung looks back in Chapter VI, 'Confrontation with the Unconscious,' to his break with Freud and the years right before and during the First World War. Like Freud's in the 1930s, Jung's thoughts are pervaded by the impending violence he senses in the culture around him, but he focuses in this chapter on his efforts to confront the darkness that lies within himself, which seems all the more dark because he no longer believes in Freud's mapping of the unconscious. As is well known, Jung insists that the forces of the psyche are living things, and that they participate in a collective unconscious that binds the living world together, spatially and historically: for Jung, what lies beyond the clear outlines of the mind is the inherence of our small self in a much larger spiritual, archetypal world, which we connect with, precisely, through the oceanic feeling. But two things are to me remarkable in this chapter. First, Jung does not portray the oceanic merely as a positive, mystical experience, in opposition to Freud: rather, he emphasizes its power, and then asserts that it is necessary to *learn* to use it ethically. What lies beyond the pleasure principle, we might then say, is not anarchy for Jung, but responsibility. He asks, how are we to handle something that may indeed be rightly called a death instinct, yet is also a life instinct – something so obscure, so contradictory, that nonetheless cannot be dismissed.

The second remarkable thing about Jung's chapter is his exploration of how we may learn, not to understand but to use or to appreciate this inner darkness. It is largely a non-verbal, non-conceptual exploration: Jung decides to build miniature villages – cottages, squares, churches – as he did when he was a child, with materials he finds in his yard. This building game, which he does not analyse as he experiences it, leads him to have dreams of blood and destruction, whereby he abandons 'heroic idealism,' that is, the notion of a radically autonomous self, who must be protected against any encroachment. This game, of course, repeats childhood and, clearly, repeats not only creation or construction but also destruction and loss; and it does so physically, with all the intensity and silence of the body absorbed in play. As such it does not so much explain as it enacts a different sense of the self, whereby traumatic loss and narcissistic reassertion are no longer opposed. In fact, Jung comes to experience the self as a constant tension between destruction and self-dissolution or between the two poles of the oceanic, one made of violent loss and the other of essential embedded relation to a living world. The point here is that, for Jung, to experience the darkness within – and in particular to experience it with physical immediacy, before conceptual differentiations intervene – is to encounter at once violence and loss as well as spiritual solidarity and a common human mortality.[20]

What Jung adds to Freud, as we try to understand our need to relive trauma, to relive it *for itself*, and not merely to change its outcome or to transform our passivity into control, is a strong claim about its darkness, or also about the connection between phantom pain, the oceanic, and a spiritual sense of humans as connected to each other in a holistic world. Interestingly, Freud and Jung seem to agree, at least implicitly, that within trauma's darkness violence and human solidarity mingle, insofar as individual boundaries break down: but whereas Freud urges us to retreat from this unhealthy mixture, Jung urges us to confront it. And I take Jung's deepest claim to be that at the core of trauma, and of traumatic repetition, lies a confusion that is in many ways degrading, even abject, certainly dangerous, yet also sacred in its ambivalence, because this is not just where horror and common humanity coexist, but where we see the fragility of the human, that is, the constructedness of any sense of solidarity. As a result, we learn that our humanness is not a given, but a full humanity that we must still create, that we create again every time we reinvent or try to reinvent the proper relationship between violent loss and common mortality. Most interestingly as regards the Unknown Soldier, such a reinvention, in Jung, passes through the ritual of playing with construction and destruction, that is, through a personal and embodied reinvention of religious rituals' enactment of the cycle of life and death. As we are about to see, not only was loss rather than lack, and thus the oceanic, made present in the Unknown Soldier Memorial and reflections on it, but also a similar personal reinvention of ritual was in play, in contrast with state scripts.[21]

The importance of a ritual response to trauma, and in particular to the obscurity of phantom pain, was spectacularly illustrated by d'Annunzio in the very years the Unknown Soldier Memorial was being invented. I mentioned his paramilitary takeover of the city of Fiume in Part I, as an attempt to denounce Italy's 'mutilated victory,' and noted that Douhet's *Il Dovere*, which launched the Unknown Soldier Memorial idea in Italy, also claimed in November 1919 that '*il santo volto*' (the sainted face) of the Unknown Soldier, Italy's new martyr, was far more present at Fiume than in Rome. I will now explain in some detail how at Fiume d'Annunzio created a 'style' which, as Michael Ledeen has argued, was innovative because it was linked to 'no single position, no single ideology,' but rather inaugurated 'the politics of mass manipulation, the politics of myth and symbol.' Ledeen, characteristically, tends to emphasize d'Annunzio's charisma and mentions his 'use of religious symbols in traditionally secular contexts' without, however, relating this to the crisis in mourning brought about by the First World War. Claudia Salaris, in turn, rightly stresses the countercultural aspect of Fiume, especially as regards the artists, revolutionaries, and intellectuals that spent time there, but does not show how much this was a reaction to the symbolic break occasioned by the war. Conversely, Emilio Gentile's pioneering work on the religion of the nation draws out how Fascist symbolism reacted to the war, but admits that a thorough estimate of d'Annunzio's influence on it lies beyond its scope. Finally, a re-evaluation of the role of Fiume, and d'Annunzio's speeches there, in the rise of Fascism is undertaken in *Der Dichter als Kommandant: D'Annunzio erobert Fiume,* edited by Hans Ulrich Gumbrecht, Friedrich Kittler, and Bernard Siegert. Together these works provide a rich context for understanding the key link I wish to make between d'Annunzian rituals and trauma. What made d'Annunzio's example so powerful was not only personal charisma, but the creation of a new symbolic language and of rituals that drew directly from the images of decay, mutilation, phantom pain, and regeneration that I have been discussing. Moreover, a properly historicized notion of trauma such as the one I advocated above helps us understand that these symbols and rituals, for all that they were not linked to a 'single ideology,' did have ethical implications. In other words, as both Jung and Young (cited above) claim, to the extent that trauma is inevitably constructed, it furthers certain 'moral arguments' and 'narratives' over others. The appeal of d'Annunzio's Fiume came, in my view, from transforming wartime trauma and its radical breaking of norms into a 'potlach,' a redistribution of cultural and concrete wealth via an anarchic festival. Approaching the question from the point of view of trauma provides a more nuanced, because more focused, view of the vexed question of d'Annunzio's relationship to Fascism. In particular, the years of the First World War and Fiume are a moment of innovation for d'Annunzio (and for Italy and beyond), which will not continue after his fall from the window and coma in 1922 (and after Mussolini's March on Rome). This parallels not only the ossification of First World War mourning under Fascism, but also the increasing lack of interest in, not to say suppression of wartime trauma in European psychology in those years, as documented by Leys.[22]

Before the First World War, the city of Fiume was part of the Austro-Hungarian Empire, and was populated largely by Croats and ethnic Italians (it is now Rijeka, in Croatia), even as it was most often dominated by Hungarians. The war 'served to catalyse the Italian revolt that had been brewing for several decades,' and thus an Italian National Council was created there in the war's immediate aftermath. According to the Italian state, its secret 1915 Treaty of London gave Italy the rights over the city; the French, in contrast, protecting the new Yugoslavia, saw Fiume as Yugoslav. Awaiting a decision to be made at the Treaty of Versailles, the Italians, the French, and the Americans entered Fiume, which was very shortly under a de facto Italian occupation. By late 1918, a conspiracy to seize Fiume organized by nationalistic Italians and Fiuman leaders was developing, but it failed in the face of Franco-Italian conflicts exploding in the city and internationally. At the same time, as I discussed in Chapter 1 especially, Italian society was destabilized by the return of veterans who could not be reintegrated into normal peacetime lives. In particular the arditi 'put political conflict on a new plane, organizing it according to military criteria,' and the ANC made a radical set of demands in January 1919, which included rewriting the constitution, 'the abolition of the senate and the creation of a series of councils elected by the various categories of workers and management, and the reduction of military service to three months' as well as the distribution of unused land to veterans (as had been promised during the war). In this atmosphere of radical discontent 'the idea of an armed advance on Fiume was quite widespread,' and d'Annunzio emerged as the leader of the grenadiers who left Fiume on 27 August 1919 seeking help in taking over the city. D'Annunzio marched on Fiume on 12 September, benefiting from a number of famous defections from the standing Italian army as well as from a paramilitary army of veterans. He would hold the city until 29 December 1920, when five days of Italian shelling would drive him out and honour the Treaty of Rapallo that had given Fiume to Yugoslavia on 12 November 1920.[23]

The story of Fiume's occupation is a complex one, in which a great deal of political manoeuvring took place simultaneously with a fascinating countercultural experiment. Though Ledeen's claim that 'virtually the entire ritual of Fascism came from the "Free State of Fiume"' appears somewhat exaggerated, what his *D'Annunzio; the First Duce* shows very well is that 'the Fiuman adventure had an almost religious significance' not just for d'Annunzio but for a significant portion of Italian society. Thus, d'Annunzio sought and maintained the most unlikely mixture of alliances while at Fiume, with right-wing nationalists, with Mussolini and Marinetti, but more predominantly, with the Socialists and with Russia (these two later rebuffed him), and most of all with revolutionaries and anarchists of various stripes. His 'Carta del Carnaro,' the city's constitution, written in collaboration with his second-in-hand Alceste de Ambris, reflects the anarcho-syndicalist convictions of the latter, but also d'Annunzio's desire to 'transcend the traditional divisions of the political realm.' D'Annunzio was fundamentally interested in fostering a radical cultural experiment, Ledeen shows, which can be emblematized by the role of the ardito Guido Keller, who led pirate attacks on ships and other targets, which provided arms and food for the occupied city, but were more often grand gestures of defiance and ridicule modelled on d'Annunzio's flight over Vienna during the war. In particular, Keller organized the hijacking of the cargo ship *Persia*, which was heading to Russia full of arms and ammunitions for the White Army; after d'Annunzio refused to return it to Italy, the Fiumans were also, amusingly, referred to as Bolsheviks, for they had aided the Russian Revolution. Overall, Keller was emblematic of the 'youthful energies of those who chose to remain [at Fiume] [which] were directed in large part against the symbols and methods of the world of their parents and other disciplinarians. As their adoption of bizarre modes of dress and their experimentation with drugs indicate, these men were interested in new models for their behavior [...], new ideas, indeed, an entirely new world view.' Keller and d'Annunzio organized numerous festivals that were modelled on traditional religious ones, but incorporated both wartime ritual elements and a new sexual and social liberation. Keller and Giovanni Comisso created the 'Yoga' association and magazine, which was a 'union of free spirits tending towards perfection' and was radically anti-establishment:

> YOGA invites you to a new dance – light perfume and music – the excitatory elements most propitious to spiritual beauty [...]
> Dance in the abyss of the sea.
> Dance in the African forest.
> Dance beyond good and evil.[24]

More concretely, the 'Carta del Carnaro' gave communes, both Croatian and Italian, as well as workers' corporations, maximum independence within a highly democratic framework; everyone had the vote, including women (which they did not get in Italy until 1945); religious tolerance was guaranteed, and religious syncretism was strongly encouraged. In terms of actual governance of the city, Ledeen shows that d'Annunzio was lenient with discipline, especially when he took the side of both soldiers and workers against the Italian National Council, which tried unsuccessfully to maintain a more hierarchical order. De facto, Fiume saw significant sexual freedom as well as a revolt of priests that demanded the right to marry and to elect their own leaders. It did not, however, see the communes developing their own forms of governance and culture as d'Annunzio had hoped. Further, under the guidance of the poet Leon Kochnitzky, the League of Fiume or League of Oppressed Peoples was created to counter the League of Nations, seen as insufficiently radical and too linked to capitalism: it obtained support from the Dalmatians, Egyptians, Indians, and Irish, and financed a projected Balkan uprising that would liberate Slovenes and Croatians. Nonetheless, neither 'Keller's world' (as Ledeen dubs the new cultural experiment) nor the new political anarchism of d'Annunzio (with its blend of syndicalism, nationalism, and aristocracy) was able take root beyond the city itself. Ultimately, d'Annunzio's Fiume succumbed to winter and scarce supplies due to the Italian blockade. Arguably, it was never meant to last, since it was an incitation and a provocation rather than a new ideology. Until the end, d'Annunzio himself seems to have hoped that discontented and oppressed peoples in Italy and beyond would spontaneously follow his example by rising up and inventing new forms of political, social, and religious life.[25]

As regards First World War trauma, what is most interesting and most misunderstood about Fiume is its mixture of the carnivalesque with the serious, of the practical joke with anti-establishment mourning, and of idiosyncratic adventures with European and perhaps global modern anxieties. To imagine Jung's childish play with construction and destruction in his backyard as a repetition (or, in fact, prefiguration) of wartime trauma brings a certain discomfort that is alleviated by the seriousness of his reflections on violence which follow. But to imagine the mayhem of Fiume as an insightful repetition of First World War trauma pushes to the limits the importance of mimesis in traumatic repetition, that is, puts great stress on the mixture of what is understood and what is not understood in trauma. Or, in short, it forces trauma to the radical margins; and it provides a narrative about trauma that is far more about contestation than it is about reintegration into society. The obscurity Freud feared is most present here, as is the 'oceanic' that is so clearly a prelude to Fascist ritual, yet cannot be reduced to it. This oceanic obscurity and contestatory understanding of mourning are also part of the Unknown Soldier Memorial: it is not a coincidence that the Memorial was debated and designed largely while d'Annunzio was at Fiume; moreover, a comparison of its 1921 inauguration to Fascist uses of it in 1922 and later shows that the Memorial, too, began as in incitation and provocation, which was later ossified into ideology. In other words, in 1921, the Unknown Soldier could still be and was associated with the anarchy of Fiume, or also with the spectacular and subversive mimesis implied by phantom pain, but this was not so after the March on Rome in 1922.[26]

Specifically, I will examine here two key symbols of the Fiuman adventure that are also central to the Unknown Soldier Memorial and its ritual inauguration, as well as to literary texts about the experience of the First World War and the materiality of soldiers' bodies; I will further consider their transformation under Fascism. The first is the 'Banner of Randaccio,' an Italian flag that belonged to d'Annunzio, and repeatedly served as a shroud. It establishes a mimetic relationship between the soldier's body and the flag in which trauma is primarily understood via touch. The second is the Promethean or apocalyptic horse, which is associated with fire: Fiume was a 'city ablaze, consumed by its own passion,' and it was to ignite Italy and the world; d'Annunzio claimed that he 'stole the horse of the Apocalypse' when he refused to give in to the Italian state.[27] Promethean fire came to represent the ritual ingestion or incorporation of the soldier's war experience by the living: that is, living bodies ingest the trauma of war and thereby transform it, not from destructive to creative, so much as from the ashes of a frozen vision of horror into the embodied, expressionistic acting out and ritual reintegration of wartime violence. Instead of seeing Fascist rituals as more complete versions of what preceded, as has too often been the case, in focusing on these two symbolic elements, I will examine Fiume and related rituals, various texts about the Unknown Soldier, the inauguration of his Memorial, and Fascist mourning as different readings of the Unknown Soldier Memorial itself, that is, as different narratives, driven by different moral arguments, about the trauma that lies at its core.

Touching the Absent Body: The 'Banner of Randaccio'

As mentioned in Chapter 1, d'Annunzio's 1917 funeral oration for Giovanni Randaccio was one of the sources for the Unknown Soldier idea, as in it he claimed that 'Giovanni Randaccio [...] was the soldier' so that to honour his humble courage was to honour every unknown peasant-soldier. Even more important, if not for the final form of the Unknown Soldier Memorial, for its symbolic power and ritual uses, especially the more subversive or innovative ones, was the 'Banner of Randaccio.' Its story is mythical almost as much as that of the Shroud of Turin, which is one of its models. D'Annunzio first fought with Randaccio and his *Lupi* (Wolves) during the ninth battle of the Isonzo, accompanying them, in particular, in their assault on the Veliki (31 October to 4 November 1916). It was at this time that d'Annunzio first gave Randaccio and his company an Italian flag to plant upon the summit once it was taken; during the assault, d'Annunzio and Randaccio would have been killed but for a shell that did not explode; in the end, success on 1 November 1916 was illustrated in the *Domenica del Corriere* of 19 November with a drawing showing Randaccio, tattered flag wound around his chest, about to reach the summit, followed by d'Annunzio. On 2 November d'Annunzio gave Randaccio another pristine flag, to celebrate his courage and his

Figure 16. Legionnaires at Fiume. © Laura Wittman

new military decorations; but on 3 November Randaccio famously removed the azure markers indicating silver medals from his uniform, announcing to his company that since they had not received them too, he would no longer wear them. D'Annunzio imitated Randaccio, though only for the duration of his stay with his company. This too wound up in the *Domenica del Corriere.*

Less than a month later Olga Levi, d'Annunzio's Triestine lover, gave him an oversized Italian flag so that he might plant it somewhere on the summits of the battlefield, whence it might be seen from Trieste. Participating months later in the tenth battle of the Isonzo (12 May to 4 June 1917), d'Annunzio joined Randaccio's company once again, on 23 May, in the Battle of the Timavo, bringing with him the oversized flag. It was in the desperate, bloody attempt to capture 'Quota 28' of the mountain, on 28 May, that Randaccio was wounded and laid upon the flag, which d'Annunzio brought to his side; he died the same day at the Monfalcone hospital, and the bloodied flag covered his coffin at the funeral in Monfalcone, on 30 May, as celebrated in the *L'illustrazione Italiana* of 17 June and the *Domenica del Corriere* of 24 June. As mentioned earlier, the commission in charge of exhumations was depicted in newspaper accounts as especially impressed by the visible 'martyrdom' represented by a skeleton from the Timavo area; d'Annunzio was invited to participate in exhumations near the Timavo because Randaccio had fallen there; Palanti proposed a mausoleum be erected at the Timavo's sources; and finally, water from the Timavo River was used to bless the Unknown Soldier. The location thus brings together Randaccio and the Unknown Soldier. Moreover, to its strategic importance and the battles fought there is added the mythical status of the Timavo River itself: as it runs underground for over forty kilometres, its sources, eventually known to be three, were as alluring as those of the Nile. Their defence made Randaccio a mythological hero.

A month after his death Randaccio was reburied at Aquileia, in 'redeemed land' (*terra redenta*), as d'Annunzio wanted, and d'Annunzio gave his funeral oration, 'La corona del Fante' (The soldier's crown or wreath). At this reburial, the bloody tattered flag was unfurled over the coffin again: it was in this second instance, where traumatic repetition takes place and points back to an unknown perhaps unknowable origin of trauma, that the flag was symbolically detached from Randaccio, in particular, and became, instead, the shroud of *il fante* (the foot soldier) and, thus, of all who died in the First World War. It is, therefore, telling that this reburial and the reappearance of the flag generated the apocryphal story of d'Annunzio throwing another flag to an 'unknown' soldier who was rushing to the top of the Faiti in 1916; just as he was about to receive it, the soldier was shot in the forehead – as a reader of *Epoca* and the poet Ada Negri imagined the Unknown Soldier to be (see Chapter 2) – and d'Annunzio gathered his blood in the flag; thereupon both the identity of the soldier and the flag were lost in the chaos of warfare. This apocryphal story shows well how the belatedness of trauma, its emergence only in repetition, casts doubts on its originality, and even more specifically on the notion of a single origin.[28]

D'Annunzio's oration for Randaccio, published in May 1918 by the Print Secretary's Office, was widely distributed among soldiers, in Italian (in the volume *La riscossa*) and in English translation (in *The Rally*). The oration develops a complex symbolism of chorality in death based on the events mentioned above, and it also is the basis for one of d'Annunzio's most famous political speeches, delivered from the balcony of the Campidoglio on 4 May 1919 and published later that same year in the volume *Contro uno e contro tutti.* On this occasion, d'Annunzio again unfurled the oversized flag, which bore the 'blood and lymph' (*sangue e sanie*) of Randaccio, over the balcony, announcing:

> This, Romans, this, Italians, this, comrades, is the flag of this hour.
>
> The sublime image of the soldier who leaned his head upon it has remained impressed in it.
>
> And it is the image of all the dead; for all who have died for the Patria and in the Patria resemble each other.
>
> It is the shroud of sacrifice.
>
> [...] I will lower this flag, until Fiume is ours, until Dalmatia is ours.
>
> Every good citizen should lower his flag, in silence, until Fiume is ours, until Dalmatia is ours.[29]

Among many other newspapers, Douhet's *Il Dovere* celebrated d'Annunzio's gesture at the Campidoglio a few days later, on 11 May 1919. This was well before any official proposals for an Unknown Soldier Memorial, in Italy or elsewhere. It is thus particularly significant that d'Annunzio not only sees the flag as an image of 'the soldier' and thus of 'all the dead,' but that he insists on how this image was produced. It is the impression of one body, and even more one face, that leaves its trace on the fabric. As with the Shroud of Turin, the point of such an image is not visual correctness but physical continuity: we do not see suffering here, but rather touch it; and the traces of one dead body ultimately allow us to touch all the dead. This recalls Caruth's analysis of how the tension between understanding and not understanding inherent in trauma, and thus the importance of imaginative re-enactment in contrast with rational understanding, can be refigured as a tension between the visual and the tactile. Her example is the film *Hiroshima mon amour,* made in 1959 by Marguerite Duras and Alain Resnais, which puts in parallel the trauma of a young French woman who witnesses the killing of her German soldier lover towards the end of the Second World War, and the cultural trauma of the atomic bomb the Americans dropped on Hiroshima. In repeated flashback sequences (as the film is set after the end of the war), we see the young woman acting 'insane' after her lover's death: locked in the basement by her dishonoured family, she bloodies her hands, bites the walls, leaves bits of blood and flesh on them, and in other ways enacts the breakdown of her physical integrity. This breakdown is linked to her later claim that she could no longer tell the difference between her living body and the dead body of her lover; their flesh was continuous, and the distance normally established by seeing and speech was abolished. A similar tactile continuity between the living and the dead is established by the Banner of Randaccio, for as we are about to see, for d'Annunzio, via the shroud we can continue to touch the suffering of a body that is absent or has fallen to dust.[30]

D'Annunzio's unfurling of the Banner of Randaccio became central to rituals at Fiume, where it was used regularly not only to mourn and commemorate, but as a relic upon which loyalty was sworn, and as an incitation to anarchy and subversive symbolic action. Thus, on 12 September 1919 it was used to consecrate the taking of Fiume, and then on 7 October 1919 to bury the two aviators who were the first to die for the cause of the city. Later it was unfurled during a trip by d'Annunzio to Zara, asserting the spread of Fiume's 'fires.' He also brought it out during a referendum regarding Fiume's National Council, to indicate his refusal to abide by their rules. At the end of the Fiuman enterprise it was used to commemorate those who died when the city was finally shelled by Italians during the 'Bloody Christmas': first on 27 December 1920 when they mourned d'Annunzio's troops, and then, more significantly, after d'Annunzio surrendered, on 2 January 1921, and on 3 January 1921, when they mourned both d'Annunzio's troops and their Italian opponents. By November 1921, in the *Notturno,* d'Annunzio would ask, rhetorically reminding readers of the link between his earlier conception and more recent events, 'And why wasn't the coffin of the Great Sufferer [the Unknown Soldier] cloaked with the flag of the Timavo

[...]?' Randaccio's flag, known as the *lábaro del Fante* (the foot soldier's banner), but also as the *bandiera del Timavo* (flag of the Timavo), was undoubtedly d'Annunzio's most eloquent commemorative move, and the official inventors of the Unknown Soldier Memorial at once borrowed and subverted its symbolism: the idea of commemorating all the dead through the traces of a single body remained, and much of the Memorial's power derives from its physicality and its link to trauma, even as it was intellectualized and repoliticized in different directions.[31]

The association of the flag with the 'shroud of sacrifice' or the Shroud of Turin not only makes Randaccio, and with him every soldier, an *alter Christus*, it also focuses – rather uncomfortably for present sensibilities – on the fleshiness of resurrection. The shroud is meant to have enveloped the dead body of Christ after the Passion, and it bears the imprint in blood and 'hematic substances' not only of his physical features, but of his flagellation and Crucifixion. What the shroud signifies, what d'Annunzio borrows from it, is a surprisingly concrete insistence that, even as bone is what appears to endure while flesh falls away, flesh too endures – and endures in its suffering. This is in direct contrast to the blankness of bones emphasized by the official creators of the Memorial, analysed in Chapter 2. The oration for Randaccio makes the imaginative connection clear, as d'Annunzio addresses all the *fanti*:

> Your [heroism] is like your own bones, it is within you as your skeleton is; it is your internal armour, it is always there; it holds your weak flesh together and keeps it continually exposed to the most awful destruction. Your life is as the bunting of your flag, and your courage as its staff. There are hoisted flags that the wind rends and tears away; but the staff remains. [...] Heroes; and yet you seemed things rather than men; poor things like crumbled stones, disembowelled sacks, shapeless helmets, empty boxes, broken bottles [...].
>
> It is said that there is a life, and that there is a death. One lives by life and one dies by death. It would seem true. But for the soldier something exists which is not life and something which is not death; a new element, a kind of suspended limit, a kind of mysterious brink where breathing is impossible; whereon, however, he manages to breathe and often laughs and sings and does not perish, because I maintain that that element is man's infinitesimal immortality.

Flesh exposed to destruction, like a lacerated flag, like shards of stone, like a disembowelled sack, is that infinitesimal limit where immortality is still in the present of *resurgo* ('I am resurrecting') and in its uncertainty. ('*Resurgo*,' you may recall, was the inscription d'Annunzio thought would be appropriate for the Unknown Soldier Memorial, which at the end of Chapter 2 I linked to the silence, liminality, and references to Limbo that were part of the creation of the Unknown Soldier Memorial.) Not only does d'Annunzio here eschew triumphalism by depicting the soldier as a victim or *miles patiens* (the suffering soldier), as Paolo Valesio has shown, he also thematizes the difficulty of commemorating victimhood without glorifying it per se. A true memorial must seek the edge between bone and flesh, between 'the staff [that] remains' and the 'hoisted flag that the wind rends and tears away,' between spiritualization and horror.[32]

A similar point was made when d'Annunzio imagined a ruined church to be the ideal memorial, as I discussed in the previous chapter, when comparing it to the Trench of the Bayonets: in his 'La Chiesa di Doberdò,' published in *L'Ardente* of 1 July 1921, d'Annunzio thus evoked 'the battered peeling pierced helmets, one on top of the other, grey like ashes, the leather inside macerated in sweat, infused with blood.' In his diary entry of 24 September 1916, when he visited Doberdò, d'Annunzio noted: 'The christ – The shoes the helmets remains of the dead – The Christ carrying the cross – the VIth station over the main altar Leaning on the wall the stretchers spotted with blood.' This entry contracts and anticipates remarkably the power of the Banner of Randaccio a year before Randaccio's death. Helmets and shoes emphasize the pathos of bodies reduced to objects, but also insist on the importance of physical traces of suffering for they are blood-soaked 'remains.' These are associated with the Sixth Station of the Cross, when Christ is vilified on the way to his Crucifixion, which again insists on the Passion and not the Resurrection. Most telling, however, this Sixth Station is associated with the blood-stained stretchers leaning on the wall, which reminds us that this station is also that of the Veil of the Veronica. As we saw, this veil is said to have been used by Veronica to wipe Christ's face during the ascent to Calvary, when his image remained impressed upon it; the veil itself is known, metonymically, as 'the Veronica,' suggesting a transmission of suffering between flesh and cloth. In all of these images, what is once again central is the fleshiness of suffering and the need for a remembrance of it that bears its direct physical traces.[33]

This is also what d'Annunzio sought to emphasize by choosing the inscription 'Veni foras Lazare' (Lazarus, come forth) for the wreath brought to the Unknown Soldier by his Fiuman legionaries in an unscripted part of the Memorial's inaugural ceremony. In a reprise of his description of the Unknown Soldier breaking open his Memorial in the 'Comento meditato,' d'Annunzio makes this connection more explicit:

> When Lazarus raises his knee out of the tomb, he expresses himself more than if he cried out the immortal cry of his resurrection. [...] I was telling you earlier: 'I want to help the dead express themselves.'
>
> This wounded knee, this exposed shin bone, this crushed ankle, all this fleshless and impoverished pain, I raise it here with my will.
>
> There is no reason to recoil in horror. I am expressing myself. This is the strength, this is the style of my expression. [...] I will be silent now. I will not speak.

This image is in close parallel to that of the *non finito* Christ, quoted at the end of Chapter 3, whose foot is still caught in stone, but who 'hurls himself at the sky.' Now, by replacing the Resurrection of Christ with that of Lazarus, who awakens not knowing how or why, d'Annunzio emphasizes even more the impenetrability of the nexus between death and resurrection: Lazarus' flesh is not transfigured but returned to life and to suffering. Randaccio's 'resurrection' is said to be of this same sort: 'tomorrow our dead, the first among martyrs, will rise up and open the tomb of Giovanni Randaccio.' He is raised not by God or even by Christ now, but by dead soldiers; by implication, then, he returns not even to everyday life but to no man's land. This was very much how d'Annunzio also envisioned the return of his dead comrades in the *Notturno*, for instance:

> [The corpse of Giuseppe Miraglia] could not rise up even if Christ called it. [...]
>
> It is the third day. The decay continues.
>
> I feel a terrible cold in my bones. To touch death, to mould death, with a living heart!

Called by Christ like Lazarus, this time the soldier cannot even rise up; rather, his flesh continues to move 'downward,' reanimated but according to a trajectory that is the opposite of transcendence. Hence, after the 'third day,' when we would expect resurrection, 'decay continues.'[34]

In a later reflection on Lazarus, in his 1924 *Faville del maglio,* d'Annunzio pulls together all these elements in another recurrent image, that of the '*luna logora*' (rent or torn moon). In the text 'Gesù e il risuscitato,' d'Annunzio imagines that on the night of Gethsemane, Jesus is so anguished about dying that he seeks out Lazarus in order to learn what lies beyond death. Significantly, Lazarus has nothing to say, and the text ends with Jesus overwhelmed by the smell of death, as 'the strips of linen soaked in ointment now piled up gave off odour.' D'Annunzio is here reinterpreting a vision of Lazarus that developed throughout the decadent and Modernist era, in which Lazarus' silence about 'the beyond' is taken as an acknowledgment that there is no resurrection, at least in the traditional Christian sense. Thus, in d'Annunzio's brief tale, the 'rent moon' appears at the end in the darkness like a shadowy 'Veronica,' perhaps also like a deflected image of Lazarus' own face, a face of suffering that is unreadable in terms of resurrection. (This white round moon with dark patches might also be imagined as a 'lacerated' Host.) This recalls how, in the opening of 'La chiesa di Doberdò,' Saint Francis is imagined, '*lacero e logoro*' (lacerated and rent), and this, in turn, is echoed in Randaccio's funeral oration, where the flesh is like a flag that the wind '*lacera e rapisce*' (lacerates and tears away). Laceration and, in particular, the adjective *logoro*, which implies rending but also wearing away, fraying, once again conflates flesh and fabric, in a suffering that is physically transmitted rather than represented visually.[35]

This sort of suffering recalls d'Annunzio's own state as he is convalescing, writing the *Notturno* in the dark, when visionary knowledge is presented in constant confrontation with tactile non-knowledge:

> You bind my forehead, bandage my eyelids, leave me in darkness.
> But I see. I still see. Day and night, I see. [...]
> I look at the watery abscess around my swollen eye, at my worn and lifeless face, at my mouth livid and bent in sorrow, at the new white hairs in my unkempt beard, at my scrawny neck: an image of miserable sadness, fixed upon my retina permanently. [...]
> And the transformations begin.
> Human sadness has become a plastic substance. I know not what mysterious thumb incessantly models it. [...]
> I no longer feel the borders of my skeleton.

With his bandaged eyes, d'Annunzio is like his later description of Lazarus, above. And here we see how the image of suffering fixed upon his inner eye leads not to understanding but to 'plastic' 'transformations' of the 'matter' of 'human sadness.' This sort of metamorphic experience reminds us not only of Cendrars' evocation of his lost arm as 'a great blossoming flower, a red lily' (cited above), as well as Giono's evocation of cadavers that move (in Chapter 4), but brings us back to Randaccio. D'Annunzio's diaries from his time with the Lupi tell of running up against a badly buried cadaver, which he must prevent the soldiers from stepping on, saying: 'There is an arm there. Do not step on it. It points towards the enemy. The fist is almost clenched. Look. The fingers are all mangled, except for the index. Look. [...] It marks our front line [...] This dead man wants his rifle back. Do not step on him. He is coming out of his grave. Tonight he will guide the company... that martyr's arm, a dry, shrivelled bunch of bones, tendons, and cartilage, seemed to tremble, as if it too were inside of me.' In this resurrection what is emphasized again is the dead soldier's descent into the flesh, into 'bones, tendons, and cartilage'; and notably, this descent is choral or plural, in that d'Annunzio now feels this trembling dead arm to be his own as well.[36]

One last image from d'Annunzio's oration brings together all the elements we have been discussing – the physical connection of the living to wartime suffering via the 'rent' banner of Randaccio, the fleshiness of resurrection, and Lazarus' return not to transcendence but to no man's land – all these come together in the double claim: 'i fanti mordevano l'azzurro' (the foot soldiers bit into the azure), and 'their chests were penetrated by that azure of courage [...]' (avevano il petto attraversato da quell'azzurro della prodezza [...]). At the ceremony inaugurating the Unknown Soldier Memorial, this meant that the sky was reflected in the 'azure signs' worn by the 'veterans' 'in the first row.' These represented gold and silver medals received, ennobling them, just as the Unknown Soldier was elevated in status by the gold metal conferred upon him by the Italian state, as is implied also by the statue of Rome rising above his tomb. At the same time, just as Randaccio, as the oration recalls, 'out of his humble regard for the thousands and thousands of unknown heroes, [...] one day took the blue stripes off of his breast,' the Unknown Soldier symbolically gave his medal, and the pension that came with it, to the association for war orphans (as did d'Annunzio): it was evidently important that he also remain a humble *fante*. We can see already that here the azure, traditionally associated with transcendence, is inscribed on the body, with the full physicality of *mordere* (biting); moreover, the rejection of traditional transcendence is one with the refusal of honours bestowed by the state or any other external authority. Further uses of the image in the 'Comento meditato' make clear that such refusal of transcendence and authority is tied both to mutilation and to the transmission of suffering via imprint and physical continuity: 'The soldiers bit into the azure. But azure turned to red. I thought all of them had my heart as their crimson banner [...] I will let you enter my chest through this dark bruise that is like an azure opening.'[37] The wounded chest, and mutilation itself, here becomes its own 'azure sign' needing no further recognition. At the same time, azure turns to red and into the dark bruise, emphasizing the physicality of suffering as did the image of soldiers as 'disembowelled sacks.' Further, the bruise visually recalls the 'rent moon,' with its dark spots against whiteness. Phantom pain, we see here, is expressed not in imagining the return of what is lacking (the closure of the open chest, the fullness of the moon or of the face) but in the monstrous flowering of loss ('*rosseggiare*' – turning to red – evokes red cloud covering over the azure). Finally, through this 'opening,' the dead soldiers and d'Annunzio interpenetrate, and life and death contaminate each other.

Though he is writing a good four years before the Unknown Soldier Memorial was inaugurated, it is remarkable that in Ungaretti also we find images that emphasize the fleshiness of resurrection – or perhaps better, of the moments of transfiguration or respite that wartime allows. In the following two poems, which thematize the night as does d'Annunzio's *Notturno*, the infinite, which would evoke traditional transcendence, becomes a 'weak touch' first, and then seems to withdraw even further, leaving the poet only able to touch his own face. This gesture recalls that of Michel Serres in the opening of *Les cinq sens*. Significantly, Serres' meditation on touch follows a traumatic confrontation with death, which is also an extreme experience of his own embodied soul, of the inextricability of body and soul, and of the truth and mystery of mortality. For Serres, touching one's own face puts the 'self' actively on both sides of touch ('in' the hands, 'in' the face) and thus emphasizes both its unity with the body and its mobility or indeed transcendence of a specific physical location.[38]

Ungaretti's experience is similar in that touch leads him to understand his singularity yet also his absolute distance from the infinite. That is, touch – the self knowing itself as embodied but never as a fixed 'thing,' for in touching itself it is more than one thing and yet one – throws him back on human mortality in its mystery:

Still Night
Vallone, 18 April 1917

My squalid

life expands
frightened of itself

In an
infinite
that tramples me and
presses upon me with its
weak touch

Another Night
Vallone, 20 April 1917

In this darkness
with frozen
hands
I make out
my face

I see myself
deserted in boundlessness

Moreover, in another poem, the tactile becomes the basis for Ungaretti's chorality, allowing a passage from the singular 'body' who experiences sorrow, longing, and most of all mortality, as 'plastic substance' in d'Annunzio's words, to the plural 'bodies':

Melancholy
Quota Centoquarantuno, 10 May 1916

Melancholy sinks upon the body bound
to its destiny

Nocturnal surrender sinks upon bodies
by souls fully taken
in the boundless silence
that eyes do not see
yet a foreboding

Sweet surrender of bodies
Heavy with bitterness

pursed lips
in replication distant lips
cruel gratification of bodies spent
in insatiable desires [...]

The poem opens with an evocation of body and soul as inextricably yet violently bound, and expands it in the second stanza to a collective experience of 'corpi / a pien'anima presi' (bodies / by souls fully taken). This image is striking for it recalls expressions such as 'prendere colpi a pieno il viso,' as in 'colpi a pieno il viso presi' (blows by the face fully taken): this implies that the body is somehow hurled at or against the soul, that the soul is physically wounded by, in, and through the body. In the third stanza, the 'melancholy' of mortality is evoked again as a choral experience, and the violent link of body and soul is expressed in the oxymoron of 'sweet' and 'bitter.' Finally, recalling d'Annunzio's claim that the face on the Banner of Randaccio enacts a physical continuity among those who experienced the war, in Ungaretti's poem the 'lips' of wartime suffering mime the curve of 'distant lips.' Though there is no shroud here, Ungaretti's images of 'la notte violentata' (the night brutalized) and 'l'aria [...] crivellata' (the air riddled), and in general of the rocky and rent landscape of the Carso in *L'Allegria* suggests that for him, too, the First World War was associated with an awakening to his own embodied mortality that extended to others not via vision or understanding but via imprint and physical contact. Not unlike d'Annunzio's 'rent moon,' which becomes an image of the body worn away by suffering, then imprinted on Randaccio's banner, Ungaretti imagines in 'Night Gully' that his body is 'used' or worn by time, becoming both a 'crumbled/leaf' and a face 'dry/like parchment' (see Chapter 2 for the whole poem).[39]

The link between trauma and transmitting the experience of embodied mortality is not only part of how the Unknown Soldier Memorial was imagined, but was also of how it was expressed at its inauguration, in two related ways. The first has to do with the unusual presence of lacerated flags. In Italy, these included 'historic flags' among which was a 'Risorgimento relic'; as the term 'relic' implies, the connection between body and flag so eloquently established by d'Annunzio was widespread. D'Annunzio himself had wept over a lacerated flag during his early battles with the Lupi, before a different flag became Randaccio's banner, and as the imagery from his oration reveals (cited above), the mobility of the flag (as opposed to the staff) is associated with that of the flesh, indicating its 'infinitesimal immortality,' a heroism no longer heroic but inseparable from the abjection of 'poor things like crumbled stones, disembowelled sacks, shapeless helmets, empty boxes, broken bottles.' Lacerated flags, then, bear the imprint of mortality and its mystery just like Randaccio's banner. They are far more than a representation of the 'the nation' that has been wounded by war because they bring back physicality to the mystical body in that they bear traces of actual wounds and weapons. In France, seventeen lacerated flags that had been returned by the Germans were prominent in the inaugural ceremony, whereas in England, the Unknown Soldier's coffin was covered with what was known as 'the Padre's flag,' a flag used by Father David Railton to cover the altar during wartime masses, described as 'old, faded.' (Railton was credited with the idea of an Unknown Soldier Memorial by the British, see Chapter 1.) In this last case, the expression of a muted or 'lacerated' immortality is striking in its closeness to d'Annunzio's interpretations of the Banner of Randaccio. The flag, in both instances, in its decay bears the imprint of incarnation in its fleshiness and by remaining lacerated (by not being replaced with a pristine flag, as you might expect) rejects the transcendence that would normally follow and, by implication, the bellicist heroism that claims that transcendence is achieved by dying for the patria.[40]

The second way embodied mortality was present at the Unknown Soldier Memorial's inauguration was through mutilated veterans. I explained in the opening of this chapter that their participation was both carefully choreographed and seen as indispensable, and at the same time feared as unruly or excessive. In other words, as state maps show, mutilated veterans were there in a very fetishistic fashion, to point to something – yet also immediately, it was hoped, cover it over. We are now in a better position to understand that something as embodied mortality, and also as incarnation without traditional resurrection, rather than as the simple absence of a limb. D'Annunzio's deployment of the Banner of Randaccio, his oration and speeches, but also the imagery in Ungaretti's poetry, all help us to grasp that the mutilated body is one in which spirit and flesh intersect, and struggle, in one metamorphic experience of suffering. Returning to Serres' description of touching one's own face as a knowing oneself at once as one body and as a mobile multiple self, we might say that in the case of a mutilated body, part of the self now lies elsewhere: like a phantom limb, it is at once physically felt, yet mysteriously exceeds the limb that is lacking, pointing to the place where body and soul connect, as something at once concrete and imaginary, palpable yet invisible. Recalling Caruth's claim that in *Hiroshima mon amour* the disintegration of the French woman's body expresses her continuity with the trauma of her lover's death, we can also grasp that the mutilated body maintains a physical connection with wartime experience which is very different from vision or rational understanding and more like symbiosis.

Significantly, in Italy, newspapers commented on how many of the mutilated veterans present at the ceremony were blind, and one writer speculated that in their apparent calmness they were remembering battles and hospitals. This refers to d'Annunzio's blind convalescence and visions of decay in the *Notturno*, and reveals how much the mutilated veterans made the battlefield palpable – not visible or comprehensible – for spectators at the inauguration of the Unknown Soldier Memorial. The phantom limb, then, functions like the '*sangue e sanie*' on the Banner of Randaccio, it is a material trace of the flesh's suffering and confrontation with mortality. Ultimately, it says that this self is at once 'not whole' and yet 'whole' in the sense that it cannot fully know its own finitude (it cannot know what it is missing except as phantom), and yet such finitude is a primal and choral experience (the phantom limb provokes a physical reaction in viewers, it imprints us with mortality). For the state, such imprinting and chorality had to remain orderly in the sense that the distancing effect of vision was supposed to control the more dangerous contagion of physical identification. Mutilated veterans were supposed to 'escort' and 'welcome' the Unknown Soldier's coffin, but when a large number of them demanded to be in immediate physical contact with the coffin as it came off the train that carried if from Aquileia to Rome, this raised alarm. Similarly, mutilated veterans were accorded a place of high honour at the Vittoriano, but also effectively kept away from the crowd of mourners. As we will see in the very last part of this chapter, this did not, in fact, stop mourners from having an intense physical reaction to the Unknown Soldier Memorial and its mutilated materiality that these soldiers also expressed.[41]

Before I get to that, however, it is important to compare the 1921 inauguration ceremonies to the later uses of the Unknown Soldier Memorial under Fascism, when the materiality of suffering, and the embodied sense of mortality were completely and deliberately erased. In 1922, 4 November was only a few days after Mussolini's March on Rome, and accordingly he celebrated the Unknown Soldier with great pomp but without major changes with respect to the inauguration, aside from a crucial one: his presence. In his article, 'Il Duce e il Milite ignoto: Dialettica di due corpi politici,' Enrico Pozzi argues that Mussolini was absent from the 1921 inauguration in order not to be seen as marginal (he was not as important a war hero as d'Annunzio, for example, nor was he a famous mutilated veteran such as Carlo Delcroix), but also, far more importantly, because 'the Duce is the living form of the Unknown Soldier, his dynamic reincarnation,' and in 1921 there was too much emphasis on death and decay. Though Mussolini could not quite have planned this reincarnation already in November 1921, Pozzi claims, quite plausibly, that at that time the Unknown Soldier's body was a 'polymorphous cadaver' that represented a crisis (for him, in place, in social cohesion, and in psychological cohesion, to which I add the crucial elements of existential integrity and spiritual cohesion). Thus, Mussolini would only wish to be associated with the Unknown Solier in 1922 and after, when he could evoke crisis but at the same time present Fascist revolution as its solution. For Pozzi, this was, in fact, a continuation of the rhetoric of the 1921 ceremony, with the mere difference that the liberal state was replaced with Fascist authority: thus, Pozzi claims that the Unknown Soldier was from the beginning a 'blank slate' and a 'universal body,' that is, an abstraction that could be appropriated by political ideologies. I hope to have shown how while this may have been true for some state scripts, it does not correspond to popular reactions or to veterans' views, and it does not explain the Memorial's original and lasting appeal. Significantly, Pozzi erroneously asserts that 'in none of the many imaginings regarding the physical state of the Unknown Soldier is it ever said that he might be mutilated'; thus, for Pozzi, the Unknown Soldier is only a 'restoration,' a covering over of 'mutilated Victory.' As we have seen, in fact, in 1921 and earlier, the Unknown Soldier was often associated with mutilation and represented a broken, not a pristine, body and patria. Pozzi's description is far more accurate when applied to 1922 and later, when he correctly claims that 'Mussolini rehabilitates [*bonifica*] in himself the social ill that he has welcomed into himself.' More specifically, Pozzi shows that both in his public appearances at commemorations and in his biographies (Sarfatti's in particular), Mussolini increasingly presents his intact body as a replacement for or resurrection of the broken body of the Unknown Soldier. The metaphor of *bonifica* ('rehabilitation,' but also more literally, 'making good'), which refers to Mussolini's draining of marshes in the south of Italy, and to colonization of that newly fertile land by farmers that came from the war-torn north, is especially apt. Thus, while it was not possible to remove mutilated veterans from the Unknown Soldier Memorial ceremonies, given their social and symbolic standing, Mussolini began, already in 1922, to associate his vigorous body with the Unknown Soldier, in order to create a vision of wartime sacrifice that denied mortality rather than confronting it. We can also perceive here a significant difference between d'Annunzio's commemorations and Fascist ones, since d'Annunzio's mutilation or blindness in one eye was so emphasized by him and by the media, even during the Fascist era, when d'Annunzio continued to celebrate rites at his home, the Vittoriale.[42]

Other changes took place more gradually, but similarly moved in the direction of 'restoration' and *bonifica*, which also entailed a passage from heterogeneity to homogeneity, that is, to the sort of blank anonymity that early conceptions of the Memorial rejected, as I showed in Chapter 2. Thus, Maurizio Ridolfi notes that gradually all non-military participants were eliminated from the ceremony, and eventually only Fascists were allowed to attend; moreover, the celebration of the March on Rome itself, on 28 October, became the beginning of a single larger celebration that ended on 4 November with the Unknown Soldier. Mussolini's own voyage from Turin to Rome was presented as an expression of the 'eternal' voyage of the Unknown Soldier from the battlefield into the City (Mussolini here became a 'type' of the Unknown Soldier in the same way that saints are a type of Christ). Until 1925, veterans' associations, and Bonomi himself, who presided over the Memorial's creation (as Minister of War from 15 June 1920 to 2 April 1921, and as Prime Minister during inauguration), denounced this conflation and reiterated the original emphasis on mourning and suffering, rather than on rebirth. In more detail, Brice notes the same evolution as Ridolfi does, explaining how the celebration of the March on Rome came to overshadow the celebration of the Unknown Soldier. Most important, she shows that the creation in 1926 of the 'Ara dei Caduti Fascisti' (Altar to the Fascist Fallen) on the Campidoglio allowed it to become dominant in the vast ceremonies that went from 28 October to 4 November, effectively eclipsing references to the First World War and its 'mutilation' of Italy.

Finally, as Brice writes in her conclusion to her analysis of the construction and ideologies of the Vittoriano, the Monument itself evinces a conflict between a monarchic sense of ceremony and a Fascist one. She notes that the former addresses individuals who must be educated whereas the latter seeks to act on a homogeneous crowd. I believe we find a similar conflict when comparing the 1921 sense of physical continuity (and heterogeneity) between the living and the dead to the later insistence on total identification (and homogeneity) between them. Thus, as Fascism consolidated its power, during ceremonies the Unknown Soldier was surrounded by soldiers on active duty, not by veterans or by mourners, and more and more, these soldiers were seen to be like him, ready for anonymous sacrifice. One very telling example of this change is that the Memorial was depicted as commemorating not only the First World War dead, but the 'martyrs' of the Fascist revolution, as the Unknown Soldier Memorial was conflated with the Ara dei Caduti Fascisti: this meant those who died in 1921–22 confrontations, but as time went on, all those who died in Fascist colonial efforts as well, and all those who would die for Fascism in the future. Significantly, every year the ritual absolution of the Unknown Soldier continued to take place symbolically at Santa Maria degli Angeli. This could only make sense, theologically, if every year he represented more soldiers, who had not yet been absolved. Further, as an annual institution, this absolution came to guarantee the salvation of all Fascist soldiers, even before they died, collapsing the distinction not only between the First World War and later conflicts, but between *all* conflicts and *all* soldiers, past and future, thus erasing any possibility of individual moral responsibility.[43]

Mussolini's most significant appropriation of First World War mourning took place not at the Unknown Soldier Memorial, however, but in its gradual overshadowing by the Grand War Cemetery at Redipuglia. As Patrizia Dogliani has shown, this began as a wartime cemetery near the front lines; by the years 1921–22 a project for twenty-two concentric kilometres of graves was under way. At the time, local sculptors and writers collaborated on personalizing each marker with a unique decoration and epigraph, both for the known and the unknown dead. But the hugeness of the task meant that the cemetery remained disorganized and even in some cases was decaying throughout the 1920s. All of this changed by the 1930s, when the regime decided to 'clean up, organize, and discipline also places of worship, dismantling the original cemeteries that had been spontaneously created during and immediately after the war' in order to 'affirm and ideologize visually the overwhelming superiority of the State over the citizen.' At Redipuglia this resulted in a monument that was and still is 'a homogeneous ensemble of white marble,' in which each soldier only has the inscription '*Presente*' (Present) visible at his grave, and the names of those who were identified are listed to one side, together, and almost invisible (also there are 39,867 of them, as opposed to the 60,330 anonymous). Based on an art project for a memorial to the Fascist soldier, this design presents all the dead as identical, and identically risen, erasing all emphasis on corporeality, suffering, decay, and mortality as an individual existential experience. It also furthers the identification 'between the masses and their leader,' because buried there are 'commanders who did not die, like the soldiers, on the battlefield or as a result of war wounds, but of natural causes years after the conflict.' Once again, individual differences – and responsibilities – are erased. Moreover, Dogliani notes that while Mussolini participated in inauguration ceremonies at Redipuglia in May 1923 as a mere 'veteran' among many, by 1931 he was at the centre of ceremonies for the burial of the Duke of Aosta, which recalled the Unknown Soldier ceremonies of ten years before, but rewrote them to subsume the 'sacred body of the Savoia' to the regime. Notably, Maria Bergamas, the woman who chose the Unknown Soldier from the eleven candidates at Aquileia with a great unscripted theatrical gesture (see Part I), as well as mutilated veterans, were central to the inauguration in 1923, according to Dogliani, but were no longer present at all in 1931. Finally, in 1938, Mussolini inaugurated a second cemetery at that same location, and used the occasion to reinforce ties with Germany in his speech.[44]

If Redipuglia functions almost as a direct replacement of the Unknown Soldier Memorial, Emilio Gentile shows, in *Il culto del littorio,* how many elements of wartime mourning were reappropriated and 'cleaned up' by Fascism. Of particular relevance here is the creation of '*sagre della bandiera*' (flag fairs), which were promoted to schoolchildren, starting in 1923, as an homage to the martyrs who had died for Italy. Gentile notes that the flag was said to be a 'new Eucharist,' which is a clear imitation of, but also redirection in the sense of transcendence and not embodiment, of the lacerated flags seen at the Unknown Soldier ceremonies. Needless to say, the flags worshipped in every classroom were pristine. Similarly, the inscription 'Present' at Redipuglia was prefigured in Fascist mourning rituals, when after the name of each of the dead was read out by a leader, the crowd answered in unison, 'Present!' As Gentile writes, 'the roll call became *the Fascist ritual* par excellence, the highest aspect of their worship,' which allowed them to 'transfigure rites of death into *rites of life*.' Nothing could be further from the silence, the dismembered bodies, and contamination of living and dead evoked by mutilated veterans at the original inauguration of the Unknown Soldier Memorial, for here the living replace the dead, answering for them and, by implication, giving over their bodies to the state in exactly the same way. In sum, therefore, what these Fascist reinventions reveal, albeit through the very force of negation, is that the materiality unleashed by the First World War was dangerously unstable, indeed, 'polymorphous' in Pozzi's word or, to return to the terms used earlier in this chapter, traumatic in

Figure 17. The new cemetery at Redipuglia by Luca Spaventa, Italy (www.fotoegragia.net). © Luca Spaventa

the very disturbing sense evoked by both Jung and Leys. That is, as Jung shows, as oceanic this materiality is both ominous (making, for example, possible the transformation of the crowd into an 'oceanic' mass) and ethically demanding (as with Ungaretti's realization of our common mortality when he is 'deserted' in the 'infinite,' or d'Annunzio's experience of a 'suspended limit' in which the abject body mutilated by war recovers its dignity). Further, as Leys claims, though 'material,' this trauma does not have an external, verifiable, 'anti-mimetic' truth, for it cannot be separated from repetition and representation. There is no 'absolute' truth of the war experience that we might 'compare' both 1921 and later ceremonies to, to evaluate them: instead, we are faced with an ethical and historical choice between different repetitions of trauma, which are themselves still ongoing, ever capable of new forms.[45]

Transmitting the Experience of Death: Promethean Fire

In the last part of this chapter, I turn now to a second element of d'Annunzian rituals at Fiume – Promethean fire – in order to discuss this ethical choice. Specifically, fire was an attempt to preserve the heterogeneity of experience (common mortality, but also individual existential realizations or indeed rebellions) and the internal heterogeneity of the self (body and soul inextricably linked in incarnation, rather than the latter replacing the former via transcendence), against the push towards homogeneity brought out by Fascism, but perhaps typical of any politicization of mourning in the modern era. As we will see, texts by d'Annunzio and Cendrars, but also the unusual and spectacular presence of women at the inauguration of the Unknown Soldier Memorial, associate his embodiment or incarnation with a generation that is at once physical birth and infusion of fiery spirit into matter.

Fire is, of course, an ancient religious symbol and, in typical fashion, d'Annunzio wove together Christian and pagan symbolism in a complex mythology of Fiume as 'la città olocausta' (the holocaust city). Coming from the ancient Greek tradition, the term 'holocaust' indicates a specific sort of sacrifice, in which the victim is completely burned up, as opposed to when it becomes a shared meal with the god and fellow humans. Holocausts were thus the more extreme sort of sacrifice, intended to appease the spirits of the underworld, especially malign ones, or the ones of angry heroes. For d'Annunzio, however, though Fiume burned in order to appease angry heroes, and the demonic violence of war, it was at the same time not consumed. The reference here was Boehme's famous simile explaining the relationship of the divine to the soul in mystical experience: when filled with the spirit, the soul is like a flaming piece of iron, which seems all fire; but when the spirit is removed, the iron remains distinct and becomes black, and this indicates the freedom of will and the uniqueness of each individual. This is seen in d'Annunzio calling his followers at Fiume '*teste di Ferro*' (ironheads), but also in the symbol of the arditi that d'Annunzio used, which was an arm patch with black flames. The fire of Fiume combined the appeasement of demonic forces with the kindling of a new 'spiritual blaze,' which, as Ledeen has shown, was d'Annunzio's image for the spreading of Fiume's anarchic cultural revolution. As he claimed, 'Every insurrection is an effort of expression, an effort of creation. [...] For all veterans, carriers of the cross who have climbed their Calvary for four years, it is time to reach towards the future'; and more specifically as regards fire, 'The meaning of my enterprise and my most obstinate resistance is becoming clearer every day [...] All the desires [...] of revolt – the world over – turn towards the fires of Fiume, which send their sparks far away [...] Since last October

I have intervened directly in the Egyptian movement [...] the Croatians turn towards me, as well.' Once again, Christian and pagan mix, this time so that crucifixion becomes an ongoing provocation or 'scandal' rather than a foundation or institutionalization: Christ continues to die, like a fire that continues to burn, and there is no final *consummatum est*, which puts an end to sacrifice and also seals its meaning.[46]

The same images of fire as common sacrifice which, however, remains alive as well as subversive, heterogeneous, irreducible to a single ideology or theology recur in both Guido Keller's 'Yoga' group at Fiume and in various celebrations that took place there. Thus, in 1920 Keller wrote, 'When the redemptive mission of the holocaust succeeded, something was expected of them [the Fiuman legionnaires]. Under the ash of their involuntary physical inactivity [once the city was taken], generous sparks kindled their hearts, and [...] they gradually understood [...] that life is born from struggle, as harmony is generated from discordant sounds.' As mentioned earlier, Keller and his group led pirate attacks that were as much about theatrical gestures as they were about provisions; also, they slept outside, invented outlandish costumes and at times practiced nudism, and orchestrated events that would now be termed 'happenings.' As Ledeen writes, 'Keller was a true contemporary of the Dadaists'; at the same time, of course, he was inspired by d'Annunzio who had behaved this way already in the 1890s, and who had described the charismatic leader of an élite proposing a spiritual and aesthetic revolution in his 1900 novel *Il fuoco* . For the day of San Vito, the patron saint of Fiume, an extraordinary festival was created around the theme of fire, as San Vito was meant to have been killed by the torture of the *catasta*, that is, by being burned alive on a grate. In the poet Kochnitzky's description:

> The illuminated piazza, the banners, the great written proclamations, the boats with their beflowered lanterns [...] They danced everywhere: in the piazza, in the streets, on the dock; by day, by night, they danced and sang [...] it was an unrestrained bacchanal. To the rhythm of martial fanfares one saw soldiers, sailors, women, citizens in bohemian embraces, recapture the triple diversity of the primitive couples hailed by Aristophanes. One's gaze, where it fixed, saw a dance: of lanterns, of sparks, of stars; starving, in ruin, in anguish, perhaps on the verge of death in the flames or under a hail of grenades, Fiume, brandishing a torch, danced before the sea. In the impoverished homes of the old city, the women had removed the sacred images. The tiny lights glowed in front of the figure of Gabriele d'Annunzio. Others

> may call this *hysteria.* It is the *Bal des Ardents.* Under the gaze of the hostile and cowardly world [...] Fiume dances before death.[47]

This description, on the one hand, recalls the 'Epiphany of Fire' in d'Annunzio's novel, which is set in Venice, and proposes also the image of fire reflected in water, itself an oxymoron similar to 'Fiume olocausta,' as both insist on the 'struggle' or primeval '*polemos*' evoked by Keller. Moreover, the reference to 'hysteria' along with the image of a personified Fiume dancing before death recalls the description of Salome, as painted by Moreau, in Huysmans' *A rebours.* There, Salome is said to be the 'goddess of immortal hysteria,' in that she represents an unquenchable desire, at once sexual and metaphysical, that ultimately devours the self, and the main character in Huysmans' novel, des Esseintes. However, in d'Annunzio's *Il fuoco,* the main character, Stelio Effrena, seeks to succeed where des Esseintes failed; inspired by Nietzsche, he thus embraces desire's ultimate lack of fulfilment and espouses eternal becoming. Still, in both novels the dark power of desire is associated with feminization or indeed androgyny as a form of sexual transgression; a similar transgression is suggested by Kotchnitzky's reference to Aristophanes, and further, descriptions of Fiume attribute androgyny to both d'Annunzio and to Keller and his followers. (This was also emphasized in d'Annunzio's association of himself with Saint Sebastian, who was also given a festival at Fiume.) Moreover, the 'bacchanal' mentioned above recalls the death of Orpheus, who was said to take precedence at Fiume 'over Christ,' and whose dismemberment is presented in another Moreau painting as very similar to that of Saint John the Baptist, caused by Salome: both are beheaded. This shows the connection between fire and a Dionysian, sexually charged, feminized, dismembered, subversive, 'insane' dance, which asserts the power of movement, becoming, transformation, and the journey to the underworld and back, against the fixity of death, or of traditional Christian transcendence.[48]

The refusal of dogmatic notions of transcendence took one very explicit form at Fiume when a group of Capuchin priests used the occasion of the occupation to make their religious Modernism more militant. They wanted the right to marry, demanded control over their own funds, and insisted that their superiors be elected from below rather than chosen by Rome; overall, they wished to modernize the dogma and rituals of Catholicism. Though they did not make any theological statements about d'Annunzian rituals, when the Church attempted to disperse them away to various dioceses, they barricaded themselves in their monastery, hanging from it a sign reading '*Hic manebimus optime*' (We will remain here splendidly) – the same slogan d'Annunzio used against the Italian government when he took over Fiume. According to Ledeen, seven Capuchins eventually left the order and joined 'Keller's world' which, presumably, they found more suited to their beliefs.[49]

A far more symbolic rejection of both Christian and politicized notions of transcendence, which brings us back to fire once again, took place a little later in the 'adventure of the Horse of the Apocalypse.' In April 1920, a group of Keller's pranksters stole forty-six horses from an Italian military stable nearby and transported them by sea to Fiume. This was a protest against mistreatment and arrest by the Italian government of ex-legionnaires who had left the city. The government retaliated by threatening a total blockade, which was called off when d'Annunzio agreed to return the horses a few days later. However, instead of returning the well-fed healthy horses that they had stolen, the legionnaires 'led forty-six emaciated beasts from the city to the Italian boundary.' D'Annunzio gloated:

> We have committed aggression against loyal troops.
> We have stolen forty-six quadrupeds.
> We have offended Italy.
> We do not know how to think 'Italianly.'
> We are not Italians.
> We deserve only to be starved, manacled, executed.
> We shall resign ourselves.
> But I must confess that last night I stole the Horse of the Apocalypse, to add it to the forty-six quadrupeds on the criminal barge.
> He has his marvellous general's harness, and a divine thunderbolt in each holster.[50]

Yet again conflating Christian and pagan symbols, d'Annunzio claims to have stolen the divine fire of Zeus, like Prometheus did when he brought it to mankind. At the same time, he rides a horse that not only takes up the common association of the First World War with apocalyptic and other 'end of the world' imagery (see Winter), but specifically refers to the first, white, horse of the Apocalypse, who is associated with conquest, and whose rider is said to be either Christ or the Holy Spirit, come to scourge yet also to renew. As with 'Fiume olocausta,' this blend of references adds up to a highly subversive notion of incarnation as the stealing of the divine spirit from the gods, as its radical descent into apocalyptic violence yet also into creative fire. D'Annunzio's project at Fiume was very much the transformation of the former into the latter, that is, the redirection of wartime rupture into the anarchist bacchanal, or also into the Promethean creation of a new civilization (in Aeschylus, Prometheus saves man from total annihilation by Zeus, and also has secret knowledge that could lead to Zeus' downfall). As his humorous tone (above, but often seen in his Fiume statements, conflated with more sombre mourning) indicates, this was a 'comical' form of traumatic repetition – in the sense of desecration, the toppling of idols, and Nietzschean laughter.

The link between stealing fire and an incarnation that repeats in a comical, desacratory, Dionysian vein the trauma of war, refusing the fixity of transcendence in favour of mimesis and becoming, brings me back to the Unknown Soldier as he is described in texts by Cendrars and d'Annunzio. As discussed in Chapter 1, Cendrars' unfinished *La Vie et le mort du soldat inconnu* opens with a frame story in which the author himself, 'the only one among us to have fought in the war,' as his mutilated hand indicates, drunkenly makes fun of his fellow-passengers on a cruise ship, suggesting in turn that the Unknown Soldier under the Arc de Triomphe might be a German, a Jew, a 'negro,' or even a woman. This desacralizing humour (which recalls d'Annunzio's claim, 'We are not Italians,' 'We do not know how to think "Italianly"' above), ends, however, with a pantheistic vision of the night over the ocean and with the reassertion of the cycle of life and death, of 'blessing' and 'cursing,' which is tied to Cendrars' making love with Caralina (and recalls d'Annunzio's 'Bal des Ardents'). As I will now explain, in the actual story of the Unknown Soldier according to Cendrars, humour gives way to a more serious appeal to pantheism, primitivism, and the flesh in its vitality and finitude, as an alternative to the official redemption offered by the Unknown Soldier Memorial. Named only as 'l'homme' (*ecce homo*, and *alter Christus*, yet also Nietzschean), Cendrars' Unknown Soldier is a mythical Russian peasant, whose intimate, visceral connection to the earth that sustains him is established in the first chapter (this recalls d'Annunzio's image of the '*corpi di terra*' [bodies of earth] of First World War soldiers, cited in Chapter 4), as is his connection to an immemorial genealogy that goes all the way back to a godlike fiery horse (hence, Prometheus, associated with fashioning humans out of clay as well as with stealing fire):

> But the ancestor of all horses, both legendary and tame, is the stallion that carried Prometheus to the top of the Caucasus to steal fire [...].

> And the hesitant voice of the old bards became more firm as they revealed the arcane lore of the poem [...] the obscure genealogy, to proclaim in a triumphant crescendo the great deeds of the founder of the race, the first member of the tribe, the father of the clan, Uûr, the fabled stallion.

In what is perhaps one of the most telling scenes of *La vie et la mort du soldat inconnu*, 'l'homme' participates with a few of his compatriots in a ritual that prepares for battle: this consists of eating the raw flesh of the most vigorous horse in the village, who is the incarnation of the fiery horse ancestor, in order to acquire his strength. However, in keeping with Cendrars' mixture of desecration and reconsecration, this time the ritual has to be improvised on a Western battlefield:

> Behind a hill, he finds a dead horse. He couldn't say how he got there. But he's not the first. [...] They are in a hurry as their tour of duty will be long tonight. They rip off the horseshoes, tear open the skin along the belly, along the legs, hurriedly skin the beast; then they leave, the bearded one taking the skin, all rolled up and dripping [...] Now everyone is screaming as though each wanted to scream the loudest. The men have taken their knives out and each bears down on his favourite piece of meat. [...] Two men grab hold of the skull, as another digs in the ground like a dog. [...] Those who will eat together and share the horse's flesh will become brothers for life, blood brothers. [...] First they drink the meat's juice [...] when they've had their fill, each one grabs his piece of flesh, rubs it on his cheeks and, before bringing it to his mouth, gives the fire its share [...] then he begins to chew [...] Do they not seem drunk? No, no, they are simply absent, far away. Each one is visited and each man speaks to Iron [descendant of Uûr].

In parallel to d'Annunzio's conflation of pagan and Christian symbols connecting the horse, fire, and sacrifice, Cendrars appeals to the East and 'primitive,' 'barbarian,' 'mythical' ritual, in order to oppose the West and established Christianity. It is not enough now to steal the Promethean horse; rather, its flesh has to be ingested, as though physically to ingest the violence of war. Yet here, too, we have Dionysian disorder, and here, too, there is a mystical transformation or better incarnation or infusion of fiery spirit into matter, as these men who animalistically consume raw flesh and blood become 'absent,' and are 'visited.' Significantly, each has to give the fire its share, as though to acknowledge and defuse the possibility of holocaust, of total annihilation, of the sacrificed horse, but also, by implication, of the clan itself. Once again, trauma is repeated in a ritual that desecrates it in the sense of refusing to give rupture abstract or transcendent power, by bringing it back to the body, where it is 'masticated' and becomes anarchic spiritual drunkenness (as we are about to see, 'mastication' is an image also in d'Annunzio).[51]

To understand Cendrars' ritual, we may refer again to Mircea Eliade and to his discussions of horse sacrifice in Central and North Asia. Though Claude Leroy shows that Cendrars' direct source for this scene was Knut Hamsun's *Au Pays des contes*, what we find there is a description, very close indeed to Cendrars' text, but no interpretation. In contrast, Eliade, drawing on a number of anthropological sources, explains that the ritual dismemberment and eating of a horse accompanies the shaman's journey to the celestial regions where he 'learns from Bai Üglan if the sacrifice has been accepted and receives predictions concerning the weather and the coming harvest; he also learns what other sacrifice the divinity expects.' He points out that the horse is 'killed in a cruel way' and that the shaman 'cuts up pieces of the horse and distributes them to those present (who represent the spirits), who noisily eat them,' so that violence is both evoked and ritually contained. Finally, Eliade also notes that earlier versions of the ritual took place without the shaman, as 'horse sacrifice was practiced by the majority of Indo-European peoples, and always offered to a god of the sky or of the storm.' What is important then about the association of the horse sacrifice with the shaman's journey is that 'by virtue of a more concrete and morphologically richer experience [...] the shaman succeeded in ousting the earlier sacrificer from the horse sacrifice, just as Bai Üglan ousted the earlier celestial god. The sacrifice became a sort of "psychophoria" leading to a dramatic encounter between the god and the shaman and to a concrete dialogue (the shaman sometimes going so far as to imitate the god's voice).' In Cendrars' text, 'l'homme' seems to become a modern version of the shaman, in that horse sacrifice is not a form of worship of the sky or storm, but of a god who like Bai Üglan is 'less distant, less passive' in that he will defend (or not) those who sacrifice to him. Moreover, though Cendrars' text does not give us an explicit dialogue between 'l'homme' and the deity, it is clear that he undertakes a journey beyond life, for we find him at the end of the scene quoted above, overtaken with 'funereal visions that made his teeth chatter, in which he saw his son fall from a train, his white hat becoming red with blood, while snakes dripped out of his eyes and a huge fly settled on his mouth.' And this vision is prophetic, since his son has indeed gone to war as well, and will be killed though we do not find out how.[52]

Moreover, Cendrars rewrites primitive ritual 'democratically' or indeed in the anarchic direction of Fiume rituals, as now all participants become shamans, 'in a state very much like the trance experienced by mullahs or whirling dervishes.' Significantly, however, 'l'homme's' prophetic vision eventually turns into the pompous newspaper article, quoted by Cendrars in the next chapter, which claims great victories for the Russians, talking about the great valour of its 'strong mountain people': 'One of them, named it seems Ostapyth, who, according to the wounded, is the singer-in-chief of his squadron, single-handedly stopped a military train filled to capacity with fleeing German soldiers and made it return to the station, where they were all made prisoner.' As is made clear, this is no doubt a media-driven falsely heroic version of the story, as in reality, 'l'homme' in his primitive madness had taken the train for a monster. This at least is what we find out from a Captain who tells the story to his mistress with great humour at a reception a few pages later. In Cendrars' notes, we further hear that 'the Czar organized an investigation that determined this Cossack did not exist' so that the Unknown Soldier falls into a 'double anonymity.' The point, then, is that modern warfare has no place for shamanic journeys, and that prophecy can only be understood as madness, or as media hype. Nonetheless, the association Cendrars makes between the Unknown Soldier or 'l'homme,' the Promethean horse, and the shamanic journey is thereby presented as a powerful contestation of the Christian (and in Cendrars' book also very clearly bourgeois) transcendence of the official memorial. Once again, as was the case at Fiume, the Unknown Soldier's journey through death is seen to endow him with a radical subversive power that is transmitted by physical continuity, like fire itself.[53]

In d'Annunzio's *Notturno,* the shamanic journey is alluded to, and again conflated with Christian symbols, in the episode of El-Nar. This was a horse whose fiery majesty d'Annunzio remembers because of the 'fire' in his wounded eye as he convalesces, and he describes pushing El-Nar to the limit as he gallops through an immense rose garden, and becomes covered with beautiful red wounds due to the thorns. These recall the 'bruise' and the 'azure opening' on the chest quoted above, in which the physical continuity of dead and living soldiers was asserted and made palpable in the Banner of Randaccio. Also, the wounds of the abject soldiers in the ruined 'Franciscan' church at Doberdò are said to flower 'like roses.' In all three cases, d'Annunzio refers to the story of how Saint Francis, visiting the monastery at Subiaco where Saint Benedict had punished his flesh against thorn bushes, made roses bloom upon the thorns.

This is a telling image associating Franciscanism with pantheism, the rediscovery of the natural world, and embodiment as that which cannot be transcended (as he preached to the birds, wrote a prayer in praise of the elements, and was obsessed with physical death, Saint Francis was a popular, but also more than we remember, uncanonical saint). Once again the journey with the horse symbolizes a stealing of fire – now associated with blood and with the continuity between soldiers – from the transcendent realm to bring it back into the sensible and the sensual.[54]

Even more notably, given Cendrars' image of ingesting divine fire and violence as a form of incarnation, d'Annunzio revisits the image of the Banner of Randaccio and compares its power to imprint the physical experience of the descent of the spirit into the suffering soldier on all who see it to a form of 'rumination' (*masticare*). Specifically, he imagines his own journey into the darkness of convalescence to be like that of the Unknown Soldier and also like that of a new Christ, or better a modern Lazarus, as mentioned; but what is added now is an extension of this choral suffering to d'Annunzio's mother. Thus, we first find the imprint of soldiers' suffering on d'Annunzio's own body: 'those blind soldiers [...] are my brothers. [...] the miserable flesh that has no fight left in it, the worried mouth of the unseeing. [...] I was naked of any privilege, without singularity, without contour, without glory save for my humble sacrifice.' Their suffering comes together in one moment that connects the soldiers, d'Annunzio, and his mother. This is where *masticare* (chewing) is the startling image used for the embodied encounter, through others, with one's own mortality and abjection: 'The tears from my [infirm] eye and the sweat from my temples drip down to my mouth. And I sip the salty drops. And it is as though I sipped them with my mother's mouth, with that deformed mouth that weighs inside me, that suffers in my disfiguration. [...] It is not an immobile image. It moves, changes. [...] The mouth opens, and it cannot speak human speech; it can only chew on the soul, ruminate desolation.' 'Disfiguration' intensifies the 'rent moon' of Doberdò, and thus we finally find that these different mouths that silently scream suffering come together in the Veronica once again:

> It is as though I could hear within me my mother's screams, which when I was born did not penetrate my sealed ears. [...]
>
> It is the Sixth Station: the shroud of the Veronica. [...]
>
> I look at that face. [...] Horrible and sublime, truly, with a gaze that does not see me, that cannot recognize me, darkened and still, where love is nothing but nameless sadness, sadness unto death and beyond death.

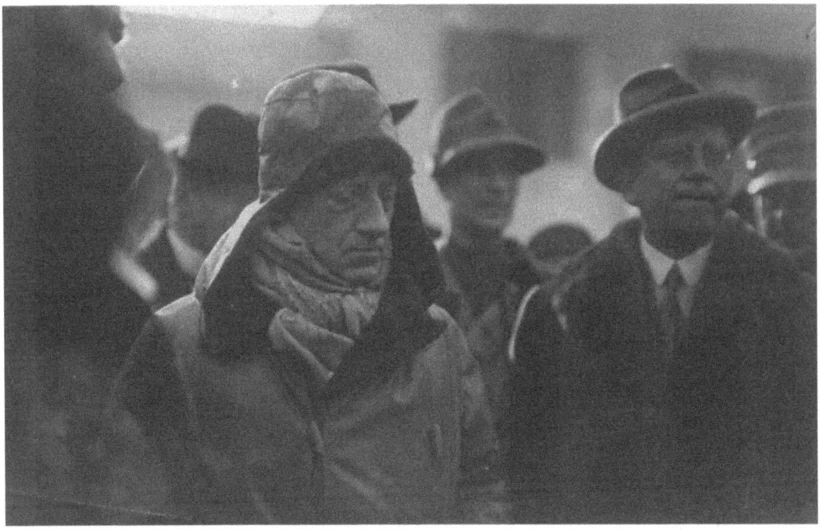

Figure 18. D'Annunzio during the First World War. © Centro Documentazione Mondadori

> My mother!
> A poor creature, humiliated, beaten, disfigured.

The Veil of the Veronica marks the physical continuity of 'nameless sadness, sadness unto death' from mother to son to soldiers; significantly, this melancholy is one that does not speak 'human speech,' but is instead a '*masticare l'anima*' (soulful rumination) in which the suffering of the flesh seems concretely to devour transcendence (the already mentioned 'the soldiers bit into the azure' here receives yet another interpretation).[55]

Yet what is most remarkable about the image of the veil is the implicit drama that it contains, and its link to birth. On the one hand, as already suggested, the veil does not merely show melancholy but it transmits it, as though taking a physical imprint from Christ / Randaccio / the Unknown Soldier and inscribing it upon the flesh of the viewer/d'Annunzio. The concrete traces of suffering – 'blood and lymph' – seem to overflow individual boundaries, contaminating all flesh with death, making the Incarnation the event of an inescapably common mortality: the cry of birth comes together with the Passion imprinted on the Veronica as the death of God marks his 'throw[ing] his lot in with humanity.' On the other hand, this identification in a shared mortality takes the form of non-recognition: d'Annunzio's mother does not recognize her son, and d'Annunzio, 'disfigur[ed],' does not recognize himself. As the theme of blindness suggests, the face on the Veronica or on Randaccio's flag seems anonymous only because we do not recognize it even as we are affected by it; indeed, the impossibility of recognizing such suffering as individualized and 'human' is the sign of how much we are affected. What anonymity signifies, then, in the Unknown Soldier Memorial is not or not only the reversal of lack of identity into choral identification, but also the drama and the horror of our shared mortality with a 'disfigured' 'creature.'[56]

This final transmutation of the Banner of Randaccio, and the association of the *masticare* and ingestion of transcendence with incarnation now seen as childbirth, and painful labour, suggests that there was a lot more to the presence of women at the Unknown Soldier Memorial's inauguration in Italy than historians usually assume. This was already suggested by Maria Bergamas' spectacular rejection of the script when she chose the Unknown Soldier claiming to have felt a tremor within her that came from within the coffin. The reference to the Veronica, with its intense physicality, and its undercurrent of sensuality (expressed also in d'Annunzio's transgressive identification of female and male faces, echoed in the placement at Fiume of images of d'Annunzio where one would normally find the Madonna), reminds us that though the 'mourning mothers' in Italy were associated most often with the Virgin, this did not necessarily mean the distant and pristine one that we might initially imagine. Indeed, the Marian cult itself has often been at odds with canonical Catholicism, because at its roots it reveres in Mary the *theotokos*, the 'God-bearer [who] was to assert that humanity was not some temporary veil which the Logos had donned and then discarded, but that in her womb God had once and for all thrown his lot in with humanity, had joined us, holding nothing back.' Images of the *theotokos* emphasize the physical continuity between Mary and Jesus, as opposed to the separateness of the Virgin: Michelangelo's *Milan Pietà*, in which the Virgin and her son dissolve as one mortal flesh into the same stone womb would have been for d'Annunzio and his contemporaries a prime example. Latent in the appeal of the Unknown Soldier is this notion of Mary, which at its most extreme collapses God's birth with his death, that is, with his relinquishing of transcendence which makes resurrection uncertain, an event that has not yet happened, that may never happen. Though the

Figure 19. Crowds kneeling at the Unknown Soldier's passage (1921 photo). © Courtesy of Department of Special Collections and University Archives, Stanford University Libraries

more conservative press coverage emphasizes the solitude, the immobility, and the purity of the mourning mothers, newspaper accounts of all political stripes also revel in details that vie with Bergamas' gesture: the innumerable mothers 'who, in order to touch even for an instant the coffin, had themselves carried onto the chariot and then refused to get down,' also depicted kissing the coffin; the one who came on foot from Livorno to Udine to see the Unknown, too poor to take the train; the one who exhumed her son's body herself to return him to his home town.[57]

These mothers may seem overwrought and even hysterical, yet their insistence on the physicality of mourning does not stem from its initial repression and subsequent re-emergence in displaced physical symptoms, as the diagnosis of hysteria would require. Rather, since these mothers' gestures express their physical continuity with their dead sons, as opposed to a desire to replace or reproduce their sons' dead bodies, they create a cathartic expressive experience which, I believe, is at the core of the appeal of the Unknown Soldier Memorial. In this experience, trauma is not so much understood as it is 'incorporated melancholically.' I borrow this expression from Amy Hollywood's analysis of mystical experience, especially by women, in its relation to mortality, mourning, and rebirth. Rejecting Freud's view that melancholy is a failure to bring mourning to its proper end, and also a failure to bring it to proper verbalization and abstraction, which results, then, in inappropriate physical symptoms, Hollywood focuses instead on the struggle of the 'speaking body,' that is, the lived disjunction between what can be said and what can be embodied. She shows that for many women mystics this struggle is one of 'melancholic incorporation,' in which melancholy blocks the movement of mourning towards abstraction and also transcendence, by accepting the physical traces of another's death in oneself not as pathological, but as expressive and liberating. This is not unlike Jung's notion that we can come to terms with trauma not by analysis but through an expressive and physical, symbolically charged and renewing, acting out of its violence. It also recalls the images of incorporation, ingestion, and disfiguring used by d'Annunzio, Ungaretti, and Cendrars to show the violence of war impressed on their living bodies. Ultimately what this expresses is not only the continuity between death and life – captured in d'Annunzio's conflation of birth and the Passion – but also the consequent dependence of humans on each other in coming to terms with death – expressed in the Veronica. Without transcendence and abstraction, that is, without a (however imaginary) perspective from beyond life, life and death as a continuum can only be understood when we give voice to, and even more, enact and embody the death of others: when we allow ourselves to become contaminated by their deaths. The mothers' gestures, then, like the *theotokos*, wrest ritual away from the abstraction of resurrection and act out instead, as an alternative melancholic form of mourning, the body's abandonment of its claims to transcendence.[58]

6 Mourning Transcendence and Re-enchanting the Flesh

Confronting Mortality

In this last chapter I seek to bring us into the present to discuss whether the enduring appeal of the Unknown Soldier Memorial is merely nostalgia for a time that seems more whole by virtue of being distant, or whether it succeeds instead in creating a new form, or better new forms, of mourning that are dynamic and allow us to reconsider our relationship, both existential and political, to death, transcendence, and the flesh. We have seen that early in the twentieth century the Unknown Soldier evoked horrific trauma, but that it also, well beyond government predictions or ability to control, enacted a human solidarity that made strong claims for the precedence of embodied existence over ideology. Today, the 'comradeship' of the First World War, and its expression in literature as well as personal correspondence, is popular precisely as an image of something we have lost or are about to lose. A case in point is the opening of Jeremy Rifkin's *The Empathic Civilization* (2009), which evokes the 1914 Christmas truce when British and German soldiers sang carols to each other and stepped into no man's land to inhabit the same physical space for one day: 'they chose to be human,' he writes, as a prelude to wondering whether we still have this capacity, as we confront ecological disaster. Why is the First World War still a model for empathy? Is this because we have managed to forget so much of its cost, not to mention its ambiguities, of which this book has in some measure reminded us? Or is there still something positive to be learned from how the first generation of men and women to confront death on a mass scale tried to make sense of it? They saw Western modernity to be in crisis, and had the advantage of believing that it could be radically transformed, however naïve or optimistic or even dangerous that might seem to us now. They also had the advantage of shock and this reminds us that since then no Western government has allowed the truth of war to be so directly available to its constituents. As I have shown, many chose to confront the truth of death and violence directly, questioning their gradual evacuation from nineteenth-century culture. As we look back, then, does their confrontation help us with ours?[1]

To answer this question, I will examine four recent works that in very different ways discuss the Unknown Soldier, but I will also draw some parallels between them and contemporary reactions and events connected to actual Unknown Soldier Memorials in Italy, France, England, and the United States. The first two are novels that are specifically about the Unknown Soldier, one French and one Italian. Olivier Barbarant's *Douze lettres d'amour au soldat inconnu* (1993) and Gian Pietro Testa's *Io sono il milite ignoto* (2006) both imagine a dialogue between a character in the present and the Unknown Soldier. Interestingly, both wish to reclaim the Unknown Soldier as a pacifist, at least in the sense that he might have supported the First World War, before knowing what it was, but that for both authors what matters is his subsequent condemnation of war and violence, which both link to a critique of the modern West. This reclaiming confirms my own sense that the Monument is not in any essential way bellicist, and indeed enacts a refusal of heroic transcendence that governments must downplay when they use the Memorial officially and militarily. Moreover, as we will see both Barbarant and Testa are fascinated by the physicality of the Unknown Soldier, and see it as running counter to our culture's tendency to abstraction, which for both authors is politically charged, in ways related to what Michel Foucault and Giorgio Agamben call 'biopolitics,' or what I would call the commodification of bodies in our culture, with its powerful imposition of normalization on marginal people or cultures. In Testa's case especially, the novel shows that to counter such commodification it is not enough to critique transcendence, for we must also develop a new sense of the body's sacredness. Yet in both novels, this sacredness is more of a wish than an actuality: does this mean that the radiant grace called for by Tournier is impossible in our world? Or is the receptive awareness and holism he represents dangerous, for they can never be separated from religion as institution, and from the dangerous conflation of spiritual and temporal authority? And how does this relate to recent uses of the Unknown Soldier Memorial for mourning, in 2003, and also to recent suicides on the site (in one case by a soldier, in Rome)?[2]

Two other works confront these problems by dwelling on the nexus of the body, the sacred, and politics that has become so pervasive in our world. In Alda Merini's poetry, from the 1960s to the present, we revisit the physicality of medieval women mystics – a topic that has become more and more fascinating throughout the twentieth century and into this one, as we seek to counter the loss of physicality and nationalistic appropriations of the mystical body, individuated by Kantorowicz in'Pro patria mori.' But in Merini's world of the insane asylum, the sacredness of the body emerges almost always as obscenity or kitsch: not unlike Pier Paolo Pasolini's 'desperate vitality,' it cannot be 'digested' by a culture that has no place for it. How, then, do we read Merini's jarring association of her body with that of the Unknown Soldier, in a 2008 poem, a year before her death? Can re-enchantment be present ironically? These issues are all tied together, finally, in Marco Bellocchio's 2002 film *L'ora di religione.* The Vittoriano, with the Unknown Soldier Memorial at its core, is a major protagonist in this work, which explores the role of religion in contemporary Italian culture, in its relation both to trauma and to political activism. Further, as we will see, Bellocchio refers to the religious icon, citing the films of Andrei Tarkovsky, which themselves sought to recapture something of the physicality and direct link to the sacredness of the Russian icon. New media, Bellocchio seems to say, echoing issues raised by the films of Gianikian and Ricci Lucchi, need not necessarily leave the body and situatedness in a living world behind. Together, Merini and Bellocchio invite us to reconsider the mediatization of the body in our culture as something that we may not reject piecemeal, but that can be thwarted, twisted, or turned in order to 'jam the theoretical machinery,' as Irigaray puts it. More radical than Barbarant or Testa, they do not see the Unknown Soldier as untouchable in his iconic quality but rather seek to transform his body in ways reminiscent of reinventions of the Virgin in South America, where she is both sexualized and politicized. Radiant grace, they imply, is best defended by its transformation.[3]

Mourning Transcendence

The format of Barbarant's 1993 novel, *Douze lettres d'amour au soldat inconnu,* runs counter to the more common practice of attributing authoritative statements to the Unknown Soldier; coming from the realization that far too many words have been put in his mouth, it seeks to engage in a dialogue that is less one-sided. At the same time, distance is foregrounded time and again, as Barbarant's letters are compared to those sent during the First World War by people to soldiers who had died, but whose deaths were not yet known. The letters thus simultaneously bracket the Unknown Soldier's being dead in order to address him directly, and insist that the dead do not return. Similarly, the letters see the Memorial simultaneously as a monument to victims, and as an unlikely choice of subject for a contemporary pacifist liberal. More broadly, they imply that history reaches out to the past, but cannot fail to misunderstand it. At first glance, Barbarant resolves these contradictions by following the lead of d'Annunzio and Cendrars, and uses fiction to separate the body of the Unknown Soldier from the official Memorial that surrounds him: 'Even more than the dissimulation of horror, what shocks me is [...] this cheap poetry [...] in Rome or Madrid [...] the body all wrapped up with glory, the impossibility of representing the face, the skin, the beating of real flesh, with anything but a mask.' As did many contemporary mourners, Barbarant's narrator insists on physicality in contrast with glory. The refusal of the mask and the insistence on skin, which recurs throughout the letters, foregrounds touch and rejects the sort of division of body and soul enacted by Tournier's Véronique or by Comerio's film footage. Moreover, as was the case with veterans and mourners after the First World War, Barbarant does not hesitate to dwell on abjection and decay, commenting on how the twentieth century has sought to cover over the reality of death: 'I would even like you to be buried under a glass tombstone, so that we could follow your decomposition. We would be forced to plunge our gaze into yours, instead of bending over it without danger.' Two elements are important here. On the one hand, the idea of a monument that shows decomposition in real time recalls proposals for an Unknown Soldier Memorial as ruined battlefield, as both insist on the marginality of mourning. In fact, Arnold van Gennep shows in *Les rites de passage*, which I mentioned in relation to sacrifice, that the time of transition – the time of change and transformation, the real passage, after separation and before incorporation – is often linked literally or symbolically to the time of decay of the cadaver. As both Philippe Ariès and Michel Vovelle note, the privatization and related medicalization of modern death conspired to minimize this marginal time: throughout the nineteenth century, prescribed mourning periods were drastically reduced and, perhaps more important, became a personal choice and one that increasingly was to be kept in the home, hidden. Mourning time was even further reduced during the First World War in order to permit women to remarry quickly and produce more children. This continued after the war, making mourning less and less of a rite of passage, as such a rite can no longer take place when the time of outsideness is so limited and so solitary as to become incommunicable.[4]

Hence, on the other hand, Barbarant's image of being forced to look into the eyes of the Unknown Soldier is also important. It claims that it is not enough to acknowledge decay, for we must relate to it intimately, in a kind of compassionate mirroring that recalls d'Annunzio's identification of his face with that of dead soldiers. But the image of 'plung[ing] our gaze into yours' is also typical of romantic love, and this underscores the central conceit of Barbarant's novel: these are truly love letters, which only at the beginning recall the letters written by women volunteers to encourage single men at the front during the war; as we read further, they evoke in detail a growing physical intimacy between the narrator and the Unknown Soldier, which culminates in their sexual encounter and night together in Letter 5. Throughout the novel, Barbarbant connects the physicality of love and that of mourning, bringing both into the same marginal time: 'This archangel of rotten flesh and mud orders me to write to you. His lips made of almost humus are what I embrace, to welcome you.' As the reference to 'humus' suggests, as well as the oxymoronic 'archangel of rotten flesh,' this marginal time is where death and rebirth may form a cycle (the archangel is also the angel of the annunciation). To further emphasize the marginal and transgressive quality of this time, Barbarant makes his narrator a homosexual male. This is not, however, a naïve reading of the evident homoerotic undertones found in much First World War comradeship (one need only think of Giono's Joseph embracing the dying Jules, discussed in Chapter 1), for the narrator acknowledges that it is most likely that homosexuality would have been frightening if not rebarbative to the actual Unknown Soldier – a fact that he accepts as a social reality of the time. Instead of conflating himself and the Unknown Soldier in a single marginal status, the narrator establishes an exchange between the place of a homosexual (and especially of his body) in contemporary France and the place of an anonymous *poilu* (and his body) in First World War France.[5]

Specifically, they are both subject to a solitude that comes from having no acceptable social identity, from society's refusal to *see* them, which deprives them of their face, that is, their personhood, of the connection between outward appearance and inwardness: 'Please know that I, too, had no face, even before I set out to speak to you. And solitude does more than disfigure you: it nullifies you, as just happened, when I felt like the steps carrying me towards you were but fog, like I was nothing but a vague thought of being turned towards you.' This recalls d'Annunzio's image connecting his mother's face, the Veronica, and childbirth to the suffering of First World War soldiers: 'a poor creature, humiliated, beaten, disfigured.' In d'Annunzio, this was an image of loss of transcendence and of mortality shared through touch and horror. Unlike d'Annunzio, Barbarant's narrator, aware that he has not participated directly in a war, presents his disfiguration as different from the Unknown Soldier's even as both involve anonymity. 'Don't worry,' he writes, 'I will do my best not to impose my features on you.' They are invisible in different ways but what they share is the death of others experienced in their own bodies as a loss of grace – a loss of that radiant beauty Robinson must discover to survive in Tournier's *Friday*: 'Please know also that I have been dragging too many corpses along my path. Just recently, there was another one, in a hospital bed again, who just closed his eyes. [...] It's been ten years since I've had the right to lightness, to grace; I should learn to survive their loss, which is what I thought I was attempting by reinventing you.' The devastation caused by AIDS is not the same as that of the First World War, but for Barbarant in both cases society fails to mourn. This is a political failure, linked to minimizing the cost of war or of AIDS because they affect only the poor or the marginal or those who are otherwise invisible as humans – those who are included via exclusion, as Agamben writes, included as *uccidibile*, or as killable. Yet it is also an existential failure, in that to most of the living this loss of grace is invisible. In terms similar to Tournier's, Barbarant links modernity's abstraction and division of mind and body to a culture that is so used to seeing the body as a machine that it no longer sees that grace is missing from it. This is a failure the narrator recognizes also has his own: 'This attempt has failed, so blatantly tonight, and I should not have bothered you with it. Instead of extracting you from it, I am melding with your mud.'[6] This leads to a vision of horror that is again about the impossibility of seeing and thus also of being seen. The narrator describes letters from a First World War doctor in order to evoke a violence he also says he cannot imagine; as in Céline, a language that '*louvoie*' (prevaricates) indicates psychic dissociation: 'Schizophrenic [...] a doctor wrote [...] making excuses for writing [...] his sentence twisted in prevarication at this point [sa phrase à cet endroit louvoyait], lost in the details of the dusk, pushing away the intolerable memory: the container was overfull with human eyes, little marbles still shiny, like egg white, each one followed by its red and black spring.'[7] Yet, in a startling reversal, Barbarant's narrator looks to the Unknown Soldier for the grace that we have lost, for the grace that he knew but which his contemporaries who survived the war were not able to keep. The narrator admits that he is tempted to 'lend you an impossible presence,' and evokes a photograph he saw in school of a dead First World War soldier, whose face was partially covered by mud: 'We were far from the "poilu" of folklore, which in any case I never believed in, and the youthfulness of this remnant of a face almost reclaimed by dirt was a violence to anyone who dwelt upon it.' A few pages later, he describes, albeit in the mode of disavowal, the emergence of the Unknown Soldier's presence – sensuous yet spiritual – in 'minuscule epiphanies': 'No doubt my weakness causes me to invent a presence for you. It is really too easy to claim that the dead maintain themselves, persist, or gesture at us within those beauties the world distributes haphazardly [...] there is nothing behind this heavy warm light I am immersed in as though blind [...] And yet every time I have to convince myself of it again [...] It may be permitted, perhaps, for me to believe that you might have known, like me, certain summers that blind us with wheat [...] the scandal is immediately evident: we can accept the loss of beings, but not of their states.'

By refusing to equate his present perception of beauty with that of the Unknown Soldier, Barbarant's narrator rejects a consolatory transcendence that they might share; yet the parallel between the soldier's possible epiphanies, now lost, and those of the narrator is what allows them to exchange grief but also joy, and to touch a shared mortality. This becomes more explicit as the narrator cites Simone Weil in order to reject dogmatic notions of sacrifice and transcendence, yet to retain a mystical sense of shared embodied mortality: 'Less than twenty years after we began worshipping you, a woman, still young, before me wrote: "And if the Gospels omitted all mention of the Resurrection, faith would be easier for me. The Cross alone suffices." Stupid the one who would see here nothing but a taste for sacrifice, when the point was to turn right side up once again the purity that can be contained at times in faith, to show how recognizing sorrow can make Humanity less brutal.' Once again the emphasis is on exchange as opposed to equation, on the distance this young woman has to cover in her compassion in order to put purity once again 'right side up' (*à l'endroit*). I will return to this image of the body without grace as upside down or inside out; before that, it is important to note how Barbarant, like many of the First World War authors that we have considered, seeks an alternative Christology, tied to the Incarnation, in order to regain a sense of the sacredness of the flesh. We see this again in another image of the exchange he wishes to establish between his narrator and the Unknown Soldier: 'Give me a piece, even minute, of your darkness, and I will lend you mine, and it will be like the legend of the cloak torn in two, which is not enough, cannot quite protect, but begins to open up a sky.' On the one hand, this refers to Saint Martin who shared his cloak with a beggar, and reiterates the idea of 'recognizing sorrow,' or shared suffering as a basis for truer humanity. And as 'darkness' (*ténèbre*) would normally be a negative term, suggesting that the dead man has either been forgotten or relegated to infernal regions, Barbarant again stresses that this sharing suffices in and of itself and does not look to a later redemption. Hence, on the other hand, the proximity of the images of the sky opening and of the torn fabric recall the Crucifixion, and in particular the moment of Christ's death, when the curtain of the temple is torn in two and the sky turns black. This darkness and symbolic tearing of the temple are signs of a world bereft of God.[8]

In such a world, Barbarant's book insists on the dignity of shared grief. But, his claim to eschew 'a taste for sacrifice' notwithstanding, the transformation of shared grief into a renewed sense of grace remains difficult. The letters pass from the 'minuscule epiphanies' mentioned above to a description of love as the realization of the particularity and uniqueness of one person's skin and body:

> My friend, who can say what smoulders in the most banal of approaches, how skin seen a little more closely under the veil of winter light reveals a secret carried from then on, for a whole life, a secret I have submitted to and taken by necessity as one of the keys to the world, the sum of everything we call reality. [...] the sudden grace of a body that tears you apart, where it is not, or not only, about desire, but about time like a glove turned inside out [...] there is no way to think oneself from you to public life, from this ordered revolution in the self to something that surpasses the number two. And yet, I insist, it is the nape of your neck, your shoulders, a golden arm in this square of daylight [...] that oppose barbarism.

Here, the image of being inside out returns, applied to time, but now to indicate putting things back the way they should be, returning to grace – and grace is, of course, also timing. Moreover, connecting time inside out to the glove and to the image of 'a body that tears you apart' (*un corps qui vous écartèle*), which mirrors yet reverses the violence of war, the passage evokes love as extreme vulnerability, radical openness to the other, almost interiority made exterior. Ultimately this suggests that the sacredness of flesh emerges from its exposure to the other. But in Barbarant's world, as he notes above, this remains a private realization, a revolution by and for two.[9]

The cover of Gian Pietro Testa's 2006 *Io sono il milite ignoto* starts with the divorce between the official Memorial and the body of the Unknown Soldier that we found in Barbarant, for what it shows is a photograph of a battlefield burial, presumably anonymous, with a bare cross, white rocks, and a helmet – the very same elements people contrasted to the pomp of the Vittoriano in 1921. It opens with the Unknown Soldier, in the first person, noting that he did not need capital letters to keep him from running away, as he is dead because he did not run away on time. The rest of the chapter continues this first person polemic against official national glory by appealing first to Christ's cry of dereliction, to evoke soldiers' terror; this is contrasted with 'Japanese kamikazes' and 'young Islamic extremists' 'trained by the lords of brutality, by the powerful who are always capable of using lives barely just in bloom for shameful purposes.' Second, to counter the empty ideals of both nationalism and fundamentalism, the Unknown Soldier attempts to answer his own question, 'but where is the spirituality proper to a dead man?' Echoing, once again, the alternative Christology we found early in the century in connection with the Unknown Soldier, Testa has him describe his spirituality as follows:

> there is no need to trust improbable fables, myths, and legends to see the obvious: you yourselves are your conscience, you are God, matter that becomes spirit, spirit that becomes matter, and all the rest is propaganda, created by people who sell divine products daily for their own glory and their own personal interest.
>
> [...] Perhaps those mystics who had visions exaggerated, those hermits eroticized by abstinence, and perhaps those poets and philosophers gave in to excessive rhetoric, but in all of them there was a deep desire for transcendental knowledge.

Once again the emphasis is on incarnation, and mysticism is seen as the opposite of dogma, as the search for an organic link between spirit and flesh. This appeal to the mystical, in Testa as in his predecessors, is an attempt to return physicality to the *corpus mysticum* and to reclaim the body from political appropriations. Thus, in the following pages he imagines the Unknown Soldier to come from a small town 'in the mountains of Irpinia' called 'the wolf town,' in which he grows up experiencing a connection to nature very similar to Barbarant's 'summers that blind us with wheat,' since he claims that 'my eyes, back then, were saturated with yellow.' The chapter ends with the Unknown Soldier as a boy witnessing a burgeoning sexual encounter between two slightly older girls, in which he learns that sexuality has 'its own sacredness,' which, we are told, he died without experiencing directly. Close to Barbarant again, Testa here makes a parallel between society's normalization of sexuality and its failure to mourn properly, as both fail to recognize the body's simultaneously material and spiritual nature.[10]

Unlike Barbarant, however, Testa seeks to bring the exchange between Unknown Soldier and present-day issues out of the personal and into the historical and the political. In his novel, as well, the marginality of the Unknown Soldier is understood to be similar to yet different from that of another, more contemporary character: his nephew, whose life the Unknown Soldier tells us about, including his (sometimes false) ideas about his uncle who died in the war. But this character is politicized, for we find out that the family named him Giuseppe Garibaldi to show that he was destined for glory; rebelling (or perhaps, Testa suggests, acting in the true spirit of Garibaldi rather than following the official Risorgimento story), he renames himself Fraschenor. *Frasca* means 'frond' and its poeticity is reminiscent of Pascoli's bucolic world (not unlike Giono's pastoral, both more or less came to an end with the First World War), but to an Italian ear it also recalls the expression '*saltare di palo in frasca*,' which means to change the subject or to refuse to be coherent (again in a sort of poetic, inspired fashion). In turn, the suffix '-enor' points to Antenor, a Trojan elder who in the *Iliad* insists that Helen should be returned to the Greeks in order to avoid war, and this refers to Frashenor's pacifism. Overall the name suggests a wisdom that is organic in contrast with the rigidity of ideology. The character himself is a mixture of Pirandello's fictional Mattia Pascal (*Il fu Mattia Pascal*) and the real-life Pier Paolo Pasolini – who both feel themselves to have outlived their lives, and thus look back on existence and society with a perspective from death not unlike that of the Unknown Soldier. Fraschenor is not mistakenly thought to be dead, like Mattia Pascal, but rather, finds himself to have become 'transparent' or invisible to society, and in particular to his wife and family. Like Mattia Pascal he embraces his new state, discovering in it a liberation from social roles: having become transparent, he decides to take a vow of silence, and thus to disappear while remaining present. Like Pasolini, he is a left-wing intellectual who nonetheless refuses party lines, and he has a group of intellectual artist friends who are also sexually transgressive. As with Pasolini's production of 'indigestible' works, especially his last film, *Salò* (1975), so shocking as to be unassimilable to mainstream culture, Fraschenor claims that his silence is an attack on a culture in which 'we can't talk to each other' and enacts his plan to 'become no one.' Finally, both Fraschenor and his uncle the Unknown Soldier are 'heretics' in the Pasolinian sense, using blasphemy against institutional religion to reclaim a more primitive spirituality: Fraschenor's vow of silence is secular but refers to monastic and mystical silence. In Chapter 2, I explained how contemporaries thought the inscription on the Unknown Soldier's Tomb should be silent, as though to manifest his refusal of political and religious speeches appropriating his death. Via the character of Fraschenor, Testa's novel seeks to bring the political relevance of that silence closer to the present.[11]

But, as mentioned, the novel also recognizes that silence has mystical roots; hence, the last question I wish to address concerns the link between politics and spirituality that is established through the relationship between Fraschenor and his dead uncle. For if Fraschenor brings political relevance to the Unknown Soldier's silence, the Unknown Soldier brings religiosity to Fraschenor's silence. The uncle, the Unknown Soldier, is associated with Franciscan innocence early in the book, for he comes from 'the wolf town': as is custom, his father thus takes him on a rite-of-passage wolf hunt, in which he fails to be a man because he cannot shoot the wolf. This recalls, of course, the legend of Saint Francis helping the city of Gubbio by going out and befriending the wolf who had been killing people, to the point that he became the people's friend and was fed by them. As we saw, the Unknown Soldier was already associated with Saint Francis in 1921 by many contemporaries, including Ungaretti, Giono, and d'Annunzio, who saw him both as 'wise fool' and as connected to nature and pantheistic religion. With somewhat heavy irony, Testa shows us that such foolishness cannot survive in the real world, as in the chapter following the wolf hunt, we hear of the Unknown Soldier as a child unable to bear the killing of the family pig, named Francesco. Finally, towards the end of the novel, the Unknown Soldier compares his nephew's silence to that of Saint Francis even as he acknowledges that the former retreats 'not, to be sure, on mount Subasio'; both, we are told, nonetheless decry the replacement of 'love' with 'rules.' As in 1921, the comparison to Saint Francis attributes a compassion to the soldier that stems from his holistic connection to the world.[12]

Further, after his failure to become a man by shooting the wolf, the Unknown Soldier's father declares him to be the son of Don Antenore, the town's priest, by which he means he is to be sent to the seminary. (Fraschenor's name thus also refers to this refusal of manhood, but not, as we will see, to the more coercive aspects of religion.) Once there, we learn that he is subject to 'mystical ecstasies,' which are undermined by the suicide of one of his cohort. This older boy, who is scathing about institutional religion, leaves him a last message about how 'powerful lords' rule poor people's minds. As though to confirm this view of religion, the religious community pretends that he died because he was so beloved by God, and enshrines him as a saint. We then have two versions of what happens next, one we are given earlier, because it is part of Fraschenor's family lore, and one told by the Unknown Soldier himself. They are very close, but for one point. In both, the Unknown Soldier returns home from the seminary and decides to enlist in the First World War, giving up on the priesthood that he no longer believes in: without telling his family, he departs early in the morning, hanging his vestments on a tree and shooting at them with the same gun he had failed to use against the wolf. For Fraschenor, this is a heroic gesture, of absolute anarchic rebellion, even if he is saddened by his uncle's subsequent 'obedience' as a soldier. As the Unknown Soldier tells the story, his motives were more a mixture of being taken in by the image of the war as just ('I am participating in a revolution'), and of his desire to affirm himself against his 'father, that asshole' as well as all the fathers at the seminary. Departing, then, he adds, 'I left the God-question [...] open.' As he walks away, however, he experiences a silence reminiscent of his mystical ecstasies: 'I realized that silence is not just an auditory category but a palpable physical phenomenon, I felt I could touch it, the silence of that moment was made of thousands of sounds and colours, all different, and the rustling of the leaves above me so light so light.' Moreover this image anticipates the very end of novel, when Testa imagines the Unknown Soldier's death and returns to the alternative Christology of the beginning:

> I got caught in the barbed wire, the steel nails cut through my uniform and tore my skin [...] arms wide as in a crucifix, head hanging. [...] I saw my blood down there mixing with the mud. After a few minutes the anxiety left me [...] Don't save my soul, I prayed as I was dying, I don't want you to save my soul, give me back my body with which I can live, love, be loved, the soul doesn't exist without the body [...] My Lord, if you exist [...] give me back the dignity of my body [...].
>
> I became light as a feather, by now I really felt like a poor flag blowing at every wind. [...] I was a light Christ on the cross.

The image of the 'poor flag' echoes Railton's 'Padre's Flag' (restored to its original place in Westminster the same year Testa's novel was published) and also d'Annunzio's 'Banner of Randaccio,' linked to the 'rent moon,' itself a reference to crucifixion without resurrection. Moreover, the prayer addressed by the Unknown Soldier to a God whose existence he doubts recalls the end of Ugo Foscolo's *Ultime lettere di Iacopo Ortis*, and thus refers to our lack of connection to our dead ancestors and to our exile from death itself. All this foregrounds suffering and undercuts the image of lightness as final liberation: though the more cruel 'anxiety' ceases, the soldier is still waiting that dignity be restored to his body. We see here that Testa's novel succeeds in revising the politics of the Unknown Soldier Memorial, perhaps more than Barbarant did, but that he cannot imagine a resacralization of the flesh except in the mode of nostalgia. His Unknown Soldier's spirituality pits the body against ideology and all the fathers, biological, religious, and political, who teach brutality, and he transmits this sort of rebellion to his nephew. But the dignity of the body itself is not really present in the novel except as anger against sexual repression.[13]

In different ways, both novels show how difficult it is to question religious absolutes and political authoritarianism while at the same time affirming a pacifism based on the fundamental sacredness of the flesh. In this, they anticipate and comment on recent events and, in particular, once again, the 2003 mourning in Italy for the dead of Nasiriyah, where we find a widespread desire to express 'solidarity' and specifically to 'touch' both the dead and the mourners, combined with general dismay at how official ceremonies fail to live up to this impulse. Specifically, in 2003 as in the novels of Testa and Barbarant, bodies were opposed to the artificial pomp surrounding them. One minute of silence was observed, even on television, which some commentators saw as a tremendous assertion of Berlusconi's media power; this, more than recalling the respectfully silent ceremonies of 1921, thus ironically acknowledged the silence society makes around violence, illness, and death, as seen in the two novels. One 2003 journalist even dared to note that many of the so-called heroes of Nasiriyah might have joined the military for financial or other personal reasons, taking up the importance given in both novels to the Unknown Soldier's lack of bellicism, and to his individual history. In the major dailies, however, no one noted that the inclusion among the dead officially mourned as 'soldiers' of the documentary film-maker and peace activist Stefano Rolla, far from generously extending state honours, was a highly effective way to turn his activism into another form of 'sacrifice' to the state. There was dissent, however, about state appropriation and control of families' mourning, whose 'unseemly' eruptions of grief could not be totally suppressed. Polemics erupted over the large number of seats (about half of the total) assigned to political figures in the religious ceremony held for the nineteen dead at San Paolo; last-minute changes were made at least to assign family members front seats, which they had not had. The Italian newspapers revelled with a mixture of irony, pleasure, and guilt on fainting women, wounded veterans, handicapped children, tearful prime ministers, and a father vocally outraged at church and state robbing him of his front-row seat. This is very close to the way our two novelists oppose their interpersonal vision of the Unknown Soldier to that of 'politics' and 'big business.' A few insightful 2003 editorials take up the central issue. 'Exposure to death' and 'unity in mourning,' we are warned by Francesco Pardi in one, 'must not become acritical consent' to a simple univocal meaning of those deaths. Gianni Vattimo insists in another that the 'martyrdom' of the nineteen, like martyrdom in general, 'has never proven the truth of any cause.' Finally, Vittorio Zucconi comments on the main news channel's use of Samuel Barber's *Adagio for Strings* as background music, noting that it was also important in Oliver Stone's *Platoon*, where it underscores useless massacres in Indochina: he asks, is this an involuntary admission that 'our soldiers too [...] have been sacrificed for nothing?' While violent death is most concretely and undeniably real, then, they insist it provides no discursive, abstract certainty, and as a result it is difficult to control the power of mourning. Edmondo Berselli thus comments on the book *Due nazioni*, about Italy's long-standing divisions, and notes that mourning unites, but leaves it unclear who will benefit from this sense of unity, that is, who will be able to capitalize on it politically. Similar issues come up in commentary around the case of Nicola Farfaglia, a soldier posted to guard the Unknown Soldier Memorial, who used his loaded weapon to kill himself while on duty on 18 January 2000. The authorities immediately claimed his suicide was due to a 'failed romance,' whereas his family refused this view, citing his letters where he complained about the dehumanizing effects of military service; psychologists commented that guard duty at the Unknown Soldier Memorial was too demoralizing. It is striking that overall Farfaglia's gesture, like Barbarant's encounter with the Unknown Soldier, was not able to reach 'public life.'[14]

Modernity and the Mystical Body

Does this mean mourning, in its existential dimension and search for a reaffirmation of life, must remain relegated to the private realm, where

Western modernity increasingly pushes it, according to Ariès and Vovelle? Two last evocations of the Unknown Soldier Memorial seek to counter this privatization by meditating explicitly on the mediatization of the body in our culture. In a short series of recently published poems for Pope John Paul II, *Santo Padre Divina Poesia* (2009), Alda Merini rather jarringly compares her body to that of the Unknown Soldier:

> [...] I love
> my body that bears me
> to the altar of the patria, like a dead soldier.

I will seek to elucidate this image by considering, first, the role of the body in relation to sacrifice in the poetry of Merini and, in particular, by asking to what sort of 'patria,' death, or home the body carries her. As we will see, more radical than the feminists who imagined the 'Unknown Wife' or 'Unknown Mother' of the Unknown Soldier in a pacifist vein, but along the same lines, Merini wants to feminize the body in the Memorial in order to question male, Western, rational notions of transcendence. In this she follows medieval women mystics, but adds to them a suffered exploration of her sexuality and maternity. Second, I will briefly address the problem of kitsch and desecration in Merini's poetry. For, as she associates a more embodied spirituality with her experience of many years in various insane asylums, at a time when electroshock treatment was normal, she can only present her spirituality as 'obscene' or demonic in the eyes of the world, able to be expressed only with irony, often in a tone of provocation. Hence, her feminization of the Unknown Soldier has a lyrical side, yet it must also be read as a sort of insult to the well-thinking reader.[15]

Merini published her first four volumes of poetry in the 1950s, and we find in them already a spirituality defined by contradiction:

> You are He with two faces: one of light,
> pasture for blessed souls,
> and one of dark,
> indefinite, in which
> the greater part of souls are submerged, crashing
> against the persistent
> enemy shadow: and they go, in that dark,
> hands outstretched as though blind ...

This is lived by the poet as an impossible attempt to reconcile within her own body these two aspects of the divine, also imagined as chaos and formed life:

> I will lie in the unformed
> myself melded, melted into the dark,
> as much as it is possible for me, formed and alive,
> to return to chaos …

In 1961, Merini, who had already had various bouts with mental illness, was interned in the Paolo Pini mental hospital, where she remained almost without interruption until 1972. During this time she did not publish, but continued to write. After she left the hospital, she struggled still with mental illness – she is now thought to have been bipolar – but also began to publish again, becoming quite prolific for the following thirty years until her death in 2009 (coincidentally her state funeral in Milan was on 4 November). 'Madness' became a recurrent motif in her work, where it is self-consciously used to describe her marginalization by society and at the same time her anguish and spiritual travail. Especially after 2000, her poetry acquired a more overt mystical bent in the sense that it referred to Christian models, including Jesus, Mary, and Saint Francis, even as she also continued to write about her sexuality, her body, and her experience of mental illness. As the title of her 2003 collection, *La carne degli angeli*, indicates, however, these themes are not easily separable from each other, and as I will suggest below, Merini's divine remained to the end double, one of suffered descent into the flesh and another equally suffered ascent out of insanity.[16]

To interpret Merini's feminization of the Unknown Soldier, we must first explore the connection between embodiment and ecstasy that runs throughout her poetry. Notwithstanding her well-known ambivalence towards her own biological mother, one of her most typical descriptions of mystical ecstasy is addressed to the mother:

> I too ate the apple
> of your omnipresence
> and I re-emerged
> void of knowledge,
> [...]
> and the calyx of your life
> grazed all the roses.
> Now you have become confused

with the obscure arguments of the lyre
[...]
[...] I will seek you,
I will unearth your ferment,
mother [...]

Here, a more canonical ecstasy as non-knowledge and omnipresence gives way to a confusion, an obscurity, a mysterious brush with roses that is also chthonic ferment. Later, this ferment is associated with pregnancy and childbirth, in the words attributed by the poet to Saint Francis: 'I too must give birth to my true God.' Merini adds in her own first person voice:

I only need your womb
to know mine
and I only need my womb
to know yours [...]

We see here the most important aspect of Merini's divine: the chiasmus, by which human and God, female and male, 'soul' and 'body,' belong to each other through an exchange of contrasting characteristics rather than via mirroring or similarity.[17]

This is also clear in Merini's depictions of Jesus. On the one hand, human limitations are 'raped by the sanctity of Jesus'; more ambiguously, but in the same direction, 'man was deflowered by God in the very moment of creation.' On the other hand – on the other side of the chiasmus – Christ, too, has been violated. This is a recurrent image, most startling perhaps in Merini's poems for the Pope:

Here with convex wings we angels
of the earth long saddened
by an unresolved maternity
no longer sing at the foot
of your cross, and we watch your flesh
flapping in your belly like a wound,
torn by a thousand demons

Christ, and John Paul II as *alter Christus*, is not only violated, but his castrated belly becomes 'an unresolved maternity.' I would propose that Merini's divine is not just this feminine Christ, wounded by the terrestriality of his own incarnation, by the fallenness of his own maternity – but also the first part of the chiasmus: the woman whose flesh is wounded by the spirit, and who cannot but compare the invasion of her body by this god to a violent defloration, to a pregnancy unwanted yet desired, and even to a rape.[18]

Throughout Merini's poetry, the blooming of the rose, its opening up into colour, is more than a metaphor for spiritual blossoming, and is tied to this violent defloration: indeed, blooming seems to become itself a living oxymoron, for the rose is at once tremendously carnal and provocatively spiritual. Thus, the barriers around the asylum are 'ferocious with flowers,' but at the same time 'something grazes you that just came near, and you strangely exult, and the beloved returns, and the rose finally opens.' More pointedly, the flower 'that opens' is 'a knot / of uncertainty and fear,' the rose is 'bitter,' 'blushing,' and 'punished,' and thus also 'the pause in God.' In the end, the rose incarnates without resolving the dynamic contradiction between decay and exultation, between abjection and consecration, between the flesh crying out for spirit and the spirit demanding descent into the joy of the flesh.[19]

This blooming is for Merini like the blood of Christ, which she understands in a way that is most uncanonical, though it is still a 'sign of contradiction.' Thus, like the rose, Christ's blood is 'of all those who have been martyred for nothing' and it is a 'variation in the blood ... the red colour of martyrdom' which the poet herself cannot understand. This blood spurts forth from the impossible reconciliation of terms that nonetheless belong together: it flows from 'a land that turns into a flower / and a flower that turns into a land,' which is also the double paradox of 'virgin ... mother' and 'mother ... virgin.' What is uncanonical here is that Christ's blood does not wash away sin or mortality, but instead flows from the contestation between the material and the spiritual: this is more than a *felix culpa*, because the embodied human, the struggle of soul and flesh, becomes the very source of Christ's blood:

> so that you, newly struck
> by human love,
> might bleed for us.

This is a 'fruit of damned certainty' and, in an image that brings everything together again, it is the silence of God which is reabsorbed into the womb and in the blood, yet still blooms when 'blood [...] turns the colour of God.' As we see already in this last image, such blooming is in and of the body. Thus, the flesh is 'imprint of God *and* of all changeable things,' and yet it is also marked by 'I know not what ghost that tramples me.' And the flesh that blooms is 'an apparition that vanishes [...] the time between two stops,' radical discontinuity, and confusion. This transforms or indeed transfigures the poet into a 'divergence of suffering.' What we see here, once again, is Merini's refusal to resolve the oxymoron of 'damned certainty' or also of 'mother virgin' into either of its two terms; what is imprinted upon her, rather, is at once their unity and their divergence, and it is only from this living contradiction that something – something that does not have a name and cannot even quite be called ecstasy – may bloom.[20]

Lastly, blooming is imprinted on the flesh in that for Merini it is also birthing; yet at the same time birthing is forever caught in the contestation between material pregnancy and a pregnancy of the soul which generates 'another unborn daughter,' '[a] daughter that ha[s] no religion [...] [a daughter] that I did not give birth to.' The same contradiction is seen in one of the more famous poems of *La terra santa*, 'Il mio primo trafugamento di madre.' Here with extreme violence material conception is equated with rape, even though in *L'altra verità* Merini rewrites this scene as passionate lovemaking 'resting on roses and thorns.' But the point in both cases, I propose, is in the last line of her poem, where she asserts that she has 'concepito una genesi': she 'conceived a genesis,' in other words, the very contradiction between the material and the immaterial world, as well as their mutual struggle and fecundation and suffering. The poet thus experiences a sort of inward blooming, a descent into the womb, as the source of all 'divergence of suffering.' Taking up the image of the Pope as *alter Christus* once again, and bringing us back to the Unknown Soldier, Merini asserts the continuity between his wound and her violated body that is also between the incarnation he celebrates and the birthing she suffers:

You had rooms wherever,
a path for your pain,
   on those paths
you went to die
day after day
   I loved you because I too
have paths wide open in me where I wait
for God to pray to me to resolve life,
and in the morning when I wake I tell myself

my body
now old and tired listens
still to his voice and so I love
my body that bears me
to the altar of the patria, like a dead soldier.

We find the chiasmus between human and divine again in that it is God who prays within the poet's wounds and not the reverse. Most important, Merini associates the Unknown Soldier with this chiasmus in which the death of Jesus and the act of giving birth come together without being identical, as the oxymoron of spiritual and sensual blooming. Thus, we see that the body as 'altar' is sacrificial, but the claim that it 'bears' her brings us back to birth, and suggests that the poet now gives birth to herself.[21]

Though Merini takes up the struggle, explored in Chapter 3, to assert the inwardness and self-generative authority of the Unknown Soldier's sacrifice against ideological appropriations, the point of her poem is not to resacralize the body within the Memorial, as was the case for both Barbarant and Testa. Rather, she foregrounds her own body, and seeks to assert its paradoxical, suffered, almost obscene grace in contrast with the soldier's, which is more acceptable, but for that very reason not truly scandalous in the Christological sense. Apparently contradicting herself, but enacting her appropriation of Jesus' suffering, which she depicts in constant tension with his appropriation of hers (as her very language is pervaded by Christian notions of suffering and transcendence), she would evoke 'the suffered pains of labour [*parto*],' and assert 'I think Jesus was inferior to me in greatness and tolerance.' Ultimately, then, Merini's strategy to resacralize the body is one that provocatively recognizes, and even plays with, the fact that enchanted flesh might appear, will most likely appear, in our world, incomprehensible, distorted, and even obscene. In doing so, she echoes the connection between sexuality and theology that is made in Latin America for the purpose of political dissent: as Marcella Althaus-Reid argues regarding the sexualization of the Virgin but, even more, the feminization of Christ, 'obscenity' in this case becomes 'a way to avoid unnecessary transcendence,' a refusal to produce the sort of sacredness that can be appropriated by religious or political ideologies.[22]

In a key scene in Marco Bellocchio's 2002 *L'ora di religione,* the main character, Ernesto Picciafuoco, his brother Erminio, and their old friend Filippo Argenti are driving home in Rome, arguing about the Picciafuoco family's attempt to have Ernesto's mother officially canonized as a saint by the Catholic Church. This canonization is the central conflict of the film, as Ernesto, a confirmed atheist, rejects it as opportunistic, whereas the rest of his family members pursue it for a variety of reasons ranging from bigotry, to the desire for social advancement, to lucre. Argenti is an important part of the case in the mother's favour, as he is a war veteran who claims to have been healed of his mutilation by her miraculous power. It is suggested, but never confirmed, that his mutilation was faked, most likely to avoid combat (we never find out exactly what his wound was or on what side of combat, presumably in the Second World War, he was), as was obviously his healing (at one point, in anger, he admits he plans on making a lot of money selling his story). As they are driving, Argenti yells at Ernesto because he is not happy with the bourgeois success canonization will bring, and he is messing up the rest of the family's plans. As they near the Vittoriano, they pass a surreal procession of veterans in wheelchairs pushed by nuns and led by Jesus himself, bearing the cross. The scene is reminiscent of Fellini's oneiric, fantastical takes on Catholicism, and like those, forces us to meditate on the power of images over the unconscious and their role in religion. Moreover, given the recent argument in the car over Argenti's faking, we are left to puzzle out whether the veterans in wheelchairs are as fake as the bleeding Jesus with his cross. Having thus made us conscious of issues of representation, Bellocchio has his characters get out of the car right in front of the Unknown Soldier Memorial: Argenti demands that the three of them cross themselves and wants a photo taken with the Memorial in the background and himself at the centre, surrounded by the two brothers. Quite clearly, this is an advance publicity photo for the 'wounded war veteran miraculously healed' that associates Argenti with the Unknown Soldier who was, as we know, often seen as saint or Christ figure.

Thus far, we might think that for Bellocchio the Vittoriano and the Unknown Soldier Memorial simply represent the collusion between the Vatican and the state, echoed in the family's collusion to get social standing out of canonization; and Ernesto Picciafuoco is the hero who sees through all this, showing us that atheism is the only answer. But in reality the film is far from this simple. We are warned by the name of Filippo Argenti, which is taken from Dante's *Inferno* VIII. Filippo Argenti is punished for his anger; when Dante encounters him, as he was a political enemy, he expresses anger that is praised by Virgil. Hence, this moment in the *Divine Comedy* is taken to illustrate the difference between unjust anger, which is partisan, and just anger, which attacks injustice. But in Bellocchio's film, where there is no guide, and no God, the character of Filippo Argenti poses the question of whether this distinction is possible at all. In essence, what basis does Ernesto have for opposing his family, and most important, for claiming his version of right is any less self-interested or ideological than the others'? For Bellocchio, who has a long record of making films about political rebels, this is a film that asks what it means to be a rebel in a post-1989 context in which distinctions between left and right and party allegiances are no longer the answer. Moreover, as the film delves into what Ernesto's unconscious motivations might be and explores how he, too, might be subject to manipulation, it is also about how existential and political rebellion may or may not connect, and about that problematic 'return' of the religious in contemporary political life, which we cannot afford to ignore. Another way of putting this, which shows why this film is such an apt conclusion for this book, is that Ernesto must discover that his impulse to destroy the Vittoriano and all the values it represents would be a base anger no better than Argenti's and that he must, instead, learn to confront and to transform this purely destructive impulse. This transformation entails rediscovering the sacredness of the flesh, in a way that draws on the religious icon but, far more than other examples we have seen, also reinvents it.

In the opening scene of *L'ora di religione,* before the credits, we see a small boy, Leonardo, yelling 'Vattene!' (Get out of here!) to the sky in his backyard; when his father, Ernesto, asks him about it, he explains that if God is everywhere, he has no freedom, and therefore he is telling God to go away. We then cut to Ernesto, who is a graphic artist, working in his studio at his computer: on the screen, he is creatively defacing a drawing of the Vittoriano. The two gestures are put in parallel when Ernesto is interrupted by a priest calling him to the Vatican for an interview regarding his mother's canonization, which until now the family had kept secret from him, anticipating his objections. For Ernesto, the big question raised by canonization is how and on what basis to allow his son the freedom he asks for, that is, how to teach his son just anger or how to 'tell God to go away' in the right way. As the film unfolds, we see that to do this, Ernesto must first learn it for himself, and therefore, as Bellocchio said in an interview, he must 'add up his life story again.' Even more relevant to our discussion, Bellocchio depicts this reckoning as a 'return to memory,' that is, a confrontation with a traumatic past, which is personal, yet cannot be separated from the political. We thus find out that aside from Argenti, the mother's main claim to sainthood comes from the fact that she was murdered by Ernesto's brother Egidio, purportedly because she was imploring him to cease shouting blasphemies; this establishes her as a martyr for the faith. According to Ernesto, however, she was killed in her sleep by a mentally ill Egidio, whose illness she had failed to address properly and perhaps even caused. Egidio himself, now interned in an asylum, has refused to speak ever since. The whole family wants to persuade Egidio, or at least Ernesto, to go with their version of the death, since divergent narratives from different brothers would obviously look bad.[23]

As the film progresses, we realize that, as we have learned to expect from trauma, it is difficult and perhaps impossible to recover its original truth: Ernesto's brothers may be liars, but this does not make Ernesto's version transparent, and Egidio's mental illness, just like its obverse, Argenti's cure, remain unverifiable, indicating that both family trauma and war trauma can never be fully, rationally explained. Bellocchio never shows us the mother's death on screen, even in the flashbacks and dream sequences he provides. Similarly, I would suggest, as Ernesto's on-screen defacing of the Vittoriano progresses to the crashing down of the whole Monument, we see a figure rising above the ruins (I will get back to this momentarily), but we do not see the destruction of the Unknown Soldier Memorial itself. I already mentioned how Bellocchio satirizes the official resurrection of the Unknown Soldier via the surreal veterans' procession and Argenti's identification with them. He also satirizes the official martyrdom of the mother, again foregrounding the slippery quality of representation, whence trauma can never be fully detached. Thus, the family's view is epitomized in a highly comical moment at a rich aunt's house, where they are staging photographs of the mother's murder with an actor and actress; these are to be used to create publicity for her cause. This repetition is, of course, a complete fantasy, as is underscored by Bellocchio's emphasis on fakery in this scene: the blood and the knife stand out as plastic, the improbably young dying mother talks on her cell phone during a break, and the light comes down from the heavens as in picture postcards of saints' apotheoses. The photographer insists, 'smile, smile, smile like a ... a saint.' This smile anticipates the coming scene, in which Ernesto's aunt explains that the family needs a saint for they need 'a protector, a second, a patron, a title.' She reacts to Ernesto's ironic smile by saying that this 'fuck you' smile ('*di chi vuole sempre sfottere il mondo*'), is the very same one his mother had and that he must stop pretending to be so superior. This is, of course, the opposite of a saint's smile; more important, it stresses a possible similarity between

Ernesto and his mother, in their common refusal of the 'live and let live' (and make money) values of the aunt. As Bellocchio said, commenting on the film's title for its release in English, *My Mother's Smile*, this smile is for Ernesto 'that part of himself [...] that he cannot deny, that he wants to rip from [his face].' In other words, it is the imprint of his mother on his body, and by implication also the trace of traumatic loss, which he must come to terms with and make his own.[24]

It is thus possible to interpret much of the film in terms of Ernesto's recurrent repetitions of his family's trauma, in which its personal and political dimensions are played out, as these are also variations on his opening gesture of defacing the Vittoriano. Ernesto acts out his anger and grief in a number of ways, including a rather surreal duel with the Conte Bulla, a monarchist, a long argument with the priest in charge of his mother's canonization, and a fight with his brothers in which the insane Egidio finally breaks his silence, screaming an obscenity at the Madonna. In the first case, anger seems all about the will to power, as the Conte Bulla himself argues; Ernesto wins their fencing match, but loses his point since he gives in to the will to power, as his body, shaking with anger and sweating profusely even after the fight is over, indicates. In the second, the priest wants to hear in Ernesto's voice a trembling that betrays his fear of God when he admits he did not love his mother; this leads him to voice real anger at her, for not helping Egidio with his mental illness. A psychoanalytic perspective on anger is for the moment juxtaposed to the religious one without further clarification. In the third case, especially as we are in the asylum in which Egidio is interned, it is the psychoanalytic interpretation of his attack on his mother (and all mothers, in the guise of insulting the Madonna) which is juxtaposed with Ernesto's interpretation of it as rebellion against all institutions (for Ernesto, Egidio was right to refuse his mother's Catholicism just as he is now right to refuse the 'cure' proposed by his doctors, who are on the family's side, and want Egidio to 'remember' that his mother was awake and praying when she died). As Ernesto hugs Egidio and holds him at length while he cries, we find him identifying with his 'insanity' as First World War veterans did with the Unknown Soldier's radical alienation, to counter the hypocritical 'sanity' that surrounds them.

Ernesto's most important acting out of anger and grief, however, has to do with the mysterious Diana Sereni. Ernesto first seeks her out in anger as well, for she is Leonardo's religious education teacher, and he is worried that she has forced the little boy to grapple with questions of free will that are too much for him. He is very disconcerted at what he finds when he reaches the school library. She is not only young and beautiful when he expected old and ugly, but she evades the entire question of religion by reciting a poem by Arseny Tarkvosky:

So summer has passed
Like it never had been
It is warm in the sun
But it's no consolation.

All that might have come off
Landed right in my hand
Like a five-fingered leaf
But it's no consolation.

Neither evil nor good
Were expended in vain
All was burning and glowed
But it's no consolation.

Life arched over its wing
It took care and preserved
And, in truth, I was blessed
But it's no consolation.

The leaves were not burned
The boughs were not broken
The day's rinsed like glass
But it's no consolation.[25]

As Bellocchio himself said in an interview, 'The poem that one of the characters recites, written by Tarkovsky's father [Arseny Tarkovsky, father of the film director, Andrei Tarkovsky], is a key moment in the story, because each of us should have, within ourselves, this permanent and total dissatisfaction, this search for something new and better.' And indeed, Diana Sereni becomes the emblem of Ernesto's quest for a different acting out of trauma, in which 'neither evil not good / were expended in vain,' and yet this, too, is 'no consolation,' or 'not enough' (as the Italian rendition of the poem goes), not the end of the quest. Ernesto thinks he is falling in love with Diana, though he barely knows her, and she, in turn, haunts his house when he is out, it seems because she is an aspiring artist and wants to leave her drawings there, so as to spy his unguarded reaction to them. In a long sequence that becomes more dreamlike and less realistic, as Ernesto comes home and moves about his large studio-house, she stays just one step ahead of him to remain hidden. Diana is also repeatedly juxtaposed with a bas-relief that Ernesto has in his studio: known as the 'Gradiva,' it was the subject of a 1903 novel by Wilhelm Jensen, which became famous when it was analysed by Freud in 1917 in *Delusion and Dream in Jensen's Gradiva.* For Freud, in the novel, the character Hanhold's obsession with the Gradiva bas-relief indicates unresolved feelings for a childhood playmate, Zoe, which he has repressed by becoming an archaeologist and which must now come to light, as it were, through an archaeology of the psyche. Just as Hanhold is cured of his obsession when he is able to express concrete love for Zoe, and thus become aware of the Gradiva as a fetish, Bellocchio implies that Ernesto, too, must undertake an archaeology of his psyche, to discover his own mechanisms of projection and fetishism. That it is Diana who seeks to spy on him unobserved is a reversal of the myth of Diana and Actaeon that already suggests she is a projection of Ernesto's desire more than she is a real woman, at least for the moment. Her visual association with the Gradiva and her elusiveness confirm this, implying that she is also a fetish whereby Ernesto can avoid confronting his mother's death. In all of this, Gradiva/Diana functions in ways similar to Hector, Véronique's model in Tournier's 'Les suaires' for, even as Diana is not the main character or agent, she at bottom stands for our tendency to project our fears and desires onto objects, separating them from ourselves and thus separating ourselves also into mind and body. The bas-relief, which is not quite three-dimensional, but gestures at that living three-dimensionality, is similar to the Veil of the Veronica; moreover, Bellocchio implies that the computer-animated drawing of the crumbling Vittoriano also has something of that quality, in which it appears that surface and depth are separable (as Tournier's Véronique thought) but they, in fact, are not. The question, then, is whether Ernesto will give in to projection and objectification, in his desire to lift the veil on trauma, or whether he will find a different way to relate to Diana/Gradiva.[26]

We here return to the figure rising above the crumbling Vittoriano, and see how it is both a commentary on Ernesto's psyche and on the collective national psyche that imagined the Unknown Soldier rising from his tomb. This figure is Gradiva, who is seen repeatedly walking up the steps towards the Unknown Soldier and then floating upwards over the rubble. But as we have just noted, she is also Diana. This is parallel to Jensen's character Hanhold imagining the Gradiva walking over the ruins of Pompeii. In both cases, it is visually made clear that Diana/Gradiva represents an escape from destruction, which in Freud's analysis of Jensen's novel was obviously also an escape from the destruction of the First World War. To love a real woman, Hanhold must cease projecting on the Gradiva a denial of violence, trauma, and mortality. But what of Ernesto in the film? Does he arrive at a new understanding of Diana, and Gradiva, which would also imply a new relationship to authority and to his own traumatic past? On one level, keeping the parallel with Freud and Jensen, we may observe that in a further sequence of drawings, the Vittoriano crumbles entirely, and Gradiva is transformed into a bird, perhaps a dove of peace, which flies away over it, leaving us at the end with a blank screen. This may be taken to mean Ernesto has come to understand Gradiva's phantasmatic quality and, therefore, the regressive nature of his own desire, both for Diana and to destroy the Monument; both are evasions of the nothingness at the core of trauma, which he must give up if he is to become a good father in the sense, at least, of becoming self-conscious and critical. In this interpretation, the blank screen confronts us with the absence of the transcendent meanings, personal, political, and religious, that official versions of the Unknown Soldier Memorial claimed, and it also implies that to destroy the Memorial gets us no closer to transcendence; all ideologies are equally delusional in their denial of death. Fabio Vighi draws on Slavoj Žižek to analyse *L'ora di religione* along these lines, proposing that we have in this film a revelation of the emptiness of the 'sublime object of ideology.' Citing Lacan, Vighi also claims that this is Ernesto's chance to accept the 'empty kernel of the subject,' '*his own* radical subjective inconsistency,' and, from there, 'to dare' to take action rather than merely to draw. I submit, however, that the film does more than expose emptiness or reject transcendent meanings as projections and that it does more than simply demand action: *L'ora di religione* meditates on the crucial link between action and the body, and ultimately, it claims that embodied experience is a genuine guide to effective rebellion.[27]

We see this, first, by the fact that the critique of ideology and exposure of its emptiness is attributed in the film itself not to Ernesto but to an architect whom Ernesto meets in the asylum where Egidio resides. In a scene that is highly comical for Italians who are aware of a long history of complaints about the Monument often called the '*orrendo Vittoriano*' (horrid Vittoriano) – ranging from Carlo Dossi's 1884 *I Mattoidi* to proposals for its demolition in the 1980s – the architect explains that he had

had elaborate plans to bomb it. Not because it was nationalistic (nothing so base!) but because it was aesthetically such a wound to the eye as to prevent all future architects from practising their art. It was his inability to execute his plan, the architect concludes, that 'made him ill.' At face value, this dialogue shows that art is an evasion and that claims to use it for political contestation are bound to fail in the face of ideology's power to incorporate all symbolic rebellions. The architect falls ill because he halts at this realization, stopping at cynicism, instead of 'daring' action. I believe Bellocchio intends this reading but, citing Tarkovsky's poem, also tells us that it is 'not enough.' We must doubt the architect's capacity for perfect self-analysis not so much because he is in the asylum, as because he is a friend of Ernesto's family, and because the dialogue ends with his admission of impotence in contrast with Ernesto's recent falling in love. Moreover, his views appear to us as rationalizations in contrast with Egidio's silence and screaming that immediately follow: if the latter is not a viable form of rebellion, it is implied, the former is too gutless to be truly contestatory.[28]

Hence, second, a more embodied sense of rebellion emerges in the film via its reference to Arseny Tarkovsky's poem. This poem is also recited in the film *Stalker* (1979) by Andrei Tarkovsky (hereafter, Tarkovsky), the poet's son; as we saw above, when asked about the poem, Bellocchio referred to 'the film director's father,' pointing us to the connection between them. In *Stalker*, the characters undertake a long and perilous journey through a 'Zone' where destruction and paranormal forces are at work; presumably, we are in some post-apocalyptic scenario, but the Zone is also reminiscent of no-man's land, and of all liminal places. The characters seek a mysterious room that fulfils wishes, but when they find it, they end up remaining outside. The 'writer' says that one can never know one's innermost wishes and claims the room is a plague that could only liberate horrors (such as the wish to kill one's mother in Bellocchio's film). The 'professor' gives up on his plan to bomb the room, however, saying that his innermost self prevents him (he would be the sceptical destroyer of hope, as the architect in Bellocchio's film wished to be, but at the end he realizes cynicism is not enough). Most important, Stalker himself never goes into the room, and refuses to take his wife there, because, apparently, his vocation lies in the Zone itself, in the journey towards a room that, he asserts at the end, is more about belief and hope (and perhaps also despair) than about any certain or provable fulfilment of wishes. One interpretation of the Zone is that it is a space of temporary anarchy, not unlike d'Annunzio's Fiume, within a world of order, referred to vaguely in the film as 'the government.' Stalker implies that it is important to journey into this world, yet to attempt to fix it into a single answer or fulfilment is counterproductive and perhaps deadly. This can be read politically but, as Bellocchio's references to Gradiva and Freud suggest, it can at the same time be read psychologically: the descent into the irrational depths of the psyche is necessary but not objectifiable; nor should it be permanent. Via the poem, Bellocchio's film makes an imaginative connection between Tarkovsky's room that we never see, the blank screen left after the destruction of the Vittoriano, and the 'truth' of Ernesto's family trauma.

But Tarkovsky is not just there to warn us that we cannot stare directly at this blankness. His entire film emphasizes the importance of the journey, and it does so in a way that returns us to the question of what form rebellion can take if ideology is no longer an absolute guide. With its long, lingering, dreamlike, and surreal sequences that pore over the materiality of all the objects and people in this semi-destroyed, strangely decaying, yet also luxuriously verdant world, the film creates a sense of heightened perception, reminiscent of Ungaretti's ecstatic moments of 'harmony' during the First World War, and also remarkably close to the effect of Gianikian and Ricci Lucchi's editing of Comerio. In all three cases, renouncing the ideas or ideals that protect us from our own mortality, renouncing external truths, leads us to inhabit the present but also the body in its perpetually changing relation to the world. As Robert Bird shows, Tarkovsky used found objects and brought them together in order to make us feel not just their strangeness – to force us to look at everyday things differently – but also in an attempt to revisit the power of Russian icon paintings to transmit a way of seeing the world rather than a specific representation of it. Among the objects are the coins that used to be embedded in ancient Russian icons, but which in more modern icons are merely painted in. Bird rightly takes this to indicate Tarkovsky's awareness that cinema, as a modern medium, and an infinitely reproducible one, must find new ways to lay claim to the iconic tradition. In the case of *Stalker*, this is attempted via slowing down and refocusing our way of seeing, but also, more overtly than in Gianikian and Ricci Lucchi, by acknowledging cinema's limitations, since here, too, the coins are 'painted' rather than real. Hence, *Stalker* also says that this process is never done, when, as the characters stand outside the room that fulfils wishes, a mutilated child recites the poem by the director's father, insisting this is 'not enough.' The reality of physical suffering, it is shown here, can be transmitted, via 'iconic' cinema in this case, but it will never be done full justice by any medium. I believe this is the essential point about rebellion and trauma that Bellocchio's wishes to make by referring to *Stalker.* With Tarkovsky, he implies that we must refuse to enter the room where the purported emptiness of our wishes is revealed either in their fulfilment or lack thereof; more important, we must learn to transform the journey itself – the mimetic displacements of trauma, the obscure ways that trauma fuels rebellion – into an 'iconic' way of seeing.[29]

How is such iconic seeing presented in *L'ora di religione* and how does it connect to the resacralized body that modern readings of the Unknown Soldier Memorial seek? The reference to the icon in Tarkovsky brings us back once more to the Veil of the Veronica, and suggests that we link it to the Gradiva bas-relief, and to Diana Sereni. We have already seen that to separate Gradiva from Diana may be deadly, like entering the room that fulfils wishes in *Stalker*, or like taking away the aura from Hector in Tournier's 'Les suaires.' The film confirms this towards the very end, for Ernesto learns that he has been mistaken for the entire film: Diana is not Leonardo's religious education teacher. He is briefly tempted to think she has been planted in his life by his family, to distract him in the middle of his divorce and perhaps even to impugn his morals and, thus, his testimony. At the end, however, he decides to love her even as he has no idea who she is, at least in terms of social identity, and we understand that had he given in to suspicion he would have fallen into the same cynicism he so dislikes in his family. Most important, Bellocchio gives us a number of shared profiles in which we see Ernesto's smile (his mother's smile) undergo a transformation through its mirroring with Diana's: their last smile together, one of clear understanding, is a prelude to his decision not to take Leonardo to the Vatican, where his grandmother's case is being presented in its final version. We see here that Ernesto has transformed his relationship to his past precisely by not clinging to it or attempting to unveil it, for just as he does not know exactly who Diana is, he also accepts that his rebellion may not have entirely rational, objectifiable justifications. At the same time, this does not mean that any rebellion, any act of daring is justified. Diana's physical presence throughout the film becomes crucial here, as confirmed by Bellocchio's noting that he needed a particular sort of face, one that would radiate without being conventionally beautiful. Something like Tarkovsky's heightened perception of materiality, and of the embeddedness of the body in the world, takes place in the long sequences where Diana is present; and this is also something like the radiant grace discovered by Robinson in *Vendredi.* Accepting Diana means accepting a certain physical joy as a guide, and demanding that vulnerability to the other as a basis for shared mortality not be limited to suffering.[30]

Conclusion

Together, these four recent works that reflect on the Unknown Soldier Memorial show that to make existential anonymity into an active re-enchantment of the flesh we cannot merely reject traditional transcendence, nor simply refuse the division of mind and body, the abstraction of the *corpus mysticum*, or the normalization and violence of modern Western culture. Returning to Tournier's terms, this would be like pretending that we can just step out of 'knowledge of things *through the other*,' or rational, divided, self-consciousness, and immediately into a world of radiant grace. But *Vendredi*, already in its title, warns us that such grace must be learned, that it passes through a mourning that needs to be shared via embodied, creative ritual, through performative language, and through the other. In Tournier's novel, it is not Friday who learns from Robinson, but the reverse, for the latter must learn from the former to live symbiotically with the island. Most important, for all that he is radiant or 'solar,' Friday knows death, and it is Friday who teaches Robinson to experience rather than to objectify it. In one central example of symbolic death and rebirth, Friday sets off Robinson's stash of explosives, and both are nearly killed as the entire landscape of the island is reinvented: Friday's laughter as he emerges from the rubble teaches Robinson to see this event as the island's legitimate reassertion of its telluric power against his attempt at domination. In another instance, Friday engages in one-to-one combat with a ram, and eventually kills him, though with difficulty; later the ram's head is preserved and transformed into an Aeolian harp, and Robinson writes that he understood this to be a symbolic killing of his old self so that his new self might be born. Crucial here is the materiality of the ram's head and Friday's literal transmutation of decaying guts into an object that is at once totemic and artistic: by reworking instead of destroying the past, Friday allows Robinson to mourn an older self without disavowing it or clinging to it.[31]

Similarly, we find in our four authors different attempts to mourn transcendence itself – to let go of yet not deny that part of ourselves that seeks absolutes, that wishes to find a singular truth or a destiny within the chaos of history. Thus, they seek to go beyond the phases of anger, denial, and negotiation typical of mourning and establish a more authentic relationship with its passing away. As with the death of a person, we are confronted with an apparent reduction to 'abject thing' that is initially unacceptable, that we can only come to terms with by gradually accepting its presence in us, thereby enacting a continuity between life and death – the continuity of touch and not of rational understanding that I have associated with the Veil of the Veronica. Before we can discover the grace of embodiment, we must experience mortality, that is, through suffering, trauma, and closeness to death we must physically live out the withdrawal of transcendence from the world, as opposed to knowing it intellectually. Further, for all four of our authors, such living out must be shared, for it is through others that we can touch our death, and most of all through others that this touch becomes healing and not only traumatic. Thus, all four of them search for a symbolic form to enact this sharing, for a language, in the broadest sense, that will be more like touch and thus go beyond analytical dualisms such as reality/appearance, mind/body, self/other, consciousness/world, or form/matter.

As I suggested in the beginning of this chapter, these four authors, like many of their predecessors, turn to the Unknown Soldier Memorial because they see in it the possibility of such a shared confrontation with death, as a prelude to reaffirming the sacredness of life. In doing so they return to a struggle that was already present at the Memorial's inception, in the debates of its creators that we have evoked: is the sacredness they seek merely a nostalgic return to older heroic models of the self? Is all heroism, notwithstanding its transformations, simply a denial of death, as Ernest Becker has argued? Is it possible to offer a genuine, and not regressive alternative to the abstraction of the mystical body and its appropriation by ideology? I have shown that during the First World War and in its immediate aftermath, nostalgia vied with more creative and subversive visions of the Memorial; I have tied the latter to an open-ended notion of sacrifice, linked to a non-canonical Christology that emphasizes incarnation and the flesh, and to the physical, ritual, and symbolic acting out of suffering that disrupts redemptive structures. Do these subversive elements live on for us in more recent interpretations, and if so, how? Another way of putting this question is: since we cannot merely return to the radiant grace of Friday, do new forms of mourning also bring us to new experiences of grace?

As we have just seen, all four of our authors seek to establish a relationship to the past – to the Unknown Soldier – that is not one of identity or nostalgic appropriation, but rather of exchange. Barbarant's narrator becomes faceless in losing his friends to AIDS and can share this with the Unknown Soldier's loss of comrades. Testa's Fraschenor becomes silent, in a defiant gesture of protest at present politics, responding to the Unknown Soldier's silent rejection of traditional Christian transcendence. In Merini's case, the exchange becomes mutual wounding as well as recognition: the transcendent God, the Pope, and the Unknown Soldier infuse the flesh with a desire for redemption; but in return the woman's flesh and her womb 'wound' God with the desire to become incarnate. Finally, in Bellocchio's film, Ernesto's confrontation with the Unknown Soldier and the past it represents is mediated by Diana Sereni, who is herself ambiguously a real person and a projection of Ernesto's psyche: referring to Tarkovsky, and eschewing the room that fulfils wishes or reveals their ultimate emptiness, Ernesto opts decisively for a creative misinterpretation of the past. Knowing we do not have direct access, or even access entirely free of nostalgia, but nonetheless continuing the journey by which we allow the past to live on within us, we transform it but are also transformed by it. Most important, by referring to Tarkovsky's icons, Bellocchio implies that to avoid nostalgia is less of an intellectual and more of a formal issue: it is the forms we use to relate to the past that constitute – or not – a denial of its pastness, and of mortality; it is by forms that we can seek to go beyond analytical dualisms.

What are these forms, and how does the Unknown Soldier Memorial shape them? In Bellocchio's film, we watch Ernesto evolve from one relationship to the past to another, radically different one. First, he depicts the Vittoriano and Gradiva on screen, abstracting them until they disappear, and he can only turn away from a blank screen on which there is unbearable emptiness. But later Diana mysteriously incarnates Gradiva, and Ernesto discovers a relationship to the past that is also embodied vulnerability to the other. Nostalgia, it is implied (by the association of Ernesto's first attitude with the architect, in particular), makes the past into an object 'for' consciousness just as it appears to be an object 'inside' Ernesto's computer. In contrast, by associating the past with being vulnerable to different forms of alterity that Ernesto cannot even fully distinguish – his family trauma inscribed by his mother's smile in his own face, Diana's incarnation of his desire to rebel – Bellocchio configures the past as larger than consciousness, or indeed as the living matter whence consciousness occasionally emerges.

Something similar happens in the Unknown Soldier Memorial because of the human body at its core: as we have seen, many attempts were made to contain, to reify, or reduce that body to mere matter and to 'frame' it ideologically; but popular reactions to and fictional representations of the Memorial return again and again to that body's strange excess with respect to its frame and to its capacity to touch us. This configures the past and its materiality as living, and indeed as that *in which* we are embedded rather than vice versa. This is not to say that there is no loss, but rather that it is we who constantly pass away; that is, the past is not made up of lost objects that our consciousness can no longer grasp, it is, rather, the finite aspect of our consciousness that prevents it from encompassing all. This shift in perspective is what impels Bellocchio, but also Merini, Testa, Barbarant, and many others throughout the twentieth century and since to return to the Unknown Soldier Memorial instead of simply inventing something new. For what they find there is not an analysis, explanation, or other type of framing of mortality; rather, in its paradoxical claim to memorialize decay, or to enact and re-enact what we usually see as the temporary passage between life and death, the Memorial challenges us to experience mortality as our own, as the very shape and finitude of our consciousness.

One last remark brings us back to recent events and the question of re-enchantment. Why is the experience of mortality liberating? Why do all four of our authors in this chapter and many earlier ones associate the Unknown Soldier's touch not only with shared suffering, but also with a renewed awareness of joy and, again and again, with flesh become 'phosphorescent' or ecstatic, radiating aliveness? As we have seen, psychologists, working on trauma, PTSD, and other confrontations with death, find that being able to share the experience, and creating a human solidarity in finitude are central to healing. But what the Unknown Soldier Memorial insists upon, as so many reactions attest, is also the symbolic form of such sharing, and the shift in perspective I discussed above. What is liberating is not just the common knowledge of mortality, a mortality we would nonetheless each separately 'own.' What is liberating, rather, is the experience that mortality is something larger than the self and something that we all participate in. Here, we experience being embedded in a world that is infinitely finite, where birth, change, joy, and finitude are inseparable. Concerning veterans of recent wars, Edward Tick thus contends that we must 'surround trauma with soul.' This, he argues, entails recognizing it as a 'rite of passage,' and inventing new forms to bring it to completion, so that veterans do not remain 'stuck in the death passage.' Concretely, what does this mean as we continue to confront death at war, violence, and trauma? Two recent events are examples.[32]

The first is the return in 2006 of Father David Railton's flag, which inspired the original creation of the Unknown Soldier Memorial, to its place over the Memorial itself at Westminster:

> The flag served as Railton's altar cloth at countless services on the Western Front. [...] It was used at the last communion of a soldier sentenced to be shot at dawn, and at services before the battles of Ypres, High Wood, Passchendaele, Cambrai, behind Vimy Ridge on Easter Day 1916, and before the Battle of the Somme, when the altar was 'a bucket turned on end in the corner of an old trench' with the flag trailing in the chalk and the mud.
>
> It flew at heart-rending roll-calls after battle, when name after name was read out and answered only by silence, and was the last covering of hundreds of bodies before their burial in hastily dug shallow graves. As the flag lay enshrouding a comrade, gunfire often forced the burial party to take cover 'behind the earth they had heaped up from the grave.'
>
> The flag was with Railton at High Wood, where some units suffered 80 per cent casualties and he was burying bodies for 'the best part of three days.' There were 'hundreds and hundreds of cases,' noted Railton, where he laid to rest soldiers 'unknown to anyone except Him from whom nothing is hidden.'
>
> Literally steeped in the blood of Britain's war dead, the Padre's Flag hung over the Unknown Warrior's grave for more than 30 years but in 1953 it was removed because it was obscuring the sightlines of television cameras at the Queen's Coronation. It has hung in a side chapel, largely forgotten, ever since.

Though the return of 'the Padre's Flag' may seem a nostalgic gesture, we have seen in discussions of the Banner of Randaccio and see again in this text that the flag serves to bring back the mobility of the flesh to a memorial that risks petrification. In the passage above, the physical connection of the flag to bodies both dead and alive, both present and absent, both loyal and not, is remarkable because it reveals the desire of present viewers, who are also experiencing war, to mourn in a more embodied fashion. That this is a refusal of the state's tendency to abstract mourning is confirmed by the spontaneous ritual that was created in the small town of Wootton Bassett, in England, starting in 2007 and until the present, because the bodies of British soldiers repatriated from Iraq had to be driven through it. People stop work and line the street, sometimes waiting for three hours to salute the hearses that drive by.[33]

My second example concerns the photographs of torture victims taken at Abu Ghraib. They are extremely concrete depictions of suffering, which, though digital, still retain something of the photograph's direct physical connection to what is portrayed. The bodies they reveal thus resemble the body in the Unknown Soldier Memorial in that both confront us with the unutterable violence humans are capable of and, presumably, elicit compassion, solidarity, and a sense of shared mortality. They were suppressed by the United States Government precisely because of their graphic nature – a decision that recalls attempts to make the Unknown Soldier's suffering blank, to make his bones clean and pure, to keep mourners away from disinterred corpses. And yet, as Susan Sontag wrote in 'Regarding the Torture of Others,' 'No: the horror of what is shown in the photographs cannot be separated from the horror that the photographs were taken.' Her point is that, in their intent at least, far from eliciting compassion or horror, these photographs were taken by the torturers – and for their pleasure. But more important, she comments on their form and their use: they include executioners along with victims, and they are meant to be circulated, actively participating in a culture of 'incessant home videoing,' in which 'to live is to be photographed, to have a record of one's life,' and eroticism exists only if 'captured in digital photographs and on video.' For Sontag, this normalizes violence. Taking this point one step further, and drawing on Tournier, I would claim the photographs create a world that is 'degraded into subject,' that is, a world in which things exist only under the narrow and estranged gaze of self-consciousness. We are real only insofar as we watch ourselves being real: this normalizes not only violence, but the disappearance of embodied experience. Watching ourselves, we become like Véronique's shrouds, a multiplication of images that have been wrenched from the living body. This is the same sort of watching that Ernesto Picciafuoco does at his computer, in Bellocchio's film, which he learns to abandon in favour of a more risky exposure to a life that has not been already photographed or filmed. Iconic seeing and embodied compassion mean turning away from the fascinating image of violence in order to take action that is uncertain in its outcome, but that asserts a world view in which objects are radiant, in which matter is inwardly alive. This is how we must react not just to the photographs of Abu Ghraib, but also to the Unknown Soldier. Having seen horror, we must refuse its tendency to fascinate and privilege the gaze, and instead take action to assert our common embodiment, mortality, and grace.[34]

Notes

Introduction

1 On the Italian Unknown Soldier, see the following: Alessandro Miniero, *Da Versailles al Milite ignoto: Rituali e retoriche della vittoria in Europa (1919–1921)* (Rome: Gangemi, 2008); Lorenzo Cadeddu, *La leggenda del soldato sconosciuto all'altare della patria* (Udine: Gaspari, 2001); Vito Labita, 'Il Milite Ignoto: Dalle trincee all'altare della patria,' in C. Grottanelli and S. Bertelli, eds., *Gli occhi di Alessandro: Potere sovrano e sacralità del corpo da Alessandro Magno a Ceaucescu* (Florence: Ponte delle Grazie, 1990), 120–53.

The Italian Memorial is also discussed in Catherine Brice, *Il Vittoriano: Monumentalità pubblica e politica a Roma* (Rome: Archivio Guido Izzi, 2005); Maurizio Ridolfi, *Le feste nazionali* (Bologna: Il Mulino, 2003); Giuseppe Talamo and Marco Pizzo, eds., *Dalla vittoria al milite ignoto. 4 novembre 1918 – 4 novembre 1921* (Rome: Ministero per i beni e le attività culturali, 2002); Marco Pizzo, *Il Vittoriano: Guida storico-artistica* (Rome: Ministero per i beni e le attività culturali, 2002). Bruno Tobia describes it as 'the greatest patriotic celebration,' in *L'altare della patria* (Bologna: Il Mulino, 1998), 84.

On the French Unknown Soldier, see the following: Jean-Pascal Soudagne, *L'histoire incroyable du soldat inconnu* (Rennes: Ouest-France, 2008); Jean-Yves Le Naour, 'Le dernier voyage du soldat inconnu,' *Histoire,* no. 336 (2008): 54–5; *Le Soldat Inconnu: La guerre, la mort, la mémoire* (Paris: Gallimard, 2008); Jean-François Jagielski, *Le soldat inconnu: Invention et postérité d'un symbole* (Paris: Imago, 2005); Gérard Fonck, *Le soldat inconnu,* vol. 1, *Les démarches* (Reims: Autoédition, 2004); M. Tauriac, 'La bataille inconnue du soldat inconnu,' *Historia,* no. 474 (1986): 47–54; Paul Virilio, 'Le soldat inconnu,' *Temps Modernes* 31/360 (1976): 2334–53; Charles Vilain, *Le Soldat inconnu: Histoire et culte* (Paris: Maurice d'Hartoy, 1933).

On the British Unknown Soldier, for a mixture of history and fiction, see Neil Hanson, *Unknown Soldiers: The Story of the Missing of the First World War* (New York: Knopf, 2006). On Edwin Luytens' cenotaph and on mourning during and after the First World War, see Jay M. Winter, *Sites of Memory, Sites of Mourning: The Great War in European Cultural History* (Cambridge and New York: Cambridge University Press, 1995).

On German memorials and mourning, see George L. Mosse, *Fallen Soldiers: Reshaping the Memory of the World Wars* (New York: Oxford University Press, 1990). Mosse discusses the Tomb of the Unknown Soldier inaugurated in Berlin in 1931 (97). I will not be considering the German Unknown Soldier, in part, because that Memorial was created later, but more important, because its meanings for a nation that had lost the war were quite different from how the winners saw it.

Broader discussions of the concept include the following: Daniel Sherman, 'Bodies and Names: The Emergence of Commemoration in Interwar France,' *American Historical Review* 102/2 (1998): 443–6; Malcolm Humble, 'The Unknown Soldier and the Return of the Fallen: The Political Dimension of Mourning in German Texts from the First World War to the Present,' *Modern Language Review* 93/4 (1998): 1034–44; Annette Becker, 'Les Soldats Inconnus,' *Historiens et Geographes* 89/364 (1998): 135–9; Thomas W. Laqueur: 'Names, Bodies, and the Anxiety of Erasure,' in Theodore R. Schatzki and Wolfgang Natter, eds., *The Social and Political Body* (New York: Guilford, 1996), 123–61, and 'Memory and Naming in the Great War,' in John R. Gillis, ed., *Commemorations: The Politics of National Identity* (Princeton: Princeton University Press, 1994), 150–67; Keith Phelan Gorman, 'The Return of the Dead: Christian Images of Resurrection and the Postwar Cult of Remembrance,' *Proceedings of the Annual Meeting of the Western Society for French History* 22 (1995): 67–77; Ken S. Inglis, 'Entombing Unknown Soldiers: From London and Paris to Baghdad,' *History and Memory* 5/2 (1993): 8–31, and 'War Memorials: Ten Questions for Historians,' *Guerres Mondiales et Conflits Contemporains* 42/167 (1992): 5–21; Daniel W. Ingersoll and James N. Nickell, 'The Most Important Monument: The Tomb of the Unknown Soldier,' in Gordon Bronitsky, ed., *Mirror and Metaphor: The Material and Social Constructions of Reality* (Lanham: University Presses of America, 1987), 199–225; Claudio Canal, 'La retorica della morte: I monumenti ai caduti della Grande guerra,' *Rivista di Storia Contemporanea*, no. 4 (1982): 659–69.

Unless otherwise noted, all translations are my own, done in collaboration with Lorenzo Giachetti; for more substantial citations, the original is given in the notes. Except where stated otherwise, all emphases are in the originals.

2 On the 'mare di popolo' in Piazza Venezia, see Mario Pirani, 'Il ritorno della Patria,' *Repubblica,* 18 Nov. 2003. For 'toccare con mano quello che è successo,' see 'Un silenzioso abbraccio: "Oggi c'è lutto, non c'è spazio per le divisioni,"' *L'Unità,* 17 Nov. 2003. On the chronology of the day, see 'Chiuso il Vittoriano, 1 milione di persone in 24 ore,' *Corriere della Sera,* 18 Nov. 2003.

On instrumentalization, from various sides of the political spectrum, see Paolo Conti, 'Onore ai caduti (senza retorica),' *Corriere della Sera,* 15 Nov. 2003. Also see Francesco Merlo, 'Carabinieri d'Italia,' *Repubblica,* 18 Nov. 2003, who begins with the assertion, 'La retorica non fa bene ai carabinieri' (rhetoric is not good for policemen). Also, condemning 'consenso acritico alla politica estera' (acritical consensus for foreign policy), see Francesco Pardi, 'Morti in guerra o in missione di pace?' *L'Unità,* 17 Nov. 2003.

For the polemic around alleged comments by the Bishop of Caserta, Raffaele Nogaro, see Red, 'Nogaro; "Dico che il terrorismo non si combatte con le armi": Ma la polemica non si placa,' *L'Unità,* 18 Nov. 2003; and 'Il caso di monsignor Nogaro crea "sconcerto" in Vaticano,' *Repubblica,* 18 Nov. 2003. Nogaro issued a written statement asserting that he never intended to condemn priests who bless coffins at state funerals, but only to warn that 'bisogna fare attenzione a non esaltare il culto dei martiri e degli eroi della patria strumentalizzando la morte di questi nostri giovani per legittimare guerre ingiuste' (we must be careful not to support the cult of martyrs and heroes of the patria, which instrumentalize the death of our young people to legitimate unjust wars).

Finally on 'remorse' and 'reproach,' see *Avanti!* 25 Oct. 1921, and Vittorio Zucconi, 'Il Vietnam italiano,' *Repubblica,* 18 Nov. 2003. This was the largest military loss suffered in one day by Italy since the Second World War, according to Frank Bruni, 'Thousands of Italians in Tribute to 19 lost,' *New York Times,* 18 Nov. 2003.

3 On the refusal to 'understand,' see Claude Lanzmann and Shoshana Felman, 'The Obscenity of Understanding: An Evening with Claude Lanzmann,' in Cathy Caruth, ed., *Trauma: Explorations in Memory* (Baltimore: Johns Hopkins University Press, 1995), 200–20. On trauma 'floating' 'just below the surface' in *Shoah,* see Dana Renga, 'Staging Memory and Trauma in French and Italian Holocaust Film,' *Romanic Review* 97/3–4 (2006): 472–5.

On war, death, and the collectivity see the following: Irvin D. Yalom, *Staring at the Sun: Overcoming the Terror of Death* (San Francisco: Jossey-Bass, 2008); Marilyn Yalom, *The American Resting Place: Four Hundred Years of History through Our Cemeteries and Burial Grounds* (Boston: Houghton Mifflin,

2008); Robert Pogue Harrison, *The Dominion of the Dead* (Chicago: University of Chicago Press, 2003); Luc Capdevila and Danièle Voldman, *Nos morts: Les sociétés occidentales face aux tués de la guerre, XIXe–XXe siècles* (Paris: Payot, 2002); Katherine Verdery, *The Political Lives of Dead Bodies: Reburial and Postsocialist Change* (New York: Columbia University Press, 1999); David I. Kertzer, *Ritual, Politics, and Power* (New Haven: Yale University Press, 1989); Michel Vovelle, *La mort et l'Occident de 1300 à nos jours* (Paris: Gallimard, 1983); Edith Wyschogrod, *Spirit in Ashes: Hegel, Heidegger, and Man-Made Mass Death* (New Haven: Yale University Press, 1985), and *The Phenomenon of Death: Faces of Mortality* (New York,: Harper and Row, 1973); Philippe Ariès, *L'homme devant la mort* (Paris: Seuil, 1977); Ernest Becker, *The Denial of Death* (New York: Free Press, 1973).

A very recent study on mourning in Italy is Oliver Janz, *Das symbolische Kapital der Trauer: Nation, Religion und Familie im italienischen Gefallenenkult des Ersten Weltkriegs* (Tubingen: Niemeyer, 2009). On two recent suicides, one by a civilian, and one by a soldier on duty at the Unknown Soldier's Tomb, see the following: Maria Rosaria Spadaccino, 'Giù dal Vittoriano, la fine terribile scelta da Francesca,' *Corriere della Sera*, 2 June 2005; Clarida Salvatori, 'Ragazza anoressica suicida al Vittoriano,' *Repubblica*, 2 June 2005; Claudio Lazzaro and Carlo Macrì, 'Marinaio si uccide all' Altare della Patria,' *Corriere della Sera*, 19 Jan. 2000; 'Aperte tre inchieste, per la Procura una "delusione sentimentale"'; 'Il fratello: "No, amava troppo la vita,"' *Repubblica*, 18 Jan. 2000. A very recent suicide attempt is discussed in Augusto Parboni, 'Afgano tenta il suicidio al Vittoriano; 'Lo Straniero: "Voglio Asilo Politico in Germania,"' *Il Tempo*, 12 May 2009.

The main character in the Peter Greenaway film *Belly of an Architect* jumps off the Vittoriano to his death.

4 On post-1921 uses of the Unknown Soldier memorial in Italy, see Ridolfi, *Le feste nazionali.* On students in 1940 France, see Maxime Tandonnet, *1940, un autre 11 novembre: Étudiant de France, malgré l'ordre des autorités opprimantes, tu iras honorer le soldat inconnu* (Paris: Tallandier, 2009). On 1935 rallies by the Croix-de-Feu, see Le Naour, *Le Soldat Inconnu*, 64. A reminder of the peace rallies held in Rome earlier in 2003 is in 'Italy Holds State Funeral for 19 Troops,' *New York Times*, 18 Nov. 2003.

Coverage of the funeral for the dead of Nasiriyah evinces a certain discomfort with the Memorial's ambivalent meanings, most notably in the attempt to elide all of its divergent uses after 1921. It is asserted that 'l'unica, e ultima cerimonia funebre nel Vittoriano risale proprio al 4 novembre 1921' (the only, and last, funeral ceremony held at the Vittoriano goes back to 1921) ('Chiuso il Vittoriano,' *Corriere della Sera*, 18 Nov. 2003). In 'Il ritorno della Patria' (*Repubblica*, 18 Nov. 2003), Mario Pirani claims 'per la *seconda volta* nella storia il Vittoriano accoglie intorno a sè un mare di popolo' (for the *second time* in history the Vittoriano welcomes a sea of people) (emphasis added), even as later in the same article he acknowledges Fascist uses of the memorial.

On the protests against the Iraq war held at the Vittoriano, including one in which Greenpeace draped a banner over the monument, see the following: *Feb. 15, 2003 anti-war protest*, retrieved from http://en.wikipedia.org/wiki/Feb._15,_2003_anti-war_protest; 'Striscione di protesta al Vittoriano,' *Corriere della Sera*, 23 March 2003; Francesco Di Frischia, Ilaria Sacchettoni, and Alessandro Capponi, 'Blitz pacifista al Vittoriano,' *Corriere della Sera*, 23 March 2003. The 15 Feb. protest was the biggest rally ever held in Italy; retrieved from http://en.wikipedia.org/wiki/February_15,_2003_anti-war_protest.

On the British Memorial to Innocent Victims of Oppression, Violence, and War, see John Ezard, 'A Tribute to Civil Victims of War and Tyranny,' *Guardian*, 11 Oct. 1996. On its use to mourn after the events of 11 Sept. 2001 in New York and Washington, see Steve Boggan and Jane Hughers, 'Like a River of Grief, Widows and Children Flowed Past a Memorial to Mass Murder,' *Independent*, 30 Nov. 2001.

On the Potsdam Memorial to the Unknown Deserter, see the following: *City of Potsdam: Platz der Einheit, Bassinplatz Squares and the Dutch Quarter*, retrieved from http://www.potsdam.de/cms/beitrag/10032561/287369/1; 'Greens Win Permission to Display Monument to Unknown Deserter,' *International News*, 21 Sept. 1989; Stephen Kinzer, 'Berlin Journal; The War Memorial: To Embrace the Guilty, Too?' *Foreign Desk*, Nov. 15 1993.

On the Finnish film and novel that it is based on, see John Sundholm, '"The Unknown Soldier": Film as a Founding Trauma and National Monument,' in John Sundholm, Conny Mithander, Maria Holmgren Troy, eds., *Collective Traumas: Memories of War and Conflict in 20th-Century Europe* (Brussels: P.I.E.–Peter Lang, 2007), 111–41. There were, in fact, two films made from this novel, one in 1955 by Edvin Laine and one in 1985 by Rauni Mollberg.

On the 'Wehrmacht Exhibition' and the related film, see Gabriel Fawcett, 'The Wehrmacht Exhibition,' *History Today* 52/4 (April 2002): 2–3; Paul Brenner, *The Unknown Soldier* (review of film by Michael Verhoeven), 2007, retrieved from http://www.filmcritic.com/misc/emporium.nsf/reviews/The–Unknown–Solider.

The United States is the only country to have added more bodies to its Unknown Soldier Memorial, adding three more to the one buried in 1921:

one each after the Second World War, the Korean War, and the Vietnam War. It was only in 1988 that controversy erupted over the suggestion that the authories had a good idea of who the Vietman Unknown Soldier might be, and that DNA testing should be done; it was, amid much controversy, and he was identified as Lieutenant Michael Blassie. For the basic story, see 'From Tomb of Unknown to Grave of Lieut. Blassie,' *New York Times,* 12 July 1998. For a discussion of the philosophical implications of this identification of the Unknown Soldier, see Michael Naas, 'History's Remains: Of Memory, Mourning, and the Event,' *Research in Phenomenology* 33 (2003): 75–96. For Naas, in the wake of the dead who were unidentifiable in the World Trade Center bombings because nothing of them remained for DNA testing to take place, the whole point of the Memorial is to show both our desire for identification and the fact that we can never fully 'know' the dead, that we can never fully own their otherness. For a more political reading, see Conrad Quintyn and Sarah Wagner, 'Dismantling a National Icon,' *Anthropology Today* 50/5 (2009): 7–9. They note that 'President Obama's decision to lift the 20-year ban on photographing flag-draped coffins [...] indexes a shift in the authority over commemoration from the government to the public' (9).

American takes on the Unknown Soldier include Robert Rossen's 1934 film *The Unknown Soldier Speaks,* in which he condemns the rise of Mussolini and Hitler. For the Unknown Soldier Memorial in Iraq, see Samir al-Khalil, *The Monument: Art, Vulgarity, and Responsibility in Iraq* (Berkeley: University of California Press, 1991); Rosie DiManno, 'Hunt for the Disappeared Leads to Hellish Tunnels,' *Toronto Star,* 21 April 2003.

On Canada, see Jennifer Delisle, '"For King and Country": Nostalgia, War, and Canada's "Tomb of the Unknown Soldier,"' *Dalhousie Review* 85/1 (2005): 15–32. On New Zealand, see Helen Tunnah, 'Soldier's Name Secret Forever,' *New Zealand Herald,* 6 Nov. 2004.

For the popularity of this memorial across the world, see Inglis, 'Entombing Unknown Soldiers.' He mentions also the Memorial to the Unknown Insurgent created in Budapest in 1989. The persistent popularity of this idea is also evident in the 2005 John Hulme film, *Unknown Soldier: Searching for a Father,* which focuses on the Vietnam generation.

5 For work on the Unknown Soldier Memorial, see n1. For the notion of figural history, I draw, from among others, on the following: Paul Fussell, *The Great War and Modern Memory* (New York and London: Oxford University Press, 1975); Eric J. Leed, *No Man's Land: Combat and Identity in World War I* (Cambridge: Cambridge University Press, 1979); Benedict R. Anderson, *Imagined Communities: Reflections on the Origins and Spread of Nationalism* (London and New York: Verso, 1991); E.J. Hobsbawm, *Nations and Nationalism*

since 1780: Programme, Myth, Reality (Cambridge and New York: Cambridge University Press, 1992); Winter, *Sites of Memory, Sites of Mourning*; Carla Freccero, *Popular Culture: An Introduction* (New York: New York University Press, 1999); Ernest Gellner, *Nations and Nationalism* (Malden: Blackwell, 2006); Alberto Mario Banti and Roberto Bizzocchi, *Immagini della nazione nell'Italia del Risorgimento* (Rome: Carocci, 2002); Alberto Mario Banti, *La nazione del risorgimento: Parentela, santità e onore alle origini dell'Italia unità* (Turin: Einaudi, 2000); Catherine Brice, 'The Many Origins of the Risorgimento Mobilization in Italy: Problems of Interpretation (1790s–1860s); or, Radical Beginnings, the Resurrection of the Dead,' presented at the California Interdisciplinary Consortium for Italian Studies Annual Meeting, Stanford University. 6–7 March 2009. Roland Greene calls for 'a version of the new historicism, but one that is genuinely comparative' in order for us 'to regard the present from a historicist angle,' in 'New World Studies and the Limits of National Literatures,' in Joan Retallack and Juliana Spahr, eds., *Poetry and Pedagogy: The Challenge of the Contemporary* (New York: Palgrave Macmillan, 2006), 64–5.

For what I mean by 'figural history,' see also Laura Wittman, 'Imagined Geographies,' in 'Imagined Geographies: Italy and France,' special issue, ed. Laura Wittman, *Romanic Review* 97/3–4 (2006): 265–74.

6 Recent books on the Unknown Soldier focus almost eclusively on the early 1920s. See the following: Miniero, *Da Versailles al Milite ignoto*; Soudagne, *L'histoire incroyable du soldat inconnu*; Le Naour, *Le Soldat Inconnu*; Jagielski, *Le soldat inconnu*; Cadeddu, *La leggenda del soldato sconosciuto*. A rare book that focuses on the popular imagination in connection to the Unknown Soldier is Hanson, *Unknown Soldiers*. For how people mourning during the First World War drew on, yet criticized, both the recent and the ancient past, see Winter, *Sites*.

On Italians' claim to mourn 'also for the dead of other countries: the Iraqi dead first of all,' see 'Un silenzioso abbraccio,' *L'Unità*, 17 Nov. 2003. On 'catharsis' as what unites 1921 and recent events, in this case the 2001 reopening of the Vittoriano, see Giorgio Battistini, 'Quel soldato sconosciuto,' *Repubblica*, 30 Oct. 2001.

7 Verdery states, 'I present the politics of corpses as being less about legitimating new governments (though it can be that, too) than about cosmologies and practices relating the living and the dead'; this leads to 'analytically enlivening or re-enchanting politics' (*Political Lives of Dead Bodies*, 27).

Richard Osgood, *The Unknown Warrior: An Archaeology of the Common Soldier* (Stroud: Sutton, 2005). Osgood's study of fourteen 'unknown soldiers' from ancient Roman times up to the British Unknown Soldier buried at Westminster is based on what is known of the physical remains of anonymous soldiers, and it seeks to uncover what their daily lives and world views were; this shows how much the very concept of an Unknown Soldier Memorial implies this sort of archeaological approach to history. Osgood also shows that while the burial of a single anonymous body did not exist until after the First World War, the commemoration of the common soldier took place in Roman times, though it was rare (44).

8 On the empty tomb, see Thucydides, *History of the Peloponnesian War*, trans. Richard Crawley (Forgotten Books, 2008; retrieved from http://www.forgottenbooks.org), Book II, 34: 'In the funeral procession cypress coffins are borne in cars, one for each tribe; the bones of the deceased being placed in the coffin of their tribe. Among these is carried one empty bier decked for the missing, that is, for those whose bodies could not be recovered.'

For the ossuary and the American Civil War, see Drew Gilpin Faust, *This Republic of Suffering: Death and the American Civil War* (New York: Knopf, 2008). The marker for the unknown remains is also described in M. Yalom, *The American Resting Place*, 245.

Mark H. Dunkelman, *Gettysburg's Unknown Soldier: The Life, Death, and Celebrity of Amos Humiston* (Westport: Praeger, 1999). Dunkelman tells the story of a famous Civil War 'unknown soldier' – a man who was only identified as Amos Humiston after a month and multiple publications of his photographs in newspapers: 'He was probably the best-known common soldier and most heralded enlisted man to emerge in the lore of the battle, but his fame centered on the heart-rending circumstances of his death' (12): he was found at Gettysburg clutching the photographs of his three small children. Humiston's story made the expression 'Unknown Soldier' a common one, before the idea for a memorial existed. For the relation of names to bodies, see Sherman, 'Bodies and Names.'

The question of the bodies' integrity emerges in 2003 in the discussion of autopsies, which remains a bit unclear: 'Onore ai caduti (senza retorica),' *Corriere della Sera*, 15 Nov. 2003, states 'non ci saranno autopsie. Per rispetto e per evitare inutili e ulteriori strazi' (there will be no autopsies. Out of respect and to avoid further useless pain). It seems that the bodies were examined to some degree, however, according to 'Solidarietà ai parenti che riconoscono le salme,' *Corriere della Sera*, 17 Nov. 2003.

Silence was emphasized in numerous articles in 2003, including: 'Un silenzioso abbraccio,' *L'Unità*, 17 Nov. 2003; 'Un silenzio assordante nel mattino di Roma,' *Repubblica*, 18 Nov. 2003; 'L'Italia si ferma per i funerali: Niente spot in tv, negozi chiusi, un minuto di silenzio in uffici e scuole,' *Repubblica*, 18 Nov. 2003.

9 On medieval nobles, see Philippe Ariès, *Images de l'homme devant la mort* (Paris: Seuil, 1983), 39–42; translated by Janet Lloyd, as *Images of Man and Death* (Cambridge: Harvard University Press, 1985). On the 'distributive' body, see Laqueur, 'Names, Bodies,' 126.

10 On embodiment as suffering, joy, and authentic experience of mortality in medieval female mystics but also in twentieth-century philosophy, see Amy M. Hollywood, *Sensible Ecstasy: Mysticism, Sexual Difference, and the Demands of History* (Chicago: University of Chicago Press, 2002). She draws, from among others, on reflections on mourning and the body in Luce Irigaray, *Speculum of the Other Woman*, trans. Gillian C. Gill (Ithaca: Cornell University Press, 1985), and *Sexes and Genealogies*, trans. Gillian C. Gill (New York: Columbia University Press, 1993).

For the importance of physical acting out of mourning, and especially vocalization, see Harrison, *Dominion*, 62–71. He is commenting on Ernesto De Martino, *Morte e pianto rituale: Dal lamento funebre antico al pianto di Maria* (Turin: Bollati Boringhieri, 2000). See also Ernesto De Martino, *La terra del rimorso: Contributo a una storia religiosa del Sud* (1961; new ed., Milan: Il saggiatore, 2008), translated by Dorothy Louise Zinn, as *The Land of Remorse: A Study of Southern Italian Tarantism* (London: Free Association, 2005).

For the idea that physical pain is an initially inconceivable experience, whose entry into language and ideology is a flowering of the imagination that also has essential ethical responsibilities, see Elaine Scarry, *The Body in Pain: The Making and Unmaking of the World* (New York: Oxford University Press, 1987).

11 On the history of the *corpus mysticum*, see Ernst Hartwig Kantorowicz, 'Pro patria mori,' in *Selected Studies* (Locust Valley: Augustin, 1965), 308–24. This will be discussed in more detail in the coming chapters.

For the importance of heterogeneity, see Wyschogrod's concluding remarks in *Spirit in Ashes*: 'The unprecedented possibility of total annihilation by human agency [...] derails the I [...] the I is overwhelmed by the ubiquity of death and gives itself over to its own foredisclosures of mass death even when these possibilities are put out of mind [...] this new I deteriorates as a personal center [...] Split off from the I, the personal center of language and coporeality, deprived of fresh infusions from the actitivities of the I the me loses its distinguishing marks and becomes part of a totality of increasingly similar selves, homogeneous units whose meaning can be expressed in quantitative terms' (212).

Homogeneity is also what Jean-Marie Apostolidès sees as characteristic of modernity, the notion of 'modern individualism' notwithstanding; see *Héroïsme et victimisation: Une histoire de la sensibilité* (Paris: Exils, 2003).

12 Winter, *Sites.* The simplistic opposition of power versus dissent breaks down when we question the validity of separating 'high' from 'low' art. As Russell A. Berman warns, in *Cultural Studies of Modern Germany: History, Representation, and Nationhood* (Madison: University of Wisconsin Press, 1993), at 23: 'there is no reason to assume they [critical moments] are more likely in commodified aesthetics than anywhere else in commodified society.'

On the issues of pacifism versus bellicism as regards teachers and education in France after the First World War see Mona Siegel, '"History Is the Opposite of Forgetting": The Limits of Memory and the Lessons of History in Interwar France,' *Journal of Modern History* 74/3 (2002): 770–800.

This same question regarding the relationship of national identity and bellicism, asking that Italy develop a policy for withdrawing from Iraq, is posed in Edmondo Berselli, 'Il cordoglio e il consenso,' *Repubblica,* 20 Nov. 2003.

13 Emilio Gentile, *Il culto del littorio: La sacralizzazione della politica nell'Italia Fascista* (Bari: Laterza, 1994), 35, 307; translated by Keith Botsford, as *The Sacralization of Politics in Fascist Italy* (Cambridge: Harvard University Press, 1996).

On Mussolini's body, see Enrico Pozzi, 'Il Duce e il Milite ignoto: Dialettica di due corpi politici,' *Rassegna Italiana di Sociologia* 39/3 (1998): 332–57.

For the notion that Fascism endlessly defers the meaning of wartime deaths in order to preserve its authority to confer meaning, formulated philosophically in the claim that 'il fascismo è [...] *attualità dell'aporia*' (fascism is [...] *the actualization of aporia*), see Fabio Vander, *L'estetizzazione della politica: Il fascismo come anti-Italia* (Bari: Dedalo, 2001), 19. A similar point is made by Jeffrey T. Schnapp, in *Staging Fascism: 18 BL and the Theater of Masses for Masses* (Stanford: Stanford University Press, 1996), at 77, where (citing Corrado Sofia) he describes Fascist theatre as 'actualized mystical experience.' This can be related to Fascism's drawing on Giovanni Gentile's actualism, as articulated by Claudio Fogu, in *The Historic Imaginary: Politics of History in Fascist Italy* (Toronto: University of Toronto Press, 2003), at 48 and 205, whereby a 'history belonging to the present' is enacted via the 'elimination of the medium of (historical) representation between (historical) agency and (historical) consciousness'; for Fogu, this results in a 'posthistoric' imaginary that is central to modern culture. See also Claudio Fogu, 'Fascism and Philosophy: The Case of Actualism,' *South Central Review* 23/1 (2006): 4–22.

On recent interest in Fascism, see Jeffrey T. Schnapp, 'Fascism after Fascism,' in Richard J. Golsan, ed., *Fascism's Return: Scandal, Revision, and Ideology since 1980* (Lincoln: University of Nebraska Press, 1998), 63–85.

14 On the Vietnam Memorial, see Harrison, *Dominion*, 136–41. On the problem of war memorials today and the mourning that veterans need in order to feel reintegrated into society and psychologically whole, see Edward Tick, *War and the Soul: Healing Our Nation's Veterans from Post-Traumatic Stress Disorder* (Wheaton: Quest Books, 2005); Daryl S. Paulson and Stanley Krippner, eds., *Haunted by Combat: Understanding PTSD in War Veterans, Including Women, Reservists, and Those Coming Back from Iraq* (Westport and London: Praeger Security International, 2007).

The importance of touch is found in 2003 Italian articles on mourning at the Vittoriano, e.g.: 'Solenni funerali di Stato: Istantanee di un paese in lutto,' *Repubblica*, 18 Nov. 2003; Giuseppe d'Avanzo, 'Il funerale infinito delle famiglie, il dolore di chi ha perso figli, papà e mariti dietro quelle divise,' *Repubblica*, 18 Nov. 2003. On the importance of touch in earlier literature, see Santanu Das, *Touch and Intimacy in First World War Literature* (Cambridge and New York: Cambridge University Press, 2005).

On the new decisions of the Obama administration, see Elisabeth Bumiller, 'U.S. Lifts Photo Ban on Military Coffins,' *New York Times*, 7 Dec. 2009, and James Dao, 'Families of Military Suicides Seek White House Condolences,' *New York Times*, 26 Nov. 2009.

For a classic discussion of 'just war,' with references to the first Gulf War see: Norberto Bobbio, *Il problema della guerra e le vie della pace* (Bologna: Il Mulino, 1979), and *Una guerra giusta? Sul conflitto del Golfo* (Venice: Marsilio, 1991); Norberto Bobbio and Pietro Polito, *Il terzo assente: Saggi e discorsi sulla pace e la guerra* (Milan: Sonda, 1989).

On 'just war' as a concept no longer valid for modern warfare, see Michel Serres, *La guerre mondiale* (Paris: Pommier, 2008). Wyschogrod (*Spirit in Ashes*) makes a similar point when she notes that since the First World War we have not war with 'sides' but a 'death-world' that 'is an artificially contrived return to the state of nature, in which every man is at war with every man' (131).

Introduction to Part I: The Return of the Dead

1 On changes brought by the First World War, see Capdevila and Voldman, *Nos morts*, 26–7. The original reads: 'l'augmentation du nombre de morts s'accompagna d'une plus grande prise de conscience de chacun d'entre eux. Comme si, à l'aube de la mort de masse, l'individu n'en devenait que plus précieux.' On policies, see Inglis, 'War Memorials,' 9.

2 On the dead returning to question the living, see Winter, *Sites of Memory, Sites of Mourning*, 15–28. On the 'living unknown soldier,' see Jean-Yves Le

Naour, *Le soldat inconnu vivant* (Paris: Hachette, 2002). A similar story, of an amnesiac recovering from First World War shell shock, is told in David Drury's 1998 miniseries for British television, *The Unknown Soldier.*

3 At first, Mangin was claimed to be the brother of Louise Vayssettes, and recognized also by his mother; but by 1920 it was clear this was false, as measurements and descriptions did not fit. This launched the newspaper campaign to identify him, which received almost 300 responses, many of which were rejected, but which also resulted in dozens of personal visits. In 1924 Marthe Mazat sought to have her son's *acte de decès* annulled, and the press took her side for a while, even as Louise Vayssettes still persisted. Then Lucie Lemay, in 1927, and the Monjoin family, in 1930, seemed to be more likely candidates: Lemay recognized him as her husband and the stories seemed to fit, but in the end details (such as handwriting) threw her case into question; the Monjoins claimed him as a 'fils' and 'frère,' but his brother barely recognized him; their case, based more on details of evidence than recognition, finally won in 1939, but it was immediately appealed by Lemay and a number of others. For the highlights of the Mangin story cited here, see Le Naour, *Le soldat inconnu vivant,* 195, 199. For the 'amnesiac of Collegno,' see *Caso Bruneri-Canella; smemorato di Collegno – Wikipedia,* retrieved from http://it.wikipedia.org/wiki/Caso_Bruneri-Canella; also S. Zago, G. Sartori, and G. Scarlat, 'Malingering and Retrograde Amnesia: The Historic Case of the Collegno Amnesic,' *Cortex* 40/3 (2004): 519–32. He is also mentioned in Le Naour, *Le soldat inconnu vivant,* 98.

4 Le Naour, *Le soldat inconnu vivant,* 59.

5 Winter, *Sites,* 15.

6 Ibid., 22; Le Naour, *Le soldat inconnu vivant,* 52.

7 Le Naour, ibid., 48.

8 Winter, *Sites,* 24.

9 Le Naour, *Le soldat inconnu vivant,* 48–9. He cites a few exceptions, where differentiation took place, as impeding mourning.

10 Sherman, 'Bodies and Names,' 462. He mentions that in France it was illegal for families to *prevent* the inscription of their dead's name on a town monument, as the patronymic was considered to be institutional property.

11 On simulation, see Le Naour, *Le soldat inconnu vivant,* 15, 19. On trauma, Freud, and the First World War see the following: Allan Young, *The Harmony of Illusions: Inventing Post-Traumatic Stress Disorder* (Princeton: Princeton University Press, 1995), 36–85; Ruth Leys, *Trauma: A Genealogy* (Chicago: University of Chicago Press, 2000), 18–19; Anne Harrington, *The Cure Within: A History of Mind-Body Medicine* (New York: Norton, 2008), 69–94.

For Young, a crucial element of definition emerges at this time: trauma is about memory, hence, always involved in time, always belated or 'constructed' in some way. In this sense, it is difficult to separate 'trauma' from PTSD, except historically, in the sense that the latter term emerges later and is part of a specific 1970s' defence of trauma as 'real' even if not physical. See Harrington, *The Cure Within*, 95. Harrington, instead, emphasizes that trauma is part of a narrative that she terms 'the body that speaks,' in which physical symptoms are the sign of repressed memories that need to be addressed. In both cases, we find a central problem for wartime discussion: trauma cannot be 'proven,' even as it prevents soldiers from being functional.

12 Leed, *No Man's Land*, 3–4, 211. Tick (*War and the Soul*, 77) argues that war trauma is not merely a 'stress' disorder but an identity crisis, and he asserts that modern war (starting with the American Civil War) is not psychologically sustainable for the large majority of those involved.

13 Leed, *No Man's Land*, 18, 21–2.

14 Ferruccio Vecchi, in the fall of 1918, quoted in Michael Ledeen, *D'Annunzio; the First Duce* (New Brunswisk, NJ: Transaction, 2002), 43. On the anti-Fascist *Arditi del popolo* and their relationship to the *arditi* in general see: Eros Francescangeli, *Gli Arditi del popolo* (Rome: Odradek, 2000); Andrea Staid, *Gli Arditi del popolo: La prima lotta armata contro il fascismo, 1921–1922* (Ragusa: La fiaccola, 2007).

15 Michel Foucault, *Madness and Civilization: A History of Insanity in the Age of Reason*, trans. Richard Howard (New York: Pantheon, 1965).

16 Two cartoons that refer to the soldier's estrangement from big business are the following: 'Le mie ragioni pesan più delle vostre' (My reasons are weightier than yours), which appeared in the first issue of *Il Dovere*, 27 April 1919, and 'Non ostante, vinse' (Nonetheless, he won), with the second caption 'non c'è posto per voi nel tempio della gloria!' (there is no room for you in the temple of glory!), in *Il Dovere*, 20–1 Aug. 1919.

17 In Italy, the most famous literary example of someone seeking wisdom from dead ancestors, who are also major cultural figures, is Ugo Foscolo, *Le ultime lettere di Iacopo Ortis* (Milan: Mursia, 1985). On Foscolo see the following: Carolyn Springer, '"L'urne de' forti": Foscolo's Sepulchral Archaeology,' *Quaderni d'Italianistica* 4/1 (1983): 26–46, and *The Marble Wilderness: Ruins and Representation in Italian Romanticism, 1775–1850* (Cambridge: Cambridge University Press, 1987); Massimo Riva, 'Ortis, o dell'ombra amorosa: Le "Ultime lettere" e la genesi del simbolismo sepolcrale,' *Italian Quarterly* 29/111 (1988): 15–39, and *Malinconie del moderno: Critica dell'incivilimento e disagio della nazionalità nella letteratura italiana del XIX secolo* (Ravenna: Longo, 2001); Margaret Brose, 'The Politics of Mourning in Foscolo's Dei Sepolcri,' *European Romantic Review* 9/1 (1998): 1–34.

18 Hence, Virilio, in 'Le soldat inconnu,' at 2338, argues that modern warfare – and he gives the example of Teilhard de Chardin and the First World War – creates the idea of '*l'homme inachevé*' (incomplete man) and thus makes possible the notion of an '*anthropogénèse*' (anthropogenesis).

For a complex analysis of the Great War's transformation of European, especially German, culture, see Modris Eksteins, *Rites of Spring: The Great War and the Birth of the Modern Age* (Boston: Houghton Mifflin, 1989). His vision of Lindbergh as 'the new Christ' (242–7) is, in particular, germane to the role that d'Annunzio played in Italy during the war.

19 For fever and stripping, see Le Naour, *Le soldat inconnu vivant,* 163.

20 Joris-Karl Huysmans, *A rebours* (1884; Paris: Pocket, 1999), 120.

21 For pre-war examples of funerary monuments with occult or spiritual exploration as an important component, see Ariès, *Images de l'homme devant la mort,* 179. On the connections between the mysticism and the body in Huysmans and late nineteenth-century culture, see Laura Wittman, 'Mystical Insight and Psychoanalysis in the Fin-de-Siècle Novel: Huysmans, Bourget, d'Annunzio,' *Forum Italicum* (2008): 30–51.

For the creative side of loss of identity, see Winter, *Sites,* 75ff. In the post–First World War era, Georg Groddek saw trauma as resulting from the denial of people's authentic needs or 'needs of the it' (*das Es*); the deeper identity beneath the social is clearly related to this 'it,' which is also related to Freud's 'Id'; see Harrington, *The Cure Within,* 82–5.

On the 'new Italian,' the canonical phrase is attributed to Massimo D'Azeglio, as he is most often cited: 'having made Italy, it remains to make the Italians.' In reality, D'Azeglio wrote, in *Things I Remember (I miei ricordi)* (London and New York: Oxford University Press, 1966), vol. I: 'The Italians have wanted to make a new Italy, but themselves remain the old Italians' (17). This, along with D'Azeglio in relation to Italians' creation of their own national past is discussed in Adrian Lyttelton, 'Creating a National Past: History, Myth, and Image in the Risorgimento,' in Albert Russell Ascoli and Krystyna Von Henneberg, eds., *Making and Remaking Italy: The Cultivation of National Identity around the Risorgimento* (Oxford and New York: Berg, 2001), 27–74.

22 See the canonical Cesare Lombroso, *Genio e follia* (Milan: Tip. e libreria Giuseppe Chiusi, 1864). On links between criminality, hysteria, mystical experience, and spiritual evolution see: Cristina Mazzoni, *Saint Hysteria: Neurosis, Mysticism, and Gender in European Culture* (Ithaca: Cornell University Press, 1996), and 'Mystical and Literary Texts: Meeting the Other, and Each Other, at the Borders of Language,' *Annali d'Italianistica* 25 (2007): 105–22; Nicoletta Pireddu, 'Primitive Marks of Modernity: Cultural Reconfigurations in the Franco-Italian Fin de Siècle,' *Romanic Review* 97/3–4 (2006): 371–400.

23 'Expressionism seeks "something else" *within* the figure's appearance – some recess in its surface,' according to Thomas Harrison, in *1910: The Emancipation of Dissonance* (Berkeley: University of California Press, 1996), 63. 'Not that artists had not taken a hard look at the self before, but few had attempted to penetrate its invisible unspeakable uniqueness, despoiling it of connections to the external, operative world,' wrote Harrison, and followed by a discussion of Munch and *The Scream* (164–7).

1 A Unanimous Idea

1 'Versailles non è l'eterna pace pei morti' [cartoon], *Il secolo illustrato*, no. 21 (1919). The original reads: 'Versailles non è l'eterna pace pei morti. essi sentono rimescolare le ossa da tutta l'umanità sopravissuta. [...] si schiudon le tombe e se ne levano i morti per alzarli a segnacolo di orgoglio o di maledizione. [...] poeti o sciacalli della politica, lasciate in pace i morti, com'essi invocano nella loro giusta pace sognata. e nel loro nome edificate il tempio della giustizia le cui fondamenta essi hanno cementato col sangue.'

This journal also had an earlier cartoon of a returning soldier with the following caption, very d'Annunzian in tone: 'Palmo a palmo il sacro terreno mi fu conteso per tre anni e mezzo dai nemici e palmo a palmo ora dagli amici. Comunque la vittoria deve restare. Null'altro ammetto. Perchè anch'io resto!' (Step by step I fought my enemies for three and a half years over the sacred land; now step by step I fight friends for it. In any case victory must remain. I will not accept anything else. For I too remain), 'Palmo a palmo' [cartoon], *Il secolo illustrato*, no. 9 (1919).

2 On the failure of massive national monuments, see Inglis, 'War Memorials,' 11.

3 Ariès, *L'homme devant la mort*; Winter, *Sites of Memory, Sites of Mourning*. For the importance of 'equality in death' see Inglis, 'War Memorials,' 7. For the modern notion that everyone has a memorable life, hence, one that must be somehow commemorated, see Laqueur, 'Names, Bodies,' 135ff.

On British policy and names, see Sherman, 'Bodies and Names,' 451; also Inglis, 'War Memorials,' 9. On Italian policy, see the following: Labita, 'Il Milite Ignoto,' 143; Ridolfi, *Le feste nazionali*, 68; Patrizia Dogliani, 'Redipuglia,' in Mario Isnenghi and Ersilia Alessandrone Perona, eds., *I luoghi della memoria: Simboli e miti dell'Italia unita* (Rome: Laterza, 1996), 382–3.

4 See cartoon, 'Palmo a palmo.'

5 Leed, *No Man's Land*, 211.

6 Mircea Eliade, *Shamanism: Archaic Techniques of Ecstasy* (New York: Bollingen,

1964), 63. Eliade is drawing on work on Eskimo and Siberian shamans, as well as on Buddhism. As I will explain in Part II, Cendrars drew on Siberian legend and the work of anthropologists on Siberia to create his vision of the Unknown Soldier as Russian peasant/shaman in his *La vie et la mort du Soldat Inconnu.* A similar duality is discussed by Michael P. Carroll, in *Veiled Threats: The Logic of Popular Catholicism in Italy* (Baltimore: Johns Hopkins University Press, 1996), at 146–7, where in his analysis of the cult of the dead in popular Italian Catholicism he shows that ideas of Purgatory developed by the Church merged with the notion that those who died violently – mainly from the plague or by execution (catacylsmic causes comparable to the First World War) – would have trouble completing the voyage to the underworld, and would remain stuck in an 'in-between domain,' whence they could both help and/or hurt the living.

7 Jagielski, *Le soldat inconnu,* 54 and note. 'Spontaneity,' is from a telegram to Bonomi in Ministero dell'Interno, Direzione generale di pubblica sicurezza, anno 1921, F.D.1, Affari Generali: 'Onoranze al Solato Ignoto', C4, busta 80, 1921, Archivio Centrale dello Stato, Rome (hereafter MdI, OaSI, 1921). The original reads: 'debbo segnalarti spontaneità dimostrazioni pubbliche che hanno superato nostre stesse previsioni. Tutti partiti hanno partecipato cerimonia.' For the dossier of reports from all over Italy regarding possible unrest and its causes in the Memorial's popular and populist appeal, see MdI, OaSI, 1921.

Harrison (*Dominion*) expresses the importance of a primordial or 'humic' connection to the dead, tied to the earth: 'The human finds its place *here,* in the "time of the utterable," which is the time of transience as such. The historical unfolding of what passes away comes to rest in the earth, its humus' (49), and 'Certainly it was Vico who first helped me understand how the human is bound up with the humus and why burial figures as the generative institution of human nature' (xi).

8 On Simon and entry into political debate, see Jagielski, *Le soldat inconnu,* 54–6. Simon's speech is reproduced in Fonck, *Le soldat inconnu,* vol. 1, at 26, along with a photo of what looks like an original pamphlet printing, which states that the speech took place on 26 Nov. 1916. The first printed mention of the speech, which does not cite it, that I have been able to find, is Vilain, *Le Soldat inconnu,* 52–3. Simon is also credited, based on Fonck, by Soudagne (in *L'histoire incroyable du soldat inconnu,* at 53–5), who, however, considers that the original popular idea, a pacifist and minimalist one, was transformed as it became official into something more 'illustrious.' On Simon and Railton respectively, see Miniero, *Da Versailles al Milite ignoto,* 127, 142. Miniero's source for the Railton story is *The Burial of the Unknown War-*

rior, retrieved from http://www.westernfrontassociation.com/great-war-people/memorials/400-burial-unkown.html.

Hanson's version is in his *Unknown Soldiers*, 263; he mentions that the Dean of Westminster did not credit Railton, but said it was his idea. Hanson cites archival sources as well as David Railton's 'The Origins of the Unknown Warrior,' *Our Empire* 6/8 (1931): 34–6. Interestingly, Hanson begins his entire book on the war experiences of three men, an Englishman, a German, and an American, with a brief, poignant description of Railton's discovery of the tomb to 'An Unknown British Soldier' at Armentières, confirming the importance of what I term the primal scene of battlefield burial (4).

9 Giulio Douhet, 'Per la tomba del soldato al Pantheon,' initially published in *Il Dovere*, 24 Aug. 1920, reproduced in Giulio Douhet, *Le Profezie di Cassandra: Raccolta di scritti* (Genoa: Lang e Padano, 1931), 424–33; quotes are respectively from 424, 426, 427, 431.

The originals read: '"*Il Dovere*" ha sempre affermato che il vero vincitore della guerra fu il Soldato, l'umile Soldato'; 'reggitori politici e militari incapaci'; 'non può essere materiato che dalla salma di un Soldato ignoto'; 'l'apoteosi del soldato non significa affatto l'apoteosi della guerra'; 'È nel desiderio di ognuno che i popoli si affratellino.'

Regarding Douhet's mixture of glorification and pacifism, which was uncomfortable for the state, see *Il Dovere*, 5 Nov. 1919, which contains a column that provocatively states that the 'cultivation of self-love' is now more important than 'sacrifice and Patria.' This is an example of the importance of internal rather than external definitions of sacrifice, which I will discuss in detail in Chapter 3. Douhet is credited with the 'invention' of the Memorial in Italy by Miniero (*Da Versailles*), most recently, who acknowledges the precedence of French and British discussions, and by Cadeddu (*La leggenda del soldato sconosciuto*), who does not. However, Caddeddu's history has the merit of great detail in the historical circumstances of the creation of the Italian Memorial.

10 Gabriele d'Annunzio, *Notturno* (Milan: Garzanti, 1995), 333–4. The original reads:

> Mentre in tristezza io trascrivevo l'esempio del contadino innominato che entra nel guado e s'inginocchia in mezzo alla correntìa e sacrificandosi incide il suo sacrifizio nell'acqua, dentro la basilica di Aquileia una madre dolorosa sceglieva tra le undici bare innominate quella che sta per discendere nel monumento.
>
> [...]

E perché il feretro del Grande Offeso non era ammantato dalla bandiera del Timavo, da quella che io custodisco, da quella che fu chiamata 'il sudario del sacrifizio' e 'il labaro del fante,' da quella che fu distesa sopra le casse dei miei morti di Fiume allineate in terra?

Prima v'era rimasta effigiata l'imagine di un solo eroe morto; ma ora v'è l'imagine di tutti i morti, ché tutti quelli che sono morti per la Patria e nella Patria si somigliano come Giovanni Randaccio nella sua arca di macigno somiglia al fante ignoto raccolto fra quattro assi.

11 In the Oct.–Nov.–Dec. 1921 issue of *L'Ardente* (admittedly a pro-d'Annunzio newspaper), Claudio Mariani goes so far as to claim that d'Annunzio is 'il Poeta Nazionale, è ancor più: il Poeta della Guerra Mondiale' (the National Poet, and even more: the Poet of the Great War), and also terms him 'il Gran Padre Spirituale dei combattenti d'Italia' (the Great Spiritual Father of Italian combatants). But the mainstream *L'illustrazione Italiana* says much the same, especially since almost no issue of this weekly illustrated newspaper, between Oct. and Dec. 1921, fails to praise d'Annunzio's *Notturno* or to mention his various commemorations for dead comrades and Italian soldiers.

For an analysis of the 'orphic' power of d'Annunzio's speeches, see Giancarlo Lancellotti, 'L'orazione dannunziana tra poetica e politica,' *Problemi,* nos. 114–15 (1999): 263. Contrasting Ungaretti's emphasis on suffering with d'Annunzio's passion for heroism, Silvio Bertoldi, in his chapter 'La Guerra di d'Annunzio e dei poeti,' in *Guerra: Italiani in trincea da Caporetto a Salò* (Milan: Rizzoli, 2003), at 83, gives a synthetic view of d'Annunzio's role, citing the Duke of Aosta, who claimed that a speech by d'Annunzio before combat equalled a battle won.

Mario Isnenghi, in *Il mito della grande guerra: Da Marinetti a Malaparte* (Rome and Bari: Laterza, 1973), at 398 and 19, indirectly acknowledges the importance of the *Notturno* when he explains that he excluded it from analysis due to its excessive visibility and power, which might overshadow other aspects of the war experience. I take this warning to be correct, but as Isnenghi admits, d'Annunzio was a massive presence for Italians.

For a thorough overview of d'Annunzio's work, as represented in his home, *Il Vittoriale,* see Lucia Re, 'Gabriele D'Annunzio's Theater of Memory: Il Vittoriale degli Italiani/Il teatro della memoria di Gabriele D'Annunzio; Il Vittoriale degli Italiani,' *Journal of Decorative and Propaganda Arts* 3 (1987): 6–51. For his most recent biographies, see John Robert Woodhouse, *Gabriele D'Annunzio Defiant Archangel* (Oxford and New York: Clarendon, 1998); Annamaria Andreoli, *Il vivere inimitabile: Vita di Gabriele d'Annunzio* (Milan: Mondadori, 2000). Especially useful is Ferruccio Ulivi, *La maschera senza il volto* (Naples: Istituto Suor Orsola Benincasa, 1989).

12 Paolo Valesio, *The Dark Flame* (New Haven: Yale University Press, 1992), 118–19; see also his 'Ungaretti and the Miles Patiens: Dannunzian Genealogies,' *Stanford Italian Review* 8/1–2 (1990): 103–37. On the importance of the *contadino* (peasant) as 'messenger from the Future' in d'Annunzio's postwar imagination, and especially his association with the revolutionaries of Fiume, see Raffaella Bertazzoli, 'Retorica fiumana e pensiero "notturno" nel messaggio dannunziano "Al legionario Alceste de Ambris,"' *Quaderni di lingue e letterature* 11 (1986): 363.

On the *fante contadino* as a dominant image in wartime and postwar newspapers, see Giovanni Sabbatucci, *La Stampa del combattentismo* (Bologna: Cappelli, 1980), 8–12. See also Mario Isnenghi, *Operai e contadini nella Grande guerra* (Bologna: Cappelli, 1982). A typical example is found in a cartoon in *Il Dovere*, 5 Nov. 1919, which depicts a veteran with a bayonet and an ox, with the caption 'Molto ho seminato, potrò raccogliere?' (I have sowed much, will I be able to reap?). More generally, this image refers to Nicolas Chauvin, a soldier who became the type for naïve nationalism in France's First Empire and was popularized in lithographs by Charlet; his name gives us the term 'chauvinism.' See 'Chauvin,' in the *Dictionnaire de l'Académie Française.*

13 Vittorio Martinelli, *La guerra di d'Annunzio* (Udine: Gaspari, 2001), 149–59. Miniero, *Da Versailles*, 215. The originals read: 'una persona conosciuta'; 'i caratteri del popolo italiano.'

14 The publication of *La riscossa* was done at the behest of the Duke of Aosta, who wished to have it distributed to all the soldiers of the Third Army; see Martinelli, *La guerra di d'Annunzio*, 166. For the date of the volume, see Andreoli, *Il vivere inimitabile*, 545. The actual book has no date printed in it, while the only publication data are as follows: 'Edizione fuori commercio a cura del Sottosegretariato per la Stampa,' 'Casa Editrice d'Arte Bestetti & Tuminelli Milano.' The English volume, *The Rally*, with the translation by Magda Sindici, is physically identical to the Italian one and bears the same publication data. For the quote, see Gabriele d'Annunzio, *La riscossa* (Milan: Casa Editrice d'Arte Bestetti & Tuminelli, 1918), 160–2. I use a modified version of Sindici's translation. The original reads:

> Ma non può essere mio questo giorno, compagni. Questo non è se non il giorno votivo del nostro martire d'Aquileia, non è se non il giorno sacro all'eroe del Timavo. [...] Giovanni Randaccio [...] era un figlio della terra, una creatura della zolla e del sasso, della mota e della polvere. [...] era il vero operaio della vittoria. Insomma, era il fante.

Era come voi [...] [Il suo] eroismo era un baleno. [...] Il vostro è come le vostra ossa, è l'armatura interna, è sempre là; [...]

'siete tutti eroi' vi gridava Giovanni Randaccio [...]

Voi, gente dei campi, gente dei mestieri, gente d'officina e d'officio, villani, operai, borghesi d'ogni parte e d'ogni arte, inselvatichiti come appostatori da spelonche, [. .] voi che vi accovacciate nella tana sudicia che sa di fogna e di sepolcro, [...] voi gente lorda e greve di sotterra, voi in quel punto non eravate se non fiamma celere, non eravate se non anima splendida, come in un Resuressi.

15 The diary entry is from Gabriele d'Annunzio, *Le tre redazioni di un taccuino di guerra di Gabriele d'Annunzio,* ed. Antonio Bruers (Milan: Mondadori, 1942), 64–6. This book published three versions side by side: d'Annunzio's original diary, from his archive – since published again in the *Taccuini,* Tutte le Opere di Gabriele d'Annunzio, ed. Egidio Bianchetti (Milan: Mondadori, 1965); the long passage reworking this entry from d'Annunzio's 'Licenza,' in *La Leda senza cigno, seguita da una Licenza* (Milan: Treves, 1916); and its later revision as it appears in various passages in *Cento e cento pagine dal libro segreto di Gabriele d'Annunzio tentato di morire* (Milan: Mondadori, 1935).

The passage I cite appears in the first two versions but not the third. Bruers comments thus on the overall changes in the three versions: 'La prima è di una schematica semplicità, la seconda, pur arrontondata nel contenuto e nella forma, appartiene allo stile serrato e potente del periodo di guerra; la terza rispecchia il periodo che potremmo dire benacense: stile riflesso nel quale si sovrappongono speciali preoccupazioni e preziosità linguistiche e filologiche, si moltiplicano le reminiscenze e i richiami eruditi e letterari, e i fatti sono, direi quasi, circondati, impregati da un soffuso senso del mistero che negli ultimi anni di vita tanto più si impadronì del Comandante, quanto più, per singolare reazione, egli accentuò l'ostentazaione nazionalista' (The first is schematic in its simplicity, the second, though more full in form and content, still belongs to the spare style of the war era; the third reflects the era we might call 'benacense' [when d'Annunzio lived on Lake Garda]: a self-reflective style full of linguistic and philological preoccupations, with multiple erudite and literary reminiscences, such that the facts are almost, I would say, surrounded, infused with that sense of mystery that overwhelmed the Comandante all the more in that time when, by a singular reaction, he embraced nationalism ostentatiously) (*Le tre redazioni,* 16).

The original reads: 'Siamo tutt'e due sul banco, l'uno accanto all'altro. Ci sembra che i nostri destini si leghino, si annodino. Egli è giovane, io non sono più giovane. E tutt'e due, martedì, prima di mezzogiorno, potremmo essere morti, essere un pugno di carniccio carbonizzato, qualche osso annerito, qualche cartilagine ritratta, un teschio spiaccicato con qualche dente d'oro luccicante nella poltiglia.'

For d'Annunzio's rhetoric of spiritual renewal, see the following: Paolo Valesio: 'Pax Italiae and the Literature of Politics,' *Yale Italian Studies* 2 (1978): 143–67; 'Declensions: D'Annunzio after the Sublime,' *New Literary History* 16/2 (1985): 401–15; and 'The Beautiful Lie: Heroic Individuality and Fascism,' in *Reconstructing Individualism: Autonomy, Individuality, and the Self in Western Thought* (Stanford: Stanford University Press, 1986), 163–83; Barbara Spackman: 'Il verbo (e)sangue: Gabriele D'Annunzio and the Ritualization of Violence,' *Quaderni d'Italianistica* 4/2 (1983): 218–29; and *Decadent Genealogies: The Rhetoric of Sickness from Baudelaire to d'Annunzio* (Ithaca: Cornell University Press, 1989); Lucia Re: 'Gabriele D'Annunzio's Novel Le vergini delle rocce: "Una cosa naturale vista in un grande specchio,"' *Stanford Italian Review* 3/2 (1983): 241–71; 'Per una lettura testualista del corpus dannunziano,' *Annali d'Italianistica* 5 (1987): 43–59; and 'D'Annunzio, Duse, Wilde, Bernhardt: Author and Actress between Decadence and Modernity,' in Luca Somigli and Mario Moroni, eds., *Italian Modernism: Italian Culture between Decadentism and Avant-Garde* (Toronto: University of Toronto Press, 2004), 86–129.

Thomas Harrison, 'D'Annunzio's Poetics: The Orphic Conceit,' *Annali d'Italianistica* 5 (1987): 69–70. For Harrison, d'Annunzio's rhetoric celebrates not infinite perception (as it appears to do) but language's own evocative power and the poet's ability to wield it. In this light, d'Annunzio's unfinished and diaristic works from the First World War and after read like a painful awakening of perception.

Jone Gaillard, 'La Terza Italia e il suo ultimo "vate": Gabriele d'Annunzio,' *Romance Languages Quarterly* 7 (1995): 245–56. Gaillard, in turn, condemns d'Annunzio's political writings up to the first year of the First World War as opportunistic and sadistically fascinated by blood, but notes that the war makes his writing 'more human, painful, and intimate' (255).

Alfredo Bonadeo, *D'Annunzio and the Great War* (Madison, London, and Cranbury: Fairleigh Dickinson University Press and Associated University Presses, 1995). More critical, Bonadeo acknowledges that d'Annunzio 'could be tormented by guilt and longing for transcendental values' (33), noting that, in particular, before the war and as he entered it his 'eagerness to fight' was all about his own need for spiritual redemption not Italy's (19, 82); nonetheless, he concludes that d'Annunzio's spiritual longing was mired in sensuality (39) up until his death (159). Similarly, Bonadeo sees in the *Tre redazioni* nothing more than 'the same eagerness to consign the soldiers to death' and a fascination with decay that is inseparable from exalting the death of soldiers, and his own, as cleansing (75–6).

16 One of the very few to notice any connection between Randaccio and the Unknown Soldier is Cadeddu who, in *La leggenda del soldato sconosciuto*, at 102 and 129, describes how searches for an eleventh anonymous body as a candidate to be the Unknown Soldier took place near the place where Randaccio died, and that d'Annunzio was invited to participate. Cadeddu also adds, that later, after the ceremony at Aquileia, in which the one to be taken to Rome was chosen, his coffin was carried around the church and that the procession paused near the tomb of Randaccio, which is in the Aquileia cemetery. However, Cadeddu makes no connection to d'Annunzio's speeches about and commemorations of Randaccio.

17 On exhumations, see Miniero, *Da Versailles*, 179. On d'Annunzio being invited to Rome, see Mario Bernardi, 'Gabriele d'Annunzio e il Milite Ignoto,' *Quaderni dannunziani (Gardone, Fondazione 'Il Vittoriale degli Italiani')*, nos. 22–3 (1962): 1127–32. On his invitation to participate in exhumations at the Timavo, ibid., 1132; this is also mentioned in Miniero, *Da Versailles*, 187. D'Annunzio was also invited to participate in the choosing ceremony at Aquileia (Bernardi, 1137). 'Lo spirito di d'Annunzio si aggirava intorno ai convenuti, ma il vate [...] non comparve' (d'Annunzio's spirit hovered around the participants, but the vate [...] did not appear) (Miniero, *Da Versailles*, 188).

18 Reassuring reports to the minister of the interior (all found in MdI, OaSI, 1921) include: 'Credo opportuno riferire che secondo assicuratami ora, persona degna di fede e in grado essere bene informata Gabriele D'Annunzio non parteciperà affatto cerimonie milite ignoto, nè in Aquileia, nè in Roma' (I think it appropriate to report that according to recent assurances by a trustworthy and well-informed person Gabriele d'Annunzio will not participate in any of the unknown soldier ceremonies, in Aquileia, or in Rome).

Less reassuring ones claim that the prefect has heard that

> il Comitato Centrale degli ex legionari fiumani per suggerimento dell'On. De Ambris proporrebbe in una prossima riunione l'opportunità di organizzare una grande manifestazione di ex legionari e di ex combattenti in Roma in onore di Gabriele D'Annunzio in occasione del di lui viaggio nella capitale per le onoranze al soldato ignoto fissate per il prossimo 4 novembre. E si aggiunge che alla proposta avrebbero dato la pro-

pria adesione tutte le associazioni repubblicane del Regno con l'intento di dare alla manifestazione stessa carattere antimonarchico.

(the Central Committee of ex-legionaries from Fiume, at the suggestion of the Hon. De Ambris, will propose at a coming reunion the organization of a great rally of ex-legionaries and veterans in Rome to honour Gabriele d'Annunzio on the occasion of his trip to the capital for the ceremonies honouring the unknown soldier planned for 4 November next. It was added that the proposal would get the approval of all the Republican associations of the Kingdom with the intention of giving the rally itself an anti-monarchical character.)

Another, from Trieste, adds, 'fascio triestino tiensi sempre pronto ordine D'Annunzio' (the Triestine *fascio* is ready at all times for an order from d'Annunzio).

19 'Gli assenti' (The Absent Ones) was a full-page headline in *L'illustrazione Italiana,* 13 Nov. 1921; below were parallel photos, in uniform, of d'Annunzio and General Cadorna. For the report from Naples, including the claim – revealing of the confusion and dissent of this era – that this march would be by *arditi autentici* (authentic arditi), see MdI, OaSI, 1921.

Articles blaming Italians' lack of civil disourse on d'Annunzio include: Nando Della Chiesa, 'Carabinieri,' *L'Unità,* 18 Nov. 2003; Gianni Vattimo, 'Ritrovare le parole,' *L'Unità,* 20 Nov. 2003. Notably, Vattimo also echoes d'Annunzio's claim that a 'truer' understanding of the Unknown Soldier would see him as a *povero Cristo,* the Christ of the poor, not a politicized one.

Rodinò's words to the 'Commissione Essercito e Marina Militare' are recorded by De Vecchi on 28 July 1921 in Atti parlamentari, Camera dei Deputati, legislatura XXVI, 1921–23, sessione 1921, documenti, disegni di legge e relazioni, no. 202-A, Sepoltura della salma di un soldato ignoto, Biblioteca Giovanni Spadolini, Rome (hereafter AP, 1921–23, no. 202-A SdSI). The original reads: 'omaggio a tutti i nostri morti, alla gente dei campi, alla gente dei mestieri, alla gente delle officine e degli offici [this last is in the singular in d'Annunzio; and Rodinò elminated the following 'villani'], agli operai, ai borghesi, a tutti coloro di ogni parte e di ogni arte.'

Rodinò also says his proposal follows 'il sogno del Poeta' (the Poet's dream) but does not mention d'Annunzio by name; arguably this was not necessary in 1921.

20 On Douhet, see Labita, 'Il Milite Ignoto,' 122; P.S. Meilinger, 'Giulio Douhet and Modern War,' *Comparative Strategy* 12/3 (1993): 321–38. Douhet not only admired d'Annunzio but was known for his extravagant proposals for the use of airplanes during the war.

On d'Annunzio's flight over Vienna, see Martinelli, *La guerra di d'Annunzio,* 257–85. On d'Annunzio's long love affair with airplane flight, and its association with a heroic and spiritual escape from the everyday, as well as on his First World War stunts and his report to Cadorna about the importance of air power in military strategy, see Peter Demetz, *The Air Show at Brescia, 1909* (New York: Farrar, Straus, and Giroux, 2002), 149–85.

21 Miniero, *Da Versailles,* 37, 49. On Sir Edwin Luytens, and possible causes for the greater popularity of his cenotaph, see Winter, *Sites,* 102–8.

22 Father David Railton, whom the British credit with first proposing an Unknown Soldier Memorial, prefered the term 'comrade' (see Inglis, 'Entombing Unknown Soldiers,' 15). For the the reference to how 'the Queen unveiled a memorial to the suffering of the Unknown Victim, as a counterpoint to the tomb of the Unknown Soldier in Westminster Abbey,' see Mick Hume, 'Are They Heroes, or Victims? We're Tying Ourselves into Knots with Yellow Ribbon,' *Times,* 31 March 2003.

23 On the ritualistic discourse of the trenches, compared to a rite of passage as described by van Gennep, involving 'separation' and 'transition' as two phases of 'liminality,' see Leed, *No Man's Land,* 14–15. 'Plebiscitary,' is in Labita, 'Il Milite Ignoto,' 122.

The originals read: 'carattere plebiscitario'; 'fare larga propaganda.' 'Naked and unarmed,' is from Douhet's initial proposal, also in Douhet, *Le Profezie di Cassandra,* 424. The original reads: 'gettato nudo e inerme contro il cemento armato ed il filo d'acciaio.' 'Humble bones' (*povere ossa*), is in *Corriere della Sera,* 4 Nov. 1921. On unanimity in France, see Jagielski, *Le soldat inconnu,* 79.

The originals read: 'enlever à la cérémonie proposée par le gouvernement son caractère d'unanimité'; 'la grandeur de son sacrifice anonyme'; 'dépasse nos petites querelles et nos discordes.' 'Suppression' (*soppressione*) is from Atti parlamentari, Camera dei Deputati, legislatura XXVI, 1921–23, sessione 1921, documenti, disegni di legge e relazioni, no. 202, Sepoltura della salma di un soldato ignoto, Biblioteca Giovanni Spadolini, Rome (hereafter AP, 1921–23, no. 202 SdSI); it is also in Labita, 'Il Milite Ignoto,' 36. The original reads: 'soppressione assoluta della propria individualità.'

On Diaz asking for the vote in the Italian Senate to be unanimous, as it already was in the Chamber of Deputies, see Atti Parlamentari, Senato del Regno, Legislatura XXVI, 1921, discussioni, tornata del 10 agosto 1921, Biblioteca Giovanni Spadolini, Rome (hereafter AP, Senato, 1921, 10 Aug.).

Hans Ulrich Gumbrecht, 'How (If at All) Can We Encounter What Remains Latent in Texts?' *Partial Answers* 7/1 (2009): 87–96, on the implicit and the power of latency. Hans Ulrich Gumbrecht, 'Reading for the Stimmung? About the Ontology of Literature Today,' *Boundary 2* 35/3 (2008): 213–21. Attention is required to *Stimmung* (mood), understood as inwardness but also as attention to a certain cultural climate and, most relevant for understanding the Unknown Soldier Memorial, as the awareness of being embedded in the material world (215, 221). On the marginal in literature, see Mario Moroni, *Al limite: L'idea di margine e confine nel Novecento italiano* (Florence: Monnier Università, 2007).

More generally, on the importance of deep images as opposed to explicit political discourse see the following: Banti, *La nazione del risorgimento*; and by Catherine Brice: 'Italia: Una allegoria debole? Sistema iconografico e identità nazionale nell'Italia della fine del XIX secolo,' *Memoria e Ricerca*, no. 25 (2007): 171–86; 'Building Nations, Transforming Landscape,' *Contemporary European History* 16/1 (2007): 109–19; and 'Riti della corona, riti del fascio,' in Emilio Gentile, ed., *Modernità totalitaria: Il fascismo italiano* (Bari: Laterza, 2008), 171–90.

Albert Russell Ascoli and Krystyna Von Henneberg, in the introduction to their *Making and Remaking Italy*, discuss the Risorgimento as 'vivid, potent – and changeable – legend' (9).

On the link between this history, the 'memory boom' or recent growth in studies of 'commemorative sites, rituals, and languages,' and the crucial importance of corporeality, see Stefan Goebel, 'Review Article: Beyond Discourse? Bodies and Memories of Two World Wars,' *Journal of Contemporary History* 42/2 (2007): 377–85.

On the primal scene, see Fonck, *Le soldat inconnu*, vol. 1, 3ff., where he shows photographs of improvised battlefield burials in his discussion of Simon's initial proposal.

24 As regards literature's ability to 'call context into question in order to surpass it,' see Russell A. Berman, *Fiction Sets You Free: Literature, Liberty, and Western Culture* (Iowa City: University of Iowa Press, 2007), xiv. This implies a 'reading for patterns,' and for their exceptions, which as Margaret Cohen warns, in *The Sentimental Education of the Novel* (Princeton: Princeton University Press, 1999), at 22–3, requires 'delimiting the field of inquiry.' A similar play of background and foreground is advocated in Franco Moretti, *Graphs, Maps, Trees: Abstract Models for a Literary History* (London: Verso, 2005).

This primal scene is ubiquitous in the literature of the First World War, but I will seek at various points in this book to draw out both its typical and its exceptional qualities.

Jean Giono, *Le grand troupeau* (Paris: Gallimard, 1971), 575, 576, 578. My translation is adapted from Jean Giono, *To the Slaughterhouse*, trans. Norman

Glass (London and Chester Springs: Peter Owen and Dufour, 2004), 44–6. The original reads:

> La raideur, comme un pieu, a repoussé la tête de Jules en arrière. Son regard est en haut de ses yeux comme pour regarder ses cheveux. Il a dormi les yeux ouverts. Il a mâché longtemps de l'air tout en dormant. [...] Enfin, il est parti pour les grands pays du dedans de sa tête et il s'est mis à parler. [...]
>
> [Joseph] a mis son bras sous la tête de Jules; mais on ne peut plus relever la tête de Jules; on ne peut plus le serrer contre soi avec sa chose molle d'homme qui est la vie et la chaleur, et la vérité du corps qui vit, il est dur et raide comme du bois coupé. Joseph s'allonge de tout son long à côté de Jules et alors il peut le serrer contre lui. [...]
>
> C'est seulement à la nuit que Joseph s'est lassé de tenir dans ses bras cette chose raide, froide, et morte. Il s'est dressé. Il a regardé Jules étendu.
>
> 'Eh oui! Il a dit ...'

Identification tags for soldiers had been used in the nineteenth century, notably in the American Civil War, initially at soldiers' own initiative; the Prussian army issued them at the beginning of the Franco-Prussian War in 1870, and they were nicknamed *Hundemarken* (markings on dogs). It was only as the First World War began, however, that they began to be issued systematically by all combatant nations. Nonetheless, soldiers often replaced them with medals depicting saints, which were felt to have a protective power, in contrast with the impersonal quality of military identification numbers. See Capdevila and Voldman, *Nos morts*, 44–6. It was a topos in the Italian press that the Unknown Soldier had replaced his identification with a religious medal or a scapular, see e.g., *Il momento*, 4 Nov. 1921, discussed by Labita ('Il Milite Ignoto,' 135–6).

25 'Piled up' (*entassé*) in Giono, *Le grand troupeau*, 613; hiding the body under the hedge, in ibid., 573, and *To the Slaughterhouse*, 43.

26 Giono, *Le grand troupeau*, 593, 594, 595; my translation is adapted from *To the Slaughterhouse*, 61–4. The original reads:

> Il viennent tous, ils sont tous là tout autour dans la grande salle de la ferme à l'âtre vide. Ils sont là raides et muets à veiller le corps absent. [...]
>
> 'Nous veillons le corps absent d'Arthur Amalric mort à la guerre,' déclame la vieille Marthe. 'Que chacun se recueuille dans son amitié pour celui qui était le sel de la terre ...'
>
> Elle met la main au pot [de sel]. Elle tire une poignée de sel, elle vient le mettre au centre de la table nue, elle en fait un petit tas. [...]

Le lourd silence revient.

'Oh! mon Arthur!' crie Félicie.

Elle est raide comme du bois. [...]

'Pauvre, que j'ai pas seulement été là pour te fermer les yeux! Pauvre, que tu es mort dans la terre comme une bête, tout seul!'

27 Douhet, *Le Profezie di Cassandra*, 426. The original reads: 'da quella tomba trarranno la maggiore consolazione le più infelici: quelle non sepperò più nulla del loro nato, che sembrò vanire nella bufera.' Also in Labita, 'Il Milite Ignoto,' 125; *Corriere*, in ibid., 137. The original reads: 'il milite ignoto più che l'eroe [...] il morto senza nome più che il coraggio guerriero [...] questo simbolo di povere ossa.'

28 Blaise Cendrars, *La vie et la mort du soldat inconnu* (Paris: Champion, 1995), 29–33. The original reads:

– Je vous demande pardon, Monsieur de Mesle, j'étais dans la lune, de quoi s'agit-il?

– Du Soldat Inconnu, dit de Mesle.

– Ah! l'Poilu, fis-je.

– Oui, reprit de Mesle, Monsieur de Tang prétend que 'notre' (de Mesle appuyait sur le 'notre' parce qu'il était déserteur) que 'notre' Soldat Inconnu qui respose sous l'Arc de Triomphe à Paris est un soldat allemand!

– Et pourquoi pas? Fis-je, ennuyé.

Tout le monde protesta avec indignation, mais Mr de Tang criait plus fort que tous, debout, triomphant:

– Vous le voyez bien, c'est possible! Et Monsieur Cendrars, qui est le seul parmi nous à avoir fait la guerre, ne voit pas d'objection de principe au fait, ou plutôt à la possibilité, à la vraisemblance ajouterai-je, que, si le prélèvement de six cadavres anonymes sur six champs de bataille différents [...] et la désignation aveugle de l'un d'eux [...], que, si ce prélèvement et ce choix [...] ont été effectués régulièrement, rien ne s'oppose, théoriquement, à ce que la dépouille, à laquelle le monde entier rend aujourd'hui hommage, ne soit, en réalité, la dépouille d'un soldat allemand. Et j'ose bien affirmer, objectivement, que c'est là toute la grandeur de la France, et sa noblesse, et son désintéressement, et que ce geste est bien dans la tradition chevaleresque de ce beau pays, que d'avoir eu l'audace de faire l'apothéose non seulement de l'un de ses enfants héroïques, mais encore, mais peut-être (n'admettons qu'une probabilité sur dix mille) de l'un de ses héroïques ennemis.

– Et pourquoi pas Juif? Fis-je.

Mr de Tang resta court. Il était pangermain, et […] ma remarque lui avait coupé son enthousiasme. […] Puis il sourit malicieusement, inclina son corps replet, le cassa en deux, fit un plongeon, me salua, éleva son verre et dit, en me souriant encore une fois:
– Ah, ces Français! ils ont toujours le mot pour rire. Rien n'est sacré pour eux et, subjectivement, oui, ils sont tous de la plus pure noblesse. Messieurs, Mesdames, je bois à la santé de la France! *Hoch! Prosit!*

Tout le monde se taisait. Tout le monde était gêné. […] Non je n'allais pas me chamailler avec von Tang, ni, pour faire plaisir à une poignée de nouveaux riches ou de nobles rastas voyageurs me lancer dans l'historique et faire l'apologie du Poilu. Je n'aurais pas été à la hauteur car, eux, tous, professaient un culte pour lui, le chérissaient et l'avaient adopté, le Soldat Inconnu, comme ils avaient adopté les courses, les théâtres, les boulevards de Paris, Montmartre et Montparnasse, Poincaré et Mistinguett, les *Folies Bergères*, le *Chabanais*, la coco du *Boeuf sur le Toit*, les robes de chez Patou, les parfums de chez Chanel et, comme des gogos, le gigolo de l'entresol, du sous-sol, du do-ré-mi-fa-sol du *Lido*. (Je regardais Caralina et j'avais envie de rire car, un peu d'humour ne messéant pas en amour, j'étais pour elle son petit poilu chéri, […].) […]

– Nous autres, Américains, c'était Ten Cate qui s'adressait à von Tang […], nous autres, Américains (il prononçait *Hammm'ricanne*), nous ne pouvons admettre cela, Monsieur von Tang. Le Poilu (il prononçait *la Poilioue*), le Soldat Inconnu (il prononçait *la Soldatte Inconnioue*) il (il prononçait *alle*) est aussi à New York, il (il prononçait encore *alle*) est Américain (il disait toujours *Hamm'ricanne*), je pense …

– Et pourquoi pas Nègre? Fis-je, car si je n'ai jamais su hurler avec les loups, j'ai toujours aimé à les exciter.

Cette fois-ci mon mot tomba à faux, car loin de faire enrager qui que ce soit, l'ahurissement de Ten Cate déclencha le fou-rire. Il restait là, bouche-bée …

[…]

Que la vie est belle!

J'ai fait l'amour jusqu'à l'aube.

[…]

O Monde!

Je passe ma tête par le hublot. L'océan est d'acier. Le ciel est du plus beau noir. A l'est, le vent se lève avant le jour, tout frissonne.

[…]

Et maintenant que je suis seul, je l'ai tout de même écrit, ce livre. Ce n'est pas le beau livre rêvé par Caralina, mais ce n'est pas non plus un cauchemar.

Ce n'est pas une histoire sainte.
Ce n'est pas une légende.
C'est un livre.
C'est un livre qui apprend tour à tour à maudire et à bénir la vie.
En somme, à l'aimer.
Tout simplement.

The number of French candidates is wrong (six instead of eight), and the location of the American Unknown Soldier is incorrect (New York instead of Washington): this is most likely a deliberate addition to Cendrars' satire. On Cendrars' unfinished novel, of which two chapters were published in 1929 to great acclaim, see Miriam Cendrars, *Blaise Cendrars: La vie, le verbe, l'écriture* (Paris: Denoël, 2006), 565–6.

29 On the cultural associations of the Pantheon in Paris, see Mona Ozouf, 'Le Panthéon: L'école normale des morts,' in Pierre Nora, ed., *Les lieux de mémoire* (Paris: Gallimard, 1997), vol. 1, 155–78. On the raging pubic debate in France, which ended up rejecting the Pantheon as a location for the burial of the Unknown Soldier, see Jagielski, *Le soldat inconnu*, 69–83, 115–20.

More specifically, on Cendrars' insistence that literature should be intimately involved with contemporary politics, as he decides in 1930 to become a journalist as well as a novelist, see M. Cendrars, *Blaise Cendrars*, 562. On Cendrars' journalism, see Georgiana M.M. Colvile, *Blaise Cendrars, écrivain protéiforme* (Amsterdam and Atlanta: Rodopi, 1994), 75ff. On his move to the left, notwithstanding his anti-German First World War patriotism, see Claude Leroy, *La main de Cendrars* ([Villeneuve d'Ascq]: Presses universitaires du Septentrion, 1996), 141, 261.

30 On the unrealizable yet seductive dream of androgyny as the solution to the psychological and social conflicts of the fin de siècle, see Spackman, *Decadent Genealogies*.

31 On the 'fantasme des simulateurs' (spectre of simulators), which Cendrars refers to by contrasting Mesle as 'deserter' to Cendrars' own mutilation, see Le Naour, *Le soldat inconnu vivant*, 16.

Louis-Ferdinand Céline, *Voyage au bout de la nuit* (1932; new ed., Paris: Gallimard, 2006). Both Cendrars and Céline are quite ironic about allowing the state to distinguish 'real' illness from simulation or desertion: Bardamu spends time in an asylum where some soldiers are clearly faking their illness whereas others who are already suffering from traumatic symptoms are made worse by the putative 'cures' administered there (95–104).

32 'Que la vie est belle,' repeated, and 'Orientaux,' in Cendrars, *La vie et la mort*, 43.

33 Leed (*No Man's Land*) notes that veterans were thought 'to have either

transcended purely social categories or to have fallen below them [...] they are merged with sacred figures [...] like gods or beasts' (13).

34 'Massa organica di uomini' is from Olindo Malagodi, 'Ammaestramenti alle cerimonie,' *La Tribuna*, 5 Nov. 1921. It is commented upon by Labita ('Il Milite Ignoto,' 132) and Miniero (*Da Versailles*, 234). The former points out that such unity is more of a state script than a reality whereas the latter sees this image as a prelude to Fascist rhetoric.

On the history of crowds in the nineteenth and twentieth centuries, see Jeffrey T. Schnapp, 'The Mass Panorama,' *Modernism/Modernity* 9/2 (2002): 243–81; Jeffrey T. Schnapp and Matthew Tiews, *Crowds* (Stanford: Stanford University Press, 2006); Olivier Bosc, *La foule criminelle* (Paris: Fayard, 2007). A canonical example of this image in Italian culture at the time is in d'Annunzio's 1900 novel, *Il fuoco*, which opens with the narrator giving an impassioned speech to a crowd described as a chimera and other fabled monsters. On d'Annunzio's manipulation of the crowd-leader relation, see Emilio Papa, 'Discorrendo di d'Annunzio politico,' *Studi Novecenteschi* 26/58 (1999): 275–93.

2 Identification and Chorality

1 On the abstraction of names, see Sherman, 'Bodies and Names,' 455. He cites a French senator who spoke of having known personally many of 'the names' (not the persons) whose memorial he was inaugurating. Laqueur ('Names, Bodies,' 125) notes the new and unusual epitaph chosen by Kipling from Ecclesiastes for First World War dead: 'Their names liveth evermore.' In this light, the 1919 Italian suggestion that instead of on 'cold monuments,' the names of the dead should be kept on electoral lists for one hundred years is an attempt to combat abstraction by indirectly revealing its political component (monumentalized 'names' have no vote). See *Il Dovere*, 27 April 1919, front page.

Antoine Prost, 'Le monuments aux morts: Culte républicain? Culte civique? Culte patriotique?' in Pierre Nora, ed., *Les lieux de mémoire* (Paris: Gallimard, 1997), vol. 1, 199–223. Prost notes that his survey of 'Monuments aux morts' throughout France shows a predominance of minimalism (the stela without any representational aspect), which however, is not to be equated with abstraction, as he concludes that 'L'innovation fondamentale du 11 Novembre, qui bouleverse les liturgies républicaines, c'est de célébrer non des principes mais des citoyens concrets' (205, 214).

2 On the convergence of lists of names and the anonymity see: Sherman, 'Bodies and Names,' 447, 454, 458; and Laqueur, 'Names, Bodies,' 123, 126,

who also echoes Labita on the 'chaos' of bodies (134). On suppression, see Labita, 'Il Milite Ignoto,' 136. On 'unity,' see Miniero, *Da Versailles*, 234. The original reads: 'unità di passioni, di voleri, e insomma di coscienza valida.'

3 One example from France is in Sherman, 'Bodies and Names,' 451. Two examples from Italy are in Tobia, *L'altare della patria*, 79, and one more example from France is in Sherman, 461.

4 For exhumations in France, see Jagielski, *Le soldat inconnu*, 92–3. For exhumations in France, Britain, Italy, and the United States respectively see Miniero, *Da Versailles*, 136–8, 138, 180–2, 235.

5 In the Old Testament, scattered bones represent desolation and sin; pardon comes as bones are gathered up and speak the Lord's praises (Ps. 21:14–19, Ps. 33:21, Ps. 50:9–10). In Oct. 1921 a number of decorated veterans searched eleven major Italian battle zones for appropriate remains (Tobia, *L'altare della patria*, 73).

For the Unknown Soldier commemorative medals, see *L'illustrazione Italiana*, 6 Nov. 1921, 533. One medal shows him in a pose reminiscent of Michelangelo's Vatican *Pietà*, which associates him, as was common, with Christ and resurrection but also, significantly, pauses here at the moment of suffering and death. See *Corriere della Sera*, 4 Nov. 1921, 2 Aug. 1921. See also *Avanti!* 30 Oct. 1921. Both also in Labita, 'Il Milite Ignoto,' 140. Note that 'brandelli' in the phrase 'brandelli delle sue ossa' (shredded bones) normally applies to flesh, not bone. The Socialist view is in *Ordine Nuovo*, 14 Oct. 1921.

Once again, I don't agree with Labita's implication that the Socialists are less interested in having the dead soldier 'live again' than other groups are; they merely imagine his return differently. See Labita, 'Il Milite Ignoto,' 144.

On Vernet, see Mona Siegel, 'To the Unknown Mother of the Unknown Soldier: Pacifism, Feminism, and the Politics of Sexual Difference among French Institutrices between the Wars,' *French Historical Studies* 22/3 (1999): 441, 423. Sulli is in *L'Ardente*, Aug.–Sept. 1921, 54ff.

6 Augusto Tognasso, *Ignoto Militi* (Milan: Tipografia Magnani, 1922). Tognasso was a mutilated veteran himself, and was part of the commission of soldiers who performed exhumations and chose one appropriate body from each battlefield. The book does not indicate the name of the photographer, but we gather it was Tognasso himself, perhaps in collaboration with other soldiers, during the work of the commission. This book is rare enough that only four libraries in Italy have a copy. It was reprinted in 1962 in Milan by the Tipografia Zanoli, in an 'edizione fuori commercio,' this time 'A favore della Casa militare veterani guerre naz. Umberto 1. e della erigenda Casa

del combattente in Bovisa.' On northern newspapers, see Miniero, *Da Versailles*, 134–5.

7 We find long passages of d'Annunzio's 'La chiesa di Doberdò,' a ruined church on the battlefield, with highly physical descriptions of wounded soldiers taking refuge there, in *L'Ardente*, 1 July 1921. This text will be discussed in more detail later. Another example of a newspaper defending d'Annunzio's takeover of Fiume and emphasizing the sacrifice of soldiers as unrecognized by the Italian government is in *L'Ardito*, 21 Sept. 1919.

As regards 'mutilated victory,' d'Annunzio asserted in Oct. 1918, in 'La preghiera di Sernaglia,' 'Vittoria nostra, non sarai mutilata' (Victory of ours, you will not be mutilated), in Gabriele d'Annunzio, 'Tre preghiere dinanzi agli altari disfatti,' *Prose*, 1922, *Il libro ascetico*, 647. Mark Choate, 'D'Annunzio's Political Dramas and His Idea-State of Fiume,' *Forum Italicum* 31/2 (1997): 367–88. As Choate points out, however, the French translation of d'Annunzio's three political plays, *La Gioconda, La Gloria*, and *La città morta* was entitled *Les victoires mutilées* (371, 383).

Mutilation is also a recurrent theme in d'Annunzio's 1893 *Poema Paradisiaco*, and is notable in his poem to Garibaldi, 'La Notte a Caprera.' See also Susan Bassnett, 'A Passion for Dismemberment: Gabriele D'Annunzio's Portrayals of Women,' in Michael St John, ed., *Romancing Decay: Ideas of Decadence in European Culture* (Aldershot: Ashgate, 1999), 128–40.

8 On how wishing to 'identificare il grado di eroismo dell'Ignoto' (identify the degree of heroism of the Unknown) runs against state scripts, see Labita, 'Il Milite Ignoto,' 136. He mentions the 'caduto con la palla in fronte' (fallen with a bullet in the forehead) as a reference to Edmondo De Amicis' 1886 novel *Cuore*, found in *Epoca*, 12 Aug. 1921, and in *Avanti!* 13 Aug. 1921. *Avanti!* 30 Oct. 1921, emphasized suffering. The description of the body from the Timavo is in Labita, 'Il Milite Ignoto,' 136, quoting the *Corriere della Sera*, 28 Oct. 1921, but it is also in *La Tribuna*, 27 and 30 Oct. 1921. The original reads: 'L'ultima [salma], raccolta nella zona del Timavo, era davvero l'espressione del martirio; con le sue gambe spezzate e il suo capo traversato da proiettili, essa ispirò una così profonda pietà alla Commissione ricercatrice che i suoi componenti s'inginocchiarono.'

This same passage is cited in Miniero (*Da Versailles*, 180), who cites Cadeddu (*La leggenda del soldato sconosciuto*, 42).

9 Negri's poem, in *Il secolo illustrato*, 1 Nov. 1921, imagines the soldier telling his young sister to stop wearing the clothes of mourning. I don't agree with Labita's claim that his wounds are only 'segni lontani del martirio sui quali prevale la nuova umiltà del santo' (distant signs of his martyrdom, over which the humility of the saint prevails). See Labita, 'Il Milite Ignoto,' 144.

In contrast, Gorman ('The Return of the Dead,' 72) argues that the imagery of resurrection, tied to martyrdom, abolishes the distance between past and present, prolonging mourning rather than bringing it to closure.

10 Plato, *Republic* X: 614–21.

11 In Siegel's view ('To The Unknown Mother'), family ties – as women saw them after the war – can be a pacifist questioning of a more militaristic reintegration into society; however, such militaristic reintegration also depends on the metaphor of the nation as family; as we will see, it is the visceral nature of women's sense of family ties that makes the difference.

The French children's book on the Unknown Soldier, *Il s'appelait ... le soldat inconnu* (Paris: Gallimard, 2004), by Augsute Ténor, also emphasizes family ties (romantic love, mainly) in a pacifist vein (its hero is depicted as incapable of killing a German he has taken prisoner, even though his own life is in danger).

12 For the choosing process in France, see Jagielski, *Le soldat inconnu*, 92–3. For the choosing process in France, Britain, Italy, and the United States respectively see Miniero, *Da Versailles*, 136–8, 138, 180–2, 235. In the United States, unlike all other nations, a new body was added to the Unknown Soldier Memorial after each major war, and thus further unknowns were buried after the Second World War, the Korean War, and the Vietnam War. After DNA testing made it possible to discover who had been buried after Vietnam, a controversy erupted about whether this should be done or not, and it is here that we find issues of visceral identification. For the official version of how 'DNA testing has proven that the remains which were buried in the Tomb of the Unknowns at Arlington National Cemetery are those of United States Air Force First Lieutenant Michael Joseph Blassie,' see *Tomb of the Unknowns, Arlington National Cemetery*, retrieved from http://www.arlingtoncemetery.net/tombofun.htm. Lieutenant Blassie's remains were returned to his family on Friday, 10 July 1998, and buried the next day in his hometown of St Louis, Missouri.

13 Jagielski, *Le soldat inconnu*, 96–9. Miniero, *Da Versailles*, 136–7. In France, familial relationships were evoked later, during the procession to the Arc de Triomphe, when the Unknown Soldier's corpse was accompanied by a father and mother who had lost a son, by a widow, and by a boy who had lost his father (Tauriac, 'La bataille inconnue du soldat inconnu,' 52). I translate 'un ancien poilu de deuxième classe' as a 'poor soldier': the French implies he is 'old' in the sense of having endured a lot.

On nineteenth-century spiritualism and how the dead 'would try to communicate again' see Winter, *Sites of Memory, Sites of Mourning*, 58. 'Offspring are we but of corpses,' is from Karl Kraus, *The Last Days of Mankind, 1915–*

22, cited in Winter, *Sites*, 191. This sort of haunting could be seen as the opposite extreme from the ideal discussed by Robert Harrison (*Dominion*, 90–105), in which in modernity we are at least to some extent allowed to 'choose' our own 'ancestors,' i.e., our spiritual and cultural fathers.

14 Otello Cavara, 'Una madre triestina ha scelto la salma ignota che assurgerà all'apoteosi di Roma,' *Corriere della Sera*, 29 Oct. 1921; Giuseppe Cappelletti, 'Le cerimonie per il Soldato Ignoto: La scelta della salma destinata all'Altare della Patria,' *Il popolo d'Italia*, 29 Oct. 1921; Umberto Micali, 'Una madre triestina designa ad Aquileia la salma del milite ignoto,' *La Tribuna*, 30 Oct. 1921. The original reads: 'per mezzo delle diverse venature e dei nodi del legno, per la diversa disposizione dei chiodi e per certe sfumature di colore.' See also Otello Cavara, *Il milite ignoto* (Milan: Alpes, 1922).

On the symbolic importance of Aquileia as a frontier, a 'martyred city' attacked by 'barbarians' since the times of the Roman Empire, see Paul Baxa, *Roads and Ruins: The Symbolic Landscape of Fascist Rome* (Toronto: University of Toronto Press, 2010), 16–17.

15 For women and mourning, see De Martino, *Morte e pianto rituale*, and *La terra del rimorso*. Michael P. Caroll, in *Madonnas that Maim: Popular Catholicism in Italy since the Fifteenth Century* (Baltimore: Johns Hopkins University Press, 1992), at 82, comments on De Martino and argues that essential to understanding *tarantismo* (ritual dancing after being bitten by the *taranta*) is the idea that this temporary madness is caused by a saint or a madonna; Maria Bergamas clearly played this dual role of placating yet also causing cathartic excess.

On brotherhood, Banti notes that in Italy, the Enlightenment and revolutionary (and more abstract) brotherhood found in France is never entirely separated from its Christological model, giving the example of 'Carbonari' united in conspiracy around a crucifix; he also notes the related importance of the Virgin, and the issue of pacifism and brothers beyond versus within national borders, as seen in the work of Manzoni and Pellico (*La nazione del Risorgimento*, 121–2, 126, 130–3, 134–5).

16 On birth in connection to death in turn-of-the-century Italy, see Cristina Mazzoni: 'Parturition, Parting, and Paradox in Turn-of-the-Century Italian Literature (D'Annunzio, Aleramo, Neera),' *Forum Italicum* 31/2 (1997): 343–66, and '"That in Giving Me Life, You Still Remain Alive": Fetal Beginnings and Maternal Endings at Two Centuries' Ends,' *Annali d'Italianistica* 18 (2000): 255–76.

Susan R. Grayzel, *Women's Identities at War: Gender, Motherhood, and Politics in Britain and France during the First World War* (Chapel Hill: University of North Carolina Press, 1999). Grayzel shows that in England and France women were largely excluded from commemoration after the First World War in the sense that memorials *to* them are very rare (and this reflected widespread fear regarding their newly emancipated and more sexually overt social roles), yet at the same time women were emblems of mourning that, crucially, could unify a socially divided nation by linking mourning to the universal experience of childbirth (156, 229–30, 236). She concludes that the grief of mothers was dual, however, in its power: 'On the one hand, women could represent the potent forces of peace, victory, and the nation itself. On the other hand, the images of grieving women left a powerful symbol of feminized loss and the devastating costs of war in the postwar world' (241).

17 See Winter, *Sites*, for Gance in general (15, 133); on Gance straddling the divide between experimental art and popular culture (143); 'virtues' (133), 'surrealism'(136), and 'pioneer'(143). As Winter notes, 'realism' was often seen as the only possible authentic response to the war's horror; yet both in film (e.g., Gance) and in fiction (e.g., d'Annunzio, Cendrars, Céline), authors who had direct experience understood realism to be intimately involved with a dominant ideology that they wished to subvert, and thus they developed techniques to fissure the more realistic surface of their art. This stylistic ambivalence may be seen to foster a 'between' position like the one advocated by Homi Bhaba as being 'in the midst of' and 'interrupting,' which according to Edith Wyschogrod, in *An Ethics of Remembering: History, Heterology, and the Nameless Others* (Chicago: University of Chicago Press, 1998), counters 'the fact that [in cinema] homogeneity is dictated by global technologies, information and commodity distribution systems' that 'imply that alterity is contrived, and that the subject of history as the marginal or exotic other is commodified, yet another collection of marketable images' (89).

On realism and ideology more generally, as regards the European novel, see Margaret Cohen and Christopher Prendergast, *Spectacles of Realism: Body, Gender, Genre* (Minneapolis: University of Minnesota Press, 1995), ix.

18 On soldiers dying before the film came out, 'clips,' Apollinaire, and Kardec, see Winter, *Sites*, 15, 20, 20, 21.

19 For nature's indifference, see Winter, *Sites*, 136. For the topos of barbed wire, see also Alfredo Bonadeo, *Mark of the Beast: Death and Degradation in the Literature of the Great War* (Lexington: University Press of Kentucky, 1989), 91.

20 See d'Annunzio, *Notturno*, for the reference to the Sybil (5), the shadow of *follia* (folly) (150), and 'bones' (70). The original reads: 'Ora mi sembra di percepire nella caduta lo splendore bianco delle mie ossa.'

For the biographical details, see the introduction by Pietro Ghibellini and the preface by Elena Ledda, in d'Annunzio, *Notturno*, xlvi. There were over 2,000 strips of paper (now in the Vittoriale archives). The diary of d'Annunzio's daughter Renata, who assisted him, reveals some details of his physical condition (see 357–80): he was almost completely deprived of water in order to facilitate the absorption of fluids into his eye; these were administered by injections, which caused abcesses in the eye area; until he lost the eye completely, he saw colours in continuous metamorphosis through it, as described in the book. On publication dates see d'Annunzio, *Notturno*, xlvii–li.

On 'scrittura notturna' and its introspective tone, see Giorgio Bàrberi Squarotti: 'D'Annunzio scrittore "politico,"' *D'Annunzio politico atti del Convegno (Il Vittoriale, 9–10 ottobre 1985)* (Milan: Garzanti, 1987), and *La scrittura verso il nulla: D'Annunzio* (Turin: Genesi, 1992). Regarding triumphalism, see Pericles Lewis' recent reading of the *Notturno*, in *Modernism, Nationalism, and the Novel* (Cambridge: Cambridge University Press, 2000), as the celebration of a 'myth of epiphanic plenitude' or 'fantasy of self-presence' (203), which is the model for Mussolini's identification with the nation. In contrast, Valesio's analysis of the last section of the book shows that d'Annunzio invents in it a recurrent image in postwar poetry, that of the humble suffering soldier or *miles patiens*. See Chapter 5 of Valesio's *The Dark Flame*; also see his 'Ungaretti and the Miles Patiens.'

Silone's Pietro Spina/Paolo Spada is another reinvention of the peasant-as-redeemer. See Elizabeth Leake, *The Reinvention of Ignazio Silone* (Toronto: University of Toronto Press, 2003). This same section, written in 1921, explicitly associates the miles patiens with the Unknown Soldier (and d'Annunzio had planned for the book's first day of sale to coincide with the Unknown Soldier ceremony on 4 Nov. 1921, though it ended up coming out late, on 22 Nov.), not only out of opportunism, but as I hope to show, because d'Annunzio had legitimate reasons to believe his writings were among the genealogical sources for the idea.

21 On reliving death, see d'Annunzio, *Notturno*, 75, 140. On critiquing the Unknown Soldier see d'Annunzio, 'Comento meditato a un discorso improvviso,' *Prose*, 1922, *Il libro ascetico*, 539. *Improvviso* is literally translated as 'sudden' or 'unexpected' but it is also the older form of *improvvisato* (improvised).

D'Annunzio's choice of 'Vittoriale' to name his house evidently echoes the *Vittoriano*, though it is also based on Pedro Niño's *Victorial* (Andreoli, *Il vivere inimitabile*, 601).

According to Leed (*No Man's Land*), while the fixation on death leads directly to the violence of the Second World War, it also reflects the reality of the First World War: 'No "rites of reaggregation" could efface the memory of utter defencelessness before authority and technology. No ceremonial conclusion to the war could restore the continuities it had ended, or recreate those "fictions" that had been left behind in the labyrinth of the trenches' (212–13).

On the 'terzo luogo' see d'Annunzio, 'Sette documenti d'amore,' *Prose, 1922, Il libro ascetico,* 705. Also, from Gabriele d'Annunzio, *Di me a me stesso* (Milan: Mondadori, 1990): 'v'è oltre la vita e la morte un'altra plaga dove possa abitare l'asceta? v'è quel "terzo luogo" dove io credetti respirare nella notte di Cattaro' (is there that 'third shore' that I thought to breathe in the night of Cattaro?) (241) and 'Il mio ardimento – che parve folle – non era se non l'avidità di conoscere la vita o la morte *di là dalla paura* e anche *di là dal coraggio*' (My daring – that seemed folly – was nothing but desire to know life or death *beyond fear* and even *beyond courage*) (141).

The reference to the archangel is a play on d'Annunzio's name, which literally signifies 'Gabriel the Announcer' (contrary to rumours throughout his life, his name is not a pseudonym; his father had the family name changed from Rapagnetta to d'Annunzio before Gabriele was born).

Renzo De Felice, 'D'Annunzio e la vita politica italiana dal 1918 al 1936,' *Quaderni dannunziani* 38–39 (1969): 84–102. D'Annunzio's fall remains a mysterious event. Police at the time agreed that d'Annunzio had been pushed, but it was never determined why or by whom. The only persons present in the room were Luisa Bàccara, d'Annunzio's lover from the days of Fiume, and her younger sister Jolanda (Jole), though three others were apparently nearby. Initial reports suggested that d'Annunzio was pushed for political reasons. D'Annunzio was scheduled to meet with Mussolini and Nitti on 15 Aug. (two days after his fall), a meeting in which he, at least, saw himself as *superiore pacificatore* between Socialists and Fascists, whose goal was to avoid a civil war.

It is true that left without d'Annunzio's leadership after his fall, the legionaries of Fiume partly flowed into Mussolini's Fasci, and partly disintegrated into warring left-wing factions. And it is also true that after Oct. 1922 Mussolini had d'Annunzio closely watched, and clearly feared some d'Annunzian intervention in Dalmatia. Thus, Luisa Bàccara, who apparently also continued to provide d'Annunzio with the cocaine he (and she) had become addicted to at Fiume, was accused of doing the Fascists' bidding. This is apparently what d'Annunzio's children, Mario and Renata, believed at the time; yet upon hearing this d'Annunzio banished them from his home, and continued instead to live with Luisa until his death in

1938. (There were rumours that he did, however, cease to have sexual relations with her.) This, and the fact that there is no evidence to make Luisa a Fascist spy, have led historians to prefer a second hypothesis regarding d'Annunzio's fall: it seems that he was a little too passionately interested in the young Jole, and either she, or Luisa, or both, inadvertently pushed him out of the window in an attempt to put an end to his advances. See the introduction by Pietro Ghibellini in Gabriele d'Annunzio, *Siamo spiriti azzurri e stelle: Diario inedito (17–27 agosto 1922)* (Florence: Giunti, 1995). Also, for historical details see: Andreoli, *Il vivere inimitabile*; De Felice, 'D'Annunzio e la vita politica italiana.'

For a lengthy discussion of the 'terzo luogo' and d'Annunzio's fascination with death in its more mystical and esoteric aspects, see the following: Giuseppe Pecci, *D'Annunzio e il mistero* (Milan: Pan, 1969) 146ff.; Attilio Mazza: *D'Annunzio e l'occulto* (Rome: Edizioni Mediterranee, 1995), and *D'Annunzio sciamano* (Milan: Edizioni Bietti, 2001).

The recent publication of a collection of autobiographical prose, *Il fastello della mirra*, ed. Angelo Piero Cappello (Florence: Vallecchi, 2004), which d'Annunzio had originally planned to publish in 1926, getting as far as proofs until the project was replaced with his *Opera omnia*, does not add any new text to the poet's corpus, but it is revealing in terms of what he chose for this relatively short anthology, and what titles he gave each section: most relevant here is that substantial passages from the *Notturno* are gathered in the section 'Il mutilato'; also notable is the importance given to Randaccio (237ff.) and to the Unknown Soldier (376).

22 Douhet, 'Per la tomba del soldato al Pantheon,' *Il Dovere*, 24 Aug. 1920; also in Douhet, *Le Profezie di Cassandra*, 425, 426, 428, and 426. The original reads:

> Tutto sopportò e tutto vinse il nostro Soldato. Dall'ingiuria gratuita dei giornalisti e dei politicanti e dei giornalastri che [...] cominciarono a meravigliarsi del suo valore [...], alla calunnia feroce diramata per il mondo a scarico di una terribile responsabilità. [...] Tutto sopportò e tutto vinse, da solo, non ostante. [...] Perciò al Soldato bisogna conferire il sommo onore [...]
>
> Nulla si deve sapere di Lui. Fu un soldato ucciso nel compimento del suo dovere. Questo solo si deve dire di Lui. E questo è immenso. Quell'Umile è una moltitudine innumerevole. [...]
>
> Tutti i cittadini debbono far ala alla via trionfale, undendosi in un unanime senso di elevazione ideale nel comune atto di reverenza verso il Figlio ed il Fratello di tutti, spentosi nella difesa della Madre Comune.
>
> [...] quelle che ebbero la sventura di perdere un loro nato.

23 D'Annunzio, 'Comento meditato,' 535. The original reads:

> Madre, Madre, da che oscurità debbo io rinascere? [...] Chi parlò così? Chi parla così?
>
> Non ha nome. E' senza nome. Non ha figura. E' senza figura. [...]
>
> Siamo tutti senza nome. Siamo senza figura. [...] Ma [...] riconosceremo e italianamente rimescoleremo le nostre ossa logore o stronche.

24 For the connection between modernity, the death of God, and the end of plenitude, and an art of 'disfiguring,' see Mark C. Taylor, *Disfiguring: Art, Architecture, Religion* (Chicago: University of Chicago Press, 1992).

25 D'Annunzio, *Tre redazioni*, 22, 24. The original reads:

> In taluni l'intero teschio traspare; e si pensa allo scheletro che attende entro la carne e che ne imita i gesti, ne segue le attitudini, prigioniero. Teste già toccate dalla morte, già segnate dall'Operaia indefessa. Una massa di carne da macello, un carnaio ben preparato. [...] Odo il canto della terra, odo la pulsazione assidua dei cuori che pompano il sangue del sacrifizio; odo il silenzio di sotterra e il silenzio che sta di là dall'azzurro. [...] Qualcuno si curva, mi riconosce, mi chude gli occhi. La marea si ritira di sotto alla volta del mio capo. Due sollevano il mio corpo per coricarlo nella barella.

26 D'Annunzio's identification with dead comrades is in d'Annunzio, *Notturno*, 108. The original reads: 'Non sento più i confini del mio scheletro.'

27 Giuseppe Ungaretti, *Vita d'un uomo: Tutte le poesie* (Milan: Mondadori, 1969), 520–1, 527, 518, 65. 'Slancio vitale' (*élan vital*) is a reference to Henri Bergson, whose courses Ungaretti had attended in Paris before the war. The originals read:

> Incomincio *Il Porto Sepolto*, dal primo giorno della mia vita in trincea, e quel giorno era il giorno di Natale del 1915, e io ero nel Carso, sul Monte San Michele. Ho passato quella notte coricato nel fango, di faccia al nemico [...] ero in presenza della morte, in presenza della natura, di una natura che imparavo a conoscere in un modo nuovo, un modo terribile. [...] Nella mia poesia non c'è traccia di odio per il nemico, né per nessuno: c'è la presa di coscienza della condizione umana, della fraternità degli uomini nella sofferenza, dell'estrema precarietà della loro condizione. C'è volontà d'espressione, [...] c'è esaltazione, nel *Porto Sepolto*, quell'estaltazione quasi selvaggia dello slancio vitale, dell'appetito di vivere [...] Viviamo nella contraddizione.

'mi scopro e mi identifico, dentro gli orrori della guerra, nell'*uomo di pena.*'

'*Uomo di pena*' (man of sorrows) refers to the passage in Isaiah 53:3–6 that is considered a prefiguration of Christ:

He is despised and rejected of men, a Man of sorrows, and acquainted with grief. And we hid as it were our faces from Him; He was despised, and we esteemed Him not.

Surely He hath borne our griefs and carried our sorrows; yet we did esteem Him stricken, smitten of God, and afflicted.

But He was wounded for our transgressions; He was bruised for our iniquities. The chastisement of our peace was upon Him, and with His stripes we are healed.

All we like sheep have gone astray; we have turned every one to his own way; and the Lord hath laid on Him the iniquity of us all.

'*Mattina*' is notoriously untranslatable, as its sound is as important as its meaning, and ambiguity between the speaker's active or passive role is also essential. This ambiguity is related to the idea explored in this chapter, of the Unknown Soldier's body's self-generated and not merely external meaning.

28 Vivienne Hand, 'Ambiguous Joy: Contradictions and Tensions in Giuseppe Ungaretti's L'Allegria (1914–19),' *Italianist* 16 (1996): 76–116; see also her *Mirage and Camouflage: Hiding behind Hermeticism in Ungaretti's L'Allegrià* (Leicester: Troubador, 2006).

On illumination in Ungaretti, see the following by Ernesto Livorni: 'Giuseppe Ungaretti e Dante: Dall' "effimero" all' "eterno,"' *Canadian Journal of Italian Studies* 14/42–3 (1991): 122–31; *Avanguardia e Tradizione: Ezra Pound e Giuseppe Ungaretti* (Florence: Lettere, 1998); and '"In sé da simulacro a fiamma vera/errando": Ungaretti's Poetry from Bergson to Plato,' *Romanic Review* 97/3–4 (2006): 517–54. See also Mario B. Mignone, 'L'Allegria ungarettiana come ricerca della parola sacrale,' in Florinda M. Iannace, ed., *Etica cristiana e scrittori del Novecento* (Stony Brook: Forum Italicum, 1993), 193–202; Margaret Brose, 'Metaphor and Simile in Giuseppe Ungaretti's L'Allegria,' *Lingua e Stile: Trimestrale di Linguistica e Critica Letteraria* 11 (1976): 43–73.

On Ungaretti as part of a generation that grappled with d'Annunzian aestheticism, seeking a place 'between form and flux,' see Thomas Harrison, 'Overcoming Aestheticism,' in Somigli and Moroni, eds., *Italian Modernism,* 187.

29 Giuseppe Ungaretti, *Vita d'un uomo: Saggi e interventi* (Milan: Mondadori, 1974), 130, 131. The original reads:

> L'innocenza, abbiamo saputo com'è fatta. Ci è apparsa, e ci ha tenuto sotto le sue ali più grandi, nei rivolgimenti di questi anni. Ci occupava tutta la mente. La memoria aveva gli occhi bendati, poteva dirsi abolita. Persino la nozione del tempo era nuova. Il tempo pareva eterno, non per modo di dire. Non ci è stato nascosto l'orrore dell'eternità. Non contava più che l'istinto. Si era in tale dimestichezza con la morte, che l'intero film del naufragio ci ripassava ogni momento in mente, e non c'era oggetto che non ce lo rilettesse; era, la nostra vita da capo a fondo, l'oggetto stesso sul quale cadeva il nostro sguardo. Non era la nostra, in realtà, vita più che oggettiva, il primo oggetto venuto. In quel frangente, ho capito perché il Negro fa gli occhi all'idolo con pezzetti di specchio. [...]
>
> Abbiamo dell'innocenza, non più, come nell'Ottocento, un'aspirazione filosofica, ma un'esperienza diretta, possediamo una conoscenza mistica della realtà. Credo che l'arte di domani sarà felice. A poco a poco, il dramma si scioglierà. Saranno andati in fumo anche i tentativi di affidare la parte del burattino alla memoria, e all'innocenza quella dell'oracolo. E della paurosa, e materna, innocenza, tornata nella memoria al suo posto oscuro, le lusinghe saranno vane.

30 Ungaretti, *Vita d'un uomo: Tutte le poesie*, 87. Stewart Flynn, 'Soldiers,' *Modern Poetry in Translation* 18 (2001): 185. The original reads:

> **Soldati**
> *Bosco di Courton luglio 1918*
>
> Si sta come
> d'autunno
> sugli alberi
> le foglie

31 Harrison, *Dominion*, 126, 127–33, 137, 138. For Harrison the Tomb of the Unknown Soldier makes the universal emerge from the particular, whereas in the Vietnam Memorial it is the reverse. I hope to show that while this was, indeed, how state scripts presented the Unknown Soldier Memorial, this is not how individuals reacted to it.

32 Ungaretti, *Vita d'un uomo: Tutte le poesie*, 39, 63. The two poems' translations are adapted verions of, respectively, *Giuseppe Ungaretti: Selected Poems*, trans. Andrew Frisardi (New York: Farrar, Straus, and Giroux, 2002), 31, and *A*

Major Selection of the Poetry of Giuseppe Ungaretti, trans. Diego Louis Bastianutti (Toronto: Exile Editions, 1997), 99. The originals read:

Fratelli
Mariano il 15 luglio 1916

Di che reggimento siete
fratelli?

Parola tremante
nella notte

Foglia appena nata

Nell'aria spasimante
involontaria rivolta
dell'uomo presente alla sua
fragilità

Fratelli

Dolina notturna
Napoli il 26 dicembre 1916

Il volto
di stanotte
è secco
come una
pergamena

Questo nomade
adunco
morbido di neve
si lascia
come una foglia
accartocciata

L'interminabile
tempo
mi adopera
come un
fruscio

33 Leopardi's 'L'Infinito' is one of the most famous poems of Italian Romanticism; in it, the sound of the wind in the leaves evokes precariousness and

time in contrast with the immensity of the night sky. It is part of Leopardi's *Idilli,* published in 1826 and reprinted in Giacomo Leopardi, *Poesie e prose,* ed. Rolando Damiani, Cesare Galimberti, and Mario Andrea Rigoni (Milan: Mondadori, 1987), 2 vols.

34 Ungaretti, *Vita d'un uomo: Tutte le poesie,* 57, and *Selected Poems,* 51. The original reads:

Italia
Locvizza l'1 ottobre 1916

Sono un poeta
un grido unanime
sono un grumo di sogni

Sono un frutto
d'innumerevoli contrasti d'innesti
maturato in una serra

Ma il tuo popolo è portato
dalla stessa terra
che mi porta
Italia

E in questa uniforme
di tuo soldato
mi riposo
come fosse la culla
di mio padre

35 For the association of church floors with flowing water, most famously, for d'Annunzio, in Ruskin (Saint Mark's) and Flaubert, see W.R. Lethaby, *Architecture, Mysticism and Myth* (1891; New York: George Braziller, 1975), 301–3. This association with water is seen in the Unknown Soldier ceremonies in Trieste, as reported by *L'illustrazione Italiana,* 13 Nov. 1921, where the altar was set up facing the ocean in Piazza Unità.

36 In Winter, *Sites*: '*From* art' (102), 'avoidance' (101), and 'triumphalism' (107); on the popularity of Romantic models (115). For Winter the strength of most memorials 'lay in the power of traditional languages, rituals, and forms to mediate bereavement,' since 'irony's cutting edge [...] could express anger and despair [...] but it could not heal' (107). Referring to Freud's *Mourning and Melancholia,* published in 1917, he adds that some element of traditional forms was needed to prevent mourners from being 'trapped in a forest of loss' (115).

Claudio Canal ('La retorica della morte') has argued that the 'anti-funerary structure' of First World War memorials is expressed in their predominant verticality (659). On visibility, see Miniero, *Da Versailles*, 170.

37 Jeffrey T. Schnapp, *In cima – Giuseppe Terragni per Margherita Sarfatti (Architetture della memoria nel 900)* (Milan: Marsilio, 2004), 22, 14. Mia Fuller, *Moderns Abroad: Architecture, Cities and Italian Imperialism* (London and New York: Routledge, 2007), 96–8. 'Forces of nature,' in Winter, *Sites*, 101.

Schnapp's originals read: 'il monumento moderno deve eseguire quello che per Freud è la riuscita elaborazione del lutto: mantenere in vita i morti tenendoli sepolti in eterno; ricordarli, certo, ma anche disaggregarli dalla scena contemporanea; consacrarli come quel qualcosa che è chiuso, completato, finito, avvolto e assorbito nel dinamismo del momento presente'; 'monumentalismo anti-monumentale'; 'una peculiare immaginazione archeologica'; 'chiudono la porta al passato solo per riaprirla a una archeologia delle strutture archaiche' (*In cima*, 22, 14).

38 Canal, 'La retorica della morte,' 659. Labita, 'Il Milite Ignoto,' 146–7. D'Annunzio writes of 'due vite e due ali' echoed in 'ossa' in *Notturno*, 26–7. 'Obscur' in Jagielski, *Le soldat inconnu*, 111. 'Oscuro,' in Miniero, *Da Versailles*, 222. The wreath indirectly recalls 'Viali della rimembranza,' in which trees are planted to commemorate the dead, and to reassert the cycle of death and rebirth (Canal, 'La retorica,' 661).

As Ingersoll and Nickell ('The Most Important Monument') note, the anonymity of the Unknown Soldier memorial, often emphasized by the whiteness of the monument itself, also reflects the loss of a symbolic vocabulary with which to comment on death (205); for them, symbolism is thus submerged in favour of utilitarian or technological concerns (as we will see, the Unknown soldier's rebirth was in part associated with mechanical purity).

39 For the double description of the Unknown Soldier submerged by water and inscribing his sacrifice in water in the *Notturno*, see the section on 'a unanimous idea.' For this water as an abyss, and indeed, a specifically Nietzschean existential one, see Valesio, *The Dark Flame*, 139.

The Unknown's coffin was blessed at Santa Maria degli Angeli by the Bishop of Trieste with water from the Timavo (see Tobia, *L'altare della patria*, 90). This was the same water that for d'Annunzio wet Randaccio's flag, before it was wet also by the 'acqua capitolina.' See the 'Discorso di Palazzo Marino,' given on 3 Aug. 1922, published as Gabriele d'Annunzio, 'Agli uomini milanesi per l'Italia degli italiani,' *Prose*, 1922, *Il libro ascetico*, 565.

40 D'Annunzio, *Notturno*, 69, 57, 57. The original reads:

> Così la sua morte e la mia vita sono una medesima cosa. [...]
> come il messaggio del vento che è la rapidità dell'infinito in cammino; con uno spirito senza riva, con un corpo senza forma, con un gaudio che sembra terrore, sento l'idealità del mondo. [...]
> Non scrivo su la sabbia, scrivo su l'acqua.

41 Valesio, *The Dark Flame*, 140, 118, 121ff., 140.

42 Ungaretti, *Vita d'un uomo: Tutte le poesie*, 43; the translation is amended from *A Major Selection*, 73, 75. The original reads:

> Mi tengo a quest'albero mutilato
> abbandonato in questa dolina
> che ha il languore
> di un circo
> prima o dopo lo spettacolo
> [...]
> stamani mi sono disteso
> in un'urna d'acqua
> e come una reliquia
> ho riposato
>
> L'Isonzo scorrendo
> mi levigava
> come un suo sasso
>
> Ho tirato su le mie quattr'ossa
> e me ne sono andato
> come un acrobata
> sull'acqua
> [...]
> Questa è la mia nostalgia
> che in ognuno
> mi traspare
> ora ch'è notte
> che la mia vita mi pare
> una corolla
> di tenebre

43 Ungaretti, *Vita d'un uomo: Tutte le poesie*, 170, and *Selected Poems*, 125. The original reads:

E nei vivi la strada dei defunti,
Siamo noi la fiumana d'ombre,
Sono esse il grano che ci scoppia in sogno
Loro è la lontananza che ci resta,
E loro è l'ombra che dà peso ai nomi.

44 'Death ceases to be the termination of life and becomes the sound of existential and historical time in the mind, the deepening of its distances as the latter are heard in the mind's temporal flow' (Harrison, *Dominion*, 76). Harrison thus also critiques our tendency towards petrification: 'instead of committing our love to the wind we engrave our dead in stone. Our loathing of death causes us to court its petrifications to death.' 'Tombstones,' in Steven P. Schneider, ed., *Complexities of Motion: New Essays on A.R. Ammons's Long Poems* (Madison: Fairleigh Dickinson University Press, 1999), 173.

45 Ungaretti, *Vita d'un uomo: Tutte le poesie*, 29–30, and *A Major Selection*, 57, 59. The original reads:

Oggi
come l'Isonzo
di asfalto azzurro
mi fisso
nella cenere del greto
scoperto dal sole
e mi trasmuto
in volo di nubi
[...]
Ho sulle labbra
il bacio di marmo

46 For water images as flood, apocalypse, end to all violence, in Barbusse, see Winter, *Sites*, 184. These may be related to imagery in Gance.

47 For music in France, see Miniero, *Da Versailles*, 151; in England, ibid., 161. For the choice of Palestrina in Italy, see Bernardi, 'Gabriele d'Annunzio e il Milite Ignoto,' 1128. For the Savoy tradition of off-key drums, see *L'illustrazione Italiana*, 13 Nov. 1921. On speeches, and their minimalism, see Miniero, *Da Versailles*, 151, 162.

48 Winter, *Sites*, 182. D'Annunzio's rejection of monuments is in the 1919 'L'erma bifronte,' in 'Il sudore di sangue,' *Prose, La penultima ventura* (Milan: Mondadori, 1947), 932. The original reads: 'oggi i fanti, contadini e non contadini, gridano a una voce, di sopra la terra e di sotterra: "Non vogliamo monumenti."'

49 For comparison of British, French, and American inscriptions, see Inglis, 'Entombing Unknown Soldiers,' 15–16. For abstraction, see Miniero, *Da Versailles,* 149. Aside from 'Ignoto militi,' the Italian memorial also includes, on the inside, the 'motivazione' for the Gold medal awarded to the Unknown (Pizzo, *Il Vittoriano,* 42).

For the French inscription, see *Arc de triomphe de l'Étoile – Wikipedia,* retrieved from http://fr.wikipedia.org/wiki/Arc_de_triomphe_de_l%27Étoile. For the initial plaque with 'le soldat Français,' see Miniero, *Da Versailles,* 147.

The British inscriptions include: 'A British Warrior who Fell in the Great War 1914–1918,' and 'In Christ shall all be made alive,' as well as: 'Beneath this stone rests the body / of a British Warrior / unknown by name or rank / brought from France to lie among / the most illustrious of the land / and buried here on armistice day / 11 Nov. 1920, in the presence of / His Majesty King George V / His Ministers of State / the Chiefs of His Forces / and a vast concourse of the nation. / Thus are commemorated the many / multitudes who during the great / war of 1914–1918 gave the most that / man can give life itself / for God / for King and Country / for loved ones home and empire / for the sacred cause of justice and / the freedom of the world. They buried him among the Kings because he / had done good toward God and toward / his House. / The Lord knoweth them that are his' (top); 'Unknown and yet well known, dying and behold we live' (side); 'Greater love hath no man than this' (side); 'In Christ shall all be made alive' (base). *The Unknown Warrior – Wikipedia,* retrieved from http://en.wikipedia.org/wiki/The_Unknown_Warrior. On Luytens' cenotaph all that is wrtitten is 'To the glorious dead.' *Cenotaph – Wikipedia,* retrieved from http://en.wikipedia.org/wiki/Cenotaph.

The American one reads: 'Here rests in honored glory an American soldier known but to God.' *Tomb of the Unknowns, Arlington National Cemetery,* retrieved from http://www.arlingtoncemetery.net/tombofun.htm.

50 For the 'generic body,' see Laqueur, 'Names, Bodies,' 126. For the feudal hero, and the shift from 'le' to 'un,' see Inglis, 'Entombing Unknown Soldiers,' 16, and 'War Memorials,' 11.

51 Paolo Valesio, *Ascoltare il silenzio: La retorica come teoria* (Bologna: Il Mulino, 1986), 353–64. Labita, 'Il Milite Ignoto,' 144. The verse is from *Inferno* IV, 81; I give the Singleton translation. On Limbo, see Gian Pietro Testa, *Io sono il milite ignoto* (Bologna: Minerva, 2006), 33.

52 D'Annunzio quotes himself in 'Comento meditato,' at 519, referring back to 'Agli Italiani degli Stati Uniti,' a speech delivered at Fiume in 1919. The original reads: 'Un uomo d'Italia vi manda questo silenzio come si manda

una lapide non iscolpita, dove / una sola parola sia da scolpire: la più grande.'

On silence in d'Annunzio's speeches at Fiume, see Jeffrey T. Schnapp, 'Worte des Schweigens in D'Annunzios Fiume-Reden,' in Hans Ulrich Gumbrecht, Friedrich Kittler, and Bernhard Siegert, eds., *Der Dichter als Kommandant: D'Annunzio erobert Fiume* (Munich: Fink, 1996), 133–45.

53 D'Annunzio's statement was: 'Là, nel piccolo cimitero, [...] c'è una pietra che porta incisa una sola parola: "Resurgo." [...] Non c'è monumento funebre, non c'è mausoleo, non c'è obelisco, non c'è piramide che valga quella lapide rozza con quell'unica parola. E' l'unica parola che doveva essere incisa sul sepolcro del soldato ignoto: "Risorgo"' ('Agli uomini milanesi,' 569). This was delivered 3 Aug. 1922, at Palazzo Marino, just a few days before d'Annunzio's fall.

54 Leed, *No Man's Land*, 21–2, 32–3, 167. He is referring to the soldier's lack of reintegration into society; I extend this now to his lack of reintegration into a coherent spiritual symbolism.

Rosa Tricerri, 'Il linguaggio della tradizione cristiana nelle scritture del tempo di guerra,' *Quaderni del Vittoriale: D'Annunzio e la religiosità*, no. 28 (1981): 97–109. Tricerri concludes her examination of Christ symbolism in d'Annunzio, noting that his work poses the question of whether the incarnation and the Passion 'can still be said' meaningfully (105). However, this problem is not merely d'Annunzio's, as she implies, but one that is symptomatic of Western modernity.

55 Winter, *Sites*, 136–7.

3 Sacrifice and the *Non finito*

1 On the difference between 'facts' (purportedly objective and unchanging) and 'values' (purportedly relative) as one of positioning, and on the importance of discussion and reinvention, see Bruno Latour, 'Why Has Critique Run Out of Steam? From Matters of Fact to Matters of Concern,' *Critical Inquiry* 30 (2004): 225–48.

Referring to soldiers returning to life in Gance's *J'Accuse*, Gorman ('The Return of the Dead,' 67) argues that nationalists as well as pacifists employed the same idiom of resurrection – keeping the dead alive so that they could instruct the living – and thus implies that the difference lies in the details of how this idiom was used. For a more elaborate discussion of the different views of sacrifice present in French culture at the end of the nineteenth century, see Ivan Strenski, *Contesting Sacrifice: Religion, Nationalism, and Social Thought in France* (Chicago: University of Chicago Press, 2002).

In Strenski's view, the predominance of 'annihilating' sacrifice (defined by the complete destruction of the self), which the French Revolution derived from the more intransigent side of Catholicism's vision of Jesus' sacrifice, continues during the First World War, even as a new model, mainly based on the work of Durkheim, Hubert, and Mauss, is emerging (162–3). He refers to Henri Hubert and Marcel Mauss, *Sacrifice: Its Nature and Its Function* (1898; Chicago: University of Chicago Press, 1964), and Émile Durkheim, *The Elementary Forms of the Religious Life* (1912; Oxford and New York: Oxford University Press, 2001).

For the connection between sacrifice and modern poetry, see Ernesto Livorni, 'Caino, Abele ed il sacrificio: La poesia moderna ed il sacro,' *Semicerchio: Rivista di Poesia Comparata* 19/2 (1998): 6–15. On sacrifice in d'Annunzio, see Nicoletta Pireddu: *Antropologi alla corte della bellezza: Decadenza ed economia simbolica nell'Europa fin de siècle* (Verona: Fiorini, 2002), and 'Gabriele D'Annunzio: The Art of Squandering and the Economy of Sacrifice,' in Mark Osteen, ed., *The Question of the Gift: Essays across Disciplines* (London: Routledge, 2002), 172–90.

2 A core influence on religious Modernism was William James, *The Varieties of Religious Experience: A Study in Human Nature* (New York: Modern Library, 1902). Early philosophical and historical texts include: Maurice Blondel, *Action: Essai d'une critique de la vie et d'une science de la pratique* (Paris: Alcan, 1893); Alfred Loisy, *L'Evangile et l'église* (Paris: A. Picard, 1902). Ernesto Buonaiuti was one of the more controversial figures of Modernism; he was condemned, in particular, for his *Il Programma dei modernisti* (Rome: Libreria Editrice Romana, 1907), which was immediately translated by Leslie Lilley, as *The Programme of Modernism* (New York and London: G.P. Putnam's Sons, 1908). He was one of only twelve university professors to lose their positions for refusing to swear an oath of loyalty to the Fascist regime in 1931. Giorgio Boatti, *Preferirei di no le storie dei dodici professori che si opposero a Mussolini* (Turin: Einaudi, 2001). Strenski (*Contesting Sacrifice*, 49, 53–4) points out that First World War deaths were so incomprehensible as to make even Loisy resort at times to old models of 'annihilating' sacrifice, but he also notes that writers such as Barbusse and others, during and right after the war, began to react against the association of nationalism and sacrifice that had dominated pre-war French culture.

On existentialism and the absurd, canonical works are Jean Paul Sartre, *La nausée: roman* (Paris: Gallimard, 1938), translated by Robert Baldick, as *Nausea* (Norfolk: New Directions, 1959); and Albert Camus, *L'homme révolté* (Paris: Gallimard, 1952), translated by Anthony Bower, as *The Rebel: An Essay on Man in Revolt* (New York: Vintage Books, 1956).

Giovanna Pieraccini, 'Il superuomo dannunziano come archetipo dei miti dell'ideologia di destra,' *Strumenti Critici* 10/3 (1995): 387–416. It has become common to compare d'Annunzio to Nietzsche and claim that for him the 'superuomo' takes over after the 'death of God' (394–5), but d'Annunzio's attempt to create a new vision of Christ, and its links to religious Modernism, remain little explored.

A singular case of taking d'Annunzio's search for the absolute seriously is in Massimo Bontempelli, 'D'Annunzio o del martirio,' *Pirandello Leopardi d'Annunzio* (Milan: Bompiani, 1938), 89–142. D'Annunzio's Christology is discussed also in Pecci, *D'Annunzio e il mistero,* 146ff.

3 Pius X's 1907 encyclical was an exhaustive condemnation of 'Modernism,' and one of the first uses of the term to describe a broad movement in European culture. See Pius X, *Pascendi dominici gregis: On the Doctrine of the Modernists,* 1907, retrieved from http://www.papalencyclicals.net/Pius10/p10pasce.htm.

Notions of the 'rite of passage' and of 'initiation' were discussed in nineteenth-century anthropology, but Arnold van Gennep's *Les rites de passage* (Paris: Nourry, 1909) was very influential, in particular, on understanding the war experience, as Leed argues (*No Man's Land,* 15). On the rite of passage in its link to virility, see George L. Mosse, *The Image of Man: The Creation of Modern Masculinity* (New York: Oxford University Press, 1996).

On Hubert and Mauss, as well as Durkheim, see Strenski, *Contesting Sacrifice,* 162–70. Contesting the notion of Jesus' sacrifice as a model, they also opposed the association of sacrifice and expiation (167), and favoured a view of sacrifice as 'giving of' rather than 'giving up' of self (166).

On a modern Christology that sees the dogmas and symbols of Christianity as 'heuristic fictions' that 'command: "Look at the unknown!"' see Michel Polanyi, *Personal Knowledge: Towards a Post-critical Philosophy* (London: Routledge and Kegan Paul, 1958), 199.

For a seminal overview of the difference between Christianity and Christ, see Thomas Sheehan, *The First Coming: How the Kingdom of God Became Christianity* (New York: Random House, 1986).

Both the contested role of the Church and the desire for an alternative Christology were present in Italy's 2003 mourning for the dead of Nasiriyah, Iraq, at the Unknown Soldier Memorial. Various articles thus describe the controversy that erupted when Monsignor Nogaro criticized Cardinal Ruini, who not only blessed the coffins of the dead, but said that Italy would 'win,' though 'peacefully.' See, e.g., the following: 'Il caso di monsignor Nogaro crea "sconcerto" in Vaticano,' *Repubblica,* 18 Nov. 2003; Gianluca Monastra, '"Niente odio, ma non fuggiremo" – Ruini: La missione continui; Tutta l'Italia in silenzio per le vittime,' *Repubblica,* 19 Nov. 2003; Roberto Monteforte, 'L'imbarazzo della Chiesa su Ruini combattente,' *L'Unità*, 19 Nov. 2003; Ottavio Ragione, 'Pisanu: Frasi intollerabili del vescovo di Caserta,' *Repubblica*, 18 Nov. 2003; Red, 'Nogaro: "Dico che il terrorismo,"' *L'Unità*, 18 Nov. 2003; Red, 'Un lungo, interminabile applauso; Poi Ruini: "Fronteggeremo gli assassini; Ma senza odio,"' *L'Unità*, 18 Nov. 2003; Red, 'Il vescovo di Belluno: "I nostri soldati hanno pagato le scelte di guerra,"' *L'Unità*, 19 Nov. 2003.

Christ is invoked as a figure of dissent in the following: d'Avanzo, 'Il funerale infinito delle famiglie,' *Repubblica*, 18 Nov. 2003; Merlo, 'Carabinieri d'Italia,' *Repubblica*, 18 Nov. 2003; Vattimo, 'Ritrovare le parole,' *L'Unità*, 20 Nov. 2003.

4 René Girard, *La violence et le sacré* (Paris: Grasset, 1972), 381, 386, 383, 466, 105–34; see also his *La route antique des hommes pervers* (Paris: Grasset, 1985), *Violence and the Sacred*, trans. Patrick Gregory (Baltimore: Johns Hopkins University Press, 1977), and *Job, the Victim of His People*, trans. Yvonne Freccero (Stanford: Stanford University Press, 1987).

On Girard, see Michel Deguy and Jean-Pierre Dupuy, eds., *René Girard et le problème du mal* (Paris: Grasset, 1982). The originals read: 'La crise se présente comme perte de différence entre les morts et les vivants, mélange des deux royaumes normalement séparés. C'est bien la preuve que les morts incarnent la violence [...] La victime émissaire meurt, semble-t-il, pour que la communauté menacée tout entière de mourir avec elle, renaisse à la fécondité d'un ordre culturel nouveau ou renouvelé'; 'les hommes ne sont pas capables de se réconcilier qu'au dépens d'un tiers'; 'il y a toujours mort d'homme à l'origine de l'ordre culturel.'

5 Joshua Landy, 'Nietzsche, Proust, and Will-to-Ignorance,' *Philosophy and Literature* 26/1 (2002): 1–23. Discussions about the purpose of sacrifice eclipse ones about its nature according to a 'will to ignorance' that, drawing from Nietzsche and Proust, Landy argues manifests as a relentless 'questing for information' by which people '*shield* themselves from the truth,' particularly the sort of truth that has to do with values, not facts (18). Also see Joshua Landy, *Philosophy as Fiction: Self, Deception, and Knowledge in Proust* (New York: Oxford University Press, 2004), 99.

Jean-Pierre Dupuy, 'Not to Know What One Knows: Some Paradoxes of Self-Deception,' *Diogenes* 43/169 (1995): 53–68. Dupuy shows that such self-deception can have specific ideological implications when he connects the Calvinist's 'power to choose his predestination' with capitalism (66). Also see Jean-Pierre Dupuy, *Self-Deception and Paradoxes of Rationality* (Stanford: CSLI, 1998), 12–15. Not coincidentlly, the first war of mass culture and capitalism, the First World War, can be associated with the equally paradoxical notion that the soldier 'chose his duty.' Dupuy extends this point in arguing that only through such self-deception can we be unaware of our impending self-destruction, in *Pour un catastrophisme éclairé: Quand l'impossible est certain* (Paris: Seuil, 2002), 49–52.

6 Atti Parlamentari, Senato del Regno, Legislatura XXVI, documenti del Senato, 1921, disegno di legge n. 163, 163A, Biblioteca Giovanni Spadolini, Rome (hereafter AP, Senato, 1921, nos. 163, 163A); AP, 1921–23, nos. 202 SdSI, 202-A SdSI.

D'Annunzio refers to the dead soldier's gun in *La riscossa*, 162. The originals read: 'simbolo della virtù collettiva di nostra gente'; 'la soppressione assoluta della propria individualità, [...] ammonisce, in fine, che l'essere stato italiano ed essere caduto per l'Italia è titolo bastevole per i supremi onori e la sempiterna venerazione, all'infuori di ogni altro segno identificatore.'

> omaggio a tutti i nostri morti [...]
>
> che caddero con la suprema sopportazione, e col supremo sacrificio, per il dovere umile che non ha nome e che irradia, così, senza individualità, tanta forza di luce sulla razza. [...]
>
> Io, ardito, fante, bombardiere, artigliere, ben ricordo che non v'era fucile più giusto del fucile d'un morto. Chi l'aveva raccolto non falliva il colpo. [...]
>
> Il fucile sollevato del caduto segnava la vittoria fatale.
>
> Le ossa sollevate del caduto segneranno l'inizio della vera pace.
>
> Così sarà se noi sapremo raccoglierci e meditare in conspetto della più alta espressione della maestà della morte passante sovra un convoglio funereo dalla silente trincea di guerra alla immortale grandezza dell'Urbe sotto il sole d'Italia.
>
> Noi pensiamo che una cerimonia supremamente austera debba aver luogo.

Labita ('Il Milite Ignoto,' 136–7) shows that this rhetoric of passive obedience is tied to the theories on military psychology seen in Agostino Gemelli, *Il nostro soldato, saggi di psicologia militare* (Milan: Treves: 1917); he also quotes *La Tribuna*, 1 Nov. 1921, on *dovere*.

Mario Isnenghi, *La Grande Guerra: Uomini e luoghi del '15–'18* (Turin: UTET, 2008). Isnenghi shows that dissent was more common than the official history would have it, noting also the importance of various types of neurosis (290–6). Strenski (*Contesting Sacrifice*) observes that in France the concept of *offensive à outrance* (attack to excess) found justification perhaps, in part, in notions of Bergsonian *élan vital* but far more clearly in the conservative Catholic idea of 'bloody sacrifice' that was used, in particular, to explain the death of the innocent at war as 'enduring an enforced Lent' (76–7, 66, 60–1).

As in Italy, the French military and later the state here appropriated Catholic notions of sacrifice, not always, though generally with the Church's approval (this will be discussed more later).

7 On Commissione, see Labita, 'Il Milite Ignoto,' 122–3; also Isnenghi, *La Grande Guerra*, 394ff. On prisoners, see *Il Dovere*, 4 May 1919, 2. The original reads, 'dopo un'azione fortunosa.' On 'caduti per la Patria,' see *Il Dovere*, 20–1 Aug. 1919.

Bàrberi Squarotti ('D'Annunzio scrittore "politico,"' 332) comments on d'Annunzio's wartime political writings, emphasizing that 'c'è l'assoluta autosufficienza dell'azione, l'assalto, la vittoria, c'è *il senso del sacrificio senza altra motivazione che se stesso*, c'è l'orgia dei morti e del sangue, la mistica dei feriti' (there is an absolute self-justification for action, assault, victory, there is *a sense of sacrifice with no motive other than itself*, there is an orgy of death and blood, the mysticism of the wounded; emphasis added).

For a remarkable defence of those accused of treason at Caporetto, which claims they are Italy's future, in contrast with their leaders whose ruinous strategy caused so much bloodshed, see Curzio Malaparte, 'La rivolta dei santi maledetti,' in Luigi Martellini and Giancarlo Vigorelli, eds., *Opere scelte* (1912; Milan: Mondadori, 1997). For a discussion of literature on Caporetto, see Mario Isnenghi, *I vinti di Caporetto nella letteratura di guerra* ([Padua]: Marsilio, 1967).

Giovanna Procacci, *Soldati e prigionieri italiani nella Grande guerra: Con una raccolta di lettere inedite* (Rome: Editori Riuniti, 1993). Procacci demonstrates that from the beginning of the war, morale was very poor among Italian soldiers, who did not trust their leaders. This was only made worse by Caporetto. On 30 Sept. 1917, according to Procacci's research, the Italian military counted as many as 100,000 deserters, or men who were conscripted but in some way or other had disappeared (60–3, 103). She also cites a 1917 statement by d'Annunzio in the *Corriere della Sera* accusing prisoners of being deserters; clearly, by the end of war he had changed his mind on that topic (172, 220).

8 This article from *Il Dovere*, 4–5 Sept. 1919, 2, refers to an earlier, but 'recent,' article in *Il giornale d'Italia*, which, however, I have not been able to locate. Douhet's newspaper claims that the Investigative Commission for Caporetto unearthed this story.

The original reads:

> ecco il genio possente della razza foggiare nella persona di un umile caporale un simbolo gigantesco di unica virtù. E' il caporale che, condannato innocente alla fucilazione, rinnovava il classico fulgore delle

più nobili devozioni civili, mostrando ai soldati che dovevano fucilarlo, come la salvezza e l'onore della patria si alimentino anche neccessariamente colla serena ed orgogliosa celebrazione del sacrificio personale. Narra alla Commissione d'inchiesta un ufficiale testimone: '[...] letta la sentenza, il caporale assunse un atteggiamento impassibile, sereno, come un soldato in combattimento. Appena finita la lettura della sentenza egli disse: *ritengo che i giudici hanno votato con coscienza e vado a morte con orgoglio.* Davanti al plotone di esecuzione non volle essere bendato e, rivolto ai soldati, disse: *soldati, io son stato con voi, ho combattuto con voi; mirate giusto, mirate al petto e servite sempre il vostro paese. Evviva l'Italia.* Questa è la bellezza sovrana.' [...] gli proponiamo di aprire una sottoscrizione per erigere un monumento a quell'eroe umile ed ancora ignoto.

9 *La Tribuna,* 1 Nov. 1921. The original reads: 'nello stretto adempimento del proprio dovere [...] è il sacrificio che si celebra, comunque compiuto.'

10 'Morto senza nome' (nameless dead) and 'mistero [mystery] di resti miseri e venerandi' are in *Corriere della Sera,* 4 Nov. 1921, as is, disapprovingly, 'coraggio guerriero' (warrior's courage). 'Martirio' (martyrdom) is from *La Tribuna,* 30 Oct. 1921. Both are also in Labita, 'Il Milite Ignoto,' 136–7.

11 On Sauro see *Nazario Sauro – Wikipedia,* retrieved from http://en.wikipedia.org/wiki/Nazario_Sauro. Aside from the series in *L'illustrazione Italiana,* Sauro was also celebrated all over Italy, as is detailed in MdI, OaSI, 1921.

12 Jean Giono, *Le déserteur et autres récits* (Paris: Gallimard, 1973), 3, 12, 18, 50, 71, 90. The originals read: 'un personnage de Victor Hugo [qui] sort des *Misérables* avant la lettre'; 'déserte une certaine forme de société pour aller vivre dans une autre'; 'une sorte de Patrie'; 'étranger en tout'; 'misérable et aristocrate'; 'aristocratie du coeur'; 'Quel est son crime (si crime il y a!): politique, professionnel, de droit commun?'; 'Le délit de Charles-Frédéric Brun n'est jamais amnistié par aucune loi: c'était le délit de misère, son crime était d'être misérable.'

13 On British gravediggers and 'uncorrupted soil,' see Miniero, *Da Versailles al Milite ignoto,* 161. On the French benediction, see Jagielski, *Le soldat inconnu,* 109. The original reads: 'tous les morts de la guerre.' On the Italian benediction, see Miniero, *Da Versailles,* 207. The original reads: 'tutti i parroci di Roma.'

For water from the Timavo River, see *L'illustrazione Italiana,* 6 Nov. 1921. This is also mentioned in Cadeddu (*La leggenda del soldato sconosciuto,* 124), where he writes that the water was in an urn bearing the inscription 'Imo ex corde Timavi' (From the depths of the heart, o Timavo), which was also inscribed on Randaccio's tomb.

14 Miniero, *Da Versailles*, 200; Jagielski, *Le soldat inconnu*, 108. The originals read: 'espressioni di pietà care alla sensibilità popolare'; 'l'Inconnu semble être entré rapidement dans la dimension universelle et apaisante tant souhaitée par ses "inventeurs."'

15 Capdevila and Voldman, *Nos morts*, 191, 193, 195, 199, 195. *L'illustrazione Italiana*, 13 Nov. 1921, sees all forms of silence as refusals of ideological appropriations of the war experience, be they by the state, the Church, or both together. On religious objects, see Lucio Fabi, ed., *1918. La guerra nella testa: Arte popolare, esperienza e memoria nel primo conflitto mondiale* (Trieste: Lint Editoriale, 1998), 109ff. Girard, *La violence et le sacré*, 381.

16 In addition to Modernist works mentioned above, important for Fogazzaro were also George Tyrrell, *External Religion: Its Use and Abuse* (London: Sands, 1899); Giovanni Semeria, *Venticinque anni del cristianesimo nascente* (Rome: F. Pustet, 1900). On Fogazzaro and religious Modernism, see Laura Wittman, '*Omnes velut aqua dilabimur*: Antonio Fogazzaro, the Saint, and Catholic Modernism,' in Somigli and Moroni, eds., *Italian Modernism*, 130–66.

17 For the scene in Fogazzaro's novel which culminates in the question 'si prega forse per il perdono di chi non può essere perdonato?' (does one pray for the forgiveness of someone who cannot be forgiven?), see Antonio Fogazzaro, *Il Santo* (Milan: Mondadori, 1970), 144.

18 'Silenzio laico' and 'preghiera cattolica' are in Labita, 'Il Milite Ignoto,' 134. Semeria's comments are in *Il corriere d'Italia*, 2 Nov. 1921, and also mentioned in Miniero, *Da Versailles*, 210–12. Semeria actually wrote the preface to Gemelli's book, in 1917, when the war was still new and triumphalist rhetoric not as grating; their views would diverge by 1921.

On Semeria and Modernism, and the anti-Modernist oath, see Maurilio Guasco, *Modernismo: I fatti, le idee, i personaggi* (San Paolo: Cinisello Balsamo, 1995), 126; the Pope's leniency was all the more surprising in that Semeria was thought to be one of the men who contributed anonymously to Buonaiuti's *Il Programma dei modernisti* (166). D'Annunzio's comment on Semeria is in his *Taccuini*, at 794.

Strenski (*Contesting Sacrifice*, 60–1) demonstrates that in France we find a collusion between the official Church view and the government's such that 'the secular republic was compromised by religion and religion finally co-opted by the secular republic' in a 'nationalist mystique' that was 'overwhelming' (though, as others and I too have shown, not without dissenting voices).

19 On 'liturgical union' (*unione liturgica*), see Labita, 'Il Milite Ignoto,' 134. For Marinetti see 'Guerra sola igiene del mondo,' in Filippo Tommaso Marinetti, *Teoria e invenzione futurista* (Milan: Mondadori, 1983). On Se-

meria's admiration for d'Annunzio as a leader, see Martinelli, *La guerra di d'Annunzio*, 153, 308. On it being a complaint about 'seguaci del corruttore d'Annunzio' (followers of that corruptor, d'Annunzio), see Miniero, *Da Versailles*, 210.

On Gemelli as the first and most stringent theoretician of the need for training in depersonalization to fight in the war, a training Italy did not concern itself with until as late as 1917, see Procacci, *Soldati e prigionieri italiani*, 60. Semeria's originals read: 'una glorificazione di guerrieri o l'apoteosi della guerra'; 'un flagello'; 'brutta e triste'; 'glorificheremo il soldato. [...] attraverso [...] l'anonimo gli onori vanno alla dignità non alla persona. [...] Lo glorificheremo perché è l'uomo del dovere [...] l'uomo del sacrifizio. Va ad uccidere potete dire anche del carnefice; va a morire non lo si dice che del soldato'; 'i superuomini nietzsciani [che] hanno predicato il culto della violenza. [...] patrioti molto troppo accesi [che] hanno trovato bello un male necessario.'

20 For the mixture of 'noble,' 'uplifting,' and 'tragic' see Winter, *Sites*, 82, 85. On the 'der des der,' see Capdevila and Voldman, *Nos morts*, 33. On fear of agitation in Italy, but a generally peaceful outcome, see MdI, OaSI, 1921.

21 Gentile, *Il culto del littorio*. Semeria's originals read: 'Coraggio, dovere, sacrifizio, sono tutte grandi parole cristiane, grandi cristiane realtà'; 'glorificazione'; 'apoteosi'; 'sa di pagano'; 'proprio e solo per amor di patria'; 'tornare al rito sacro'; 'Noi da Dio siamo condotti alla patria, non dalla patria a Dio e, peggio, nella patria a Dio. I nostri morti per noi sono eroi, non sono *dei* o qualcosa di simile'; 'salire [...] all'ara della patria per rinnovare nella Roma di Cristo lo spettacolo della Roma dei Cesari.' Douhet's assertion was quoted in Chapter 1; it is found in Douhet, *Le Profezie di Cassandra*, 427. The original reads: 'l'apoteosi del soldato non significa affatto l'apoteosi della guerra.'

22 The postcard is found in MdI, OaSI, 1921. The medals are in *L'illustrazione Italiana*, 6 Nov. 1921, illustrating Otello Cavara's 'Apoteosi ad Aquileia.'

23 Kantorowicz, 'Pro patria mori,' 308, 309.

24 Ibid., 312–16.

25 Miniero, *Da Versailles*, 124–5. The originals read: 'tono mistico' and 'religione della patria.' *Corriere della Sera*, 31 Oct. 1920. The original reads:

> La grande guerra ha agitato nomi e figure: eroi veri e men veri hanno luce nella memoria. Ma c'è l'Eroe senza nome e senza volto, che passò per la vie del destino e s'immerse nell'ombra, [...] stese le braccia sulla terra come se la terra fosse la sua croce [...] al Grappa segnò col suo corpo il confine sacro tra la disperazione e la speranza, [...] piegò le gi-

> nocchia sulle zolle del Friuli riconsacrato cedendo, morituro, alla vittoria la via che avanzava verso Trieste e verso Trento. Fu la volontà d'amore il suo coraggio? O fu la disciplina soltanto [...]? Non importa. [...] I nomi, i volti? Nulla, nulla di così preciso e piccolo. [...] all'oscurissimo [...] ogni particolare abolito nella incandescenza solare della gloria – la tomba in Roma, nell'Altare della Patria.

On resurrection *of* the Patria, see Mosse, *Fallen Soldiers*, 79. Douhet, *Le Profezie di Cassandra*, 434. The Vittoriano was conceived as a 'Patriotic altar' before the Unknown Soldier was buried there, but the Christological significance of the altar, and the ambiguity as to whether the altar was dedicated to the Patria, or whether the Patria was sacrificed on it, appropriating the body of the soldier, only crystallized with his burial (Pizzo, *Il Vittoriano*, 14).

On barbed wire, and recent archaeology of the trenches, see Nicholas J. Saunders, 'Material Culture and Conflict: The Great War, 1914–2003,' in his *Matters of Conflict: Material, Memory, and the First World War* (London; New York: Routledge, 2004).

26 For 'santo volto,' see *Il Dovere*, 5 Nov. 1919. For 'voce del poeta' and quotes by d'Annunzio, such as 'non piegare d'un ugna,' see the first issue, 27 April 1919. For the cartoon, and the electoral lists proposal, see the first issue as well.

27 Kantorowicz, 'Pro patria mori,' 318, 321–2.

28 Ibid., 322–4.

29 D'Annunzio, 'Comento meditato,' 539. Notably this passage is included as pages 376–7 in *Il fastello della mirra*, the selection of autobiographical prose planned for 1926, but it is not included in the sections evoking the war, rather it is moved to much later, and presented as the path towards the future of Italy; thus it is in a later section titled 'Verso l'Impero d'Italia,' and in it d'Annunzio portrays himself as an unheard and isolated prophet, making no mention of Mussolini, positive or negative. The original reads:

> Io voglio aiutare i morti a esprimersi. Anche se io debba a uno a uno prenderli in me e con la virtù del mio amore rigenerarli e inspirarli, io voglio ch'essi apprendano a esprimersi. Voglio che il soldato ignoto si esprima e ch'egli sforzi il marmo del suo sepolcro e scoperchi il suo monumento e deluda i portatori di aròmati vani.
>
> Per quanti secoli l'Italia non aveva più potuto esprimersi come nazione intiera e unanime ma soltanto in alcuna voce e in alcun segno?
>
> S'era nazionalmente espressa nel Rinascimento? S'era nazionalmente espressa nel Risorgimento?
>
> Oggi io rinserro nei musei tutti i suoi capolavori per vivificare e per

ricelebrare l'incomparabile capolavoro della sua espressione: la sua guerra.

La sua guerra non è se non uno sforzo d'espressione, quale non fu veduto mai in moltitudine d'uomini armati o disarmati.

Per ciò metto il Buonarroto ai massi del Carso, metto l'Alighiero nelle bolge del Carso. [...]

Disperatamente, e speranzosamente, dall'ombra di un argine del Piave, ci fu qualcuno che gridò: 'Che c'è da gettare alla grande fiamma? Getteremo tutto, se è necessario, anche le tavole più sacre.'

30 On the law in Foucault, see Colin Gordon, 'Introduction,' in Michel Foucault, *The Essential Foucault: Selections from Essential Works of Foucault, 1954–1984,* trans. Nikolas S. Rose, ed. Paul Rabinow and Nikolas S. Rose (New York: New Press, 2003), xxx–xxxi. In Matthew 28:1 there is an earthquake and an angel is said to roll away the stone of Christ's sepulchre so that Mary can see he is already gone. Different versions without the angel nonetheless emphasize that the risen Christ can appear and disappear, or become more or less material; see Mark 16:1, Luke 25, John 20:12. D'Annunzio's image of the peasant-soldier ploughing stones and fertilizing them with his blood is from his 'Comento meditato,' 526–9.

31 De Felice, 'D'Annunzio e la vita politica italiana.' In 'D'Annunzio nella vita politica italiana,' *Il Veltro* 32/5–6 (1988): 470. De Felice also points out that the only improvised aspect of the takeover of Fiume was d'Annunzio's leadership.

See Chapter 1 on d'Annunzio's fall, as well as De Felice. For a detailed examination of Mussolini's careful political manoeuvring of d'Annunzio, in particular, through Giuseppe Giulietti, responsible for the Federazione Italiana Lavoratori di Mare, see Guglielmo Salotti, 'I rapporti d'Annunzio-Giulietti dall' "idillio" fiumano al "pactum sine nomine,"' in Renzo De Felice, ed., *D'Annunzio politico, 1918–1938* (Rome and Bari: Laterza, 1978), 23–35. On d'Annunzio and Mussolini, see also Thomas E. Peterson, 'Schismogenesis and National Character: The D'Annunzio-Mussolini Correspondence,' *Italica* 81/1 (2004): 44–64.

For Hans Ulrich Gumbrecht, d'Annunzio's indecisiveness is inseparable from his belief in a future redemption that war sacrifice had already secured; see his 'I rendentori della vittoria: On Fiume's Place in the Genealogy of Fascism,' *Journal of Contemporary History* 31 (1996): 253–72.

32 Comments by Bruers are from d'Annunzio, *Tre redazioni,* 16. He is cited also in Chapter 2. On d'Annunzio's 'quest for a moment of pure authenticity,' different from, but related to Fascism's call for permanent revolution, see

Lewis, *Modernism, Nationalism, and the Novel*, 176. Gumbrecht, in 'I rendentori della vittoria,' at 257, 267, and 269, argues that both d'Annunzio at Fiume and the Fascist regime in the 1930s found themselves caught in a rhetoric of redemption that puts a possible redeemed state in the future and thus requires more and more acts of redemption to perpetuate itelf.

On the Fascist imagination, including the importance of blood and soil, see the following: Schnapp, *Staging Fascism*; Ruth Ben-Ghiat, *Fascist Modernities: Italy, 1922–1945* (Berkeley: University of California Press, 2001); Fogu, *The Historic Imaginary*; and by Emilio Gentile: *Le origini dell'ideologia fascista (1918–1925)* ([Rome, Bari]: Laterza, 1975), translated by Robert L. Miller, as *The Origins of Fascist Ideology, 1918–1925* (New York: Enigma, 2005), and *La grande Italia: Il mito della nazione nel XX secolo* (Rome: GLF editori Laterza, 2006). Jeffrey T. Schnapp, in *A Primer of Italian Fascism* (Lincoln: University of Nebraska Press, 2000), summarizes: 'the fascist espousal of fluidity bespeaks an essentially vitalistic concept of thought that stands in opposition to Enlightenment ideals. It repudiates all rationalistic models of behavior and politics, be they liberal-democratic or socialist, as utopistic, abstract, universalistic, out of touch' (x).

33 On Actual Idealism, see Giovanni Gentile, *Teoria generale dello spirito come atto puro* (Bari: Laterza, 1920). Gentile was to author the 'Manifesto of Fascist Intellectuals' in 1925. On Mussolini and the Unknown Soldier's body, see Pozzi, 'Il Duce e il Milite ignoto.'

34 Letter of 16 Nov. 1918 to Olga Levi, in Andreoli, *Il vivere inimitabile*, 559. The original reads: 'Essere oggi un "superstite" è per me la più grave disgrazia che potesse accadere al mio corpo e al mio spirito. [...] E poi la vittoria crudele, che recide tante cose vive e lascia me mal vivo.'

On taking one, d'Annunzio, *Notturno*, 26. The original reads: 'la morte che doveva prendere i due, ne prese uno, uno solo, contro il patto, contro l'offerta.' This is echoed in *Il Dovere*, 27 April 1919, the inaugural issue. Fiume is mentioned as a 'stato d'immolazione' (state of immolation) in d'Annunzio, *Di me a me stesso*, 136.

35 Leed, *No Man's Land*, 211.

36 Gabriele d'Annunzio, 'Messaggio del convalescente agli uomini di pena,' *Prose*, 1922, *Il libro ascetico*, 573, 576, 578, 579, 580, 581. The original reads:

> Uomini di pena, lavoratori, compagni, avete il viso pallido; e io anche ho il viso pallido, forse più di voi. Visitate l'infermo, confortate il convalescente. Tutto il sangue si raccoglie intorno al cuore fraterno?
>
> Così sia. [...]
>
> Certo avevamo dietro di noi tutti i nostri morti, [...] ma avevamo davanti a noi i nascituri, più numerosi degli uccisi. [...]

> Nei nostri corpi miseri, nelle nostre anime umili, abitano e operano le forze eterne. [...]
>
> Ho patito la più trista delle prove. Sono stato precipitato dalla rupe tarpea, [...] Ho sentito l'osso [del cranio] fendersi [...]
>
> Ho sentito passare su di me la sciagura di Caporetto; io l'ho sentita su tutto me, come se essa fosse per calpestare per infangare per sotterrare me soltanto. [...]
>
> E vero, uomini di pena, operai, marinai, contadini, è vero quel che dai Padri veggenti è sentenziato: non esservi se non un peccato mortale, un solo; [...]
>
> E mi chiedete quale? Venire a disperazione, venire a disperazione in sé medesimi, come direbbe Gregorio Magno.
>
> Di questo peccato conviene che io m'incolpi, nella mia non prona umiltà. [...]
>
> E da quella volontà di morte scampai [...]
>
> Sanguinante a piè della rupe tarpea, non ho disperato. Non dispero.

37 On the Tarpeian rock, see *Tarpeian Rock – Wikipedia,* retrieved from http://en.wikipedia.org/wiki/Tarpeian_Rock.

38 Giorgio Agamben, *Homo sacer: Il potere sovrano e la nuda vita* (Turin: Einaudi, 1995). 'Defenestrazione dalla rupe Tarpea' as a 'rito[o] di purificazione' rather than a 'pen[a] di morte in senso moderno' is a core example that Agamben gives for the constitution of 'vita sacra' (91). On 'vita sacra' as 'elemento politico originario' (109, 98). Translations are from Giorgio Agamben, *Homo Sacer: Sovereign Power and Bare Life,* trans. Daniel Heller-Roazen (Stanford: Stanford University Press, 1998), 81, 98, 89. 'Between categories' is from Leed, *No Man's Land,* 18.

The 'offensive à outrance' (attack to excess) attributed to Saint-Just, Foch, and in the First World War, Joffre, sees massive engagement of men, used as though totally expendable, as the way to win a battle.

39 Agamben, *Homo sacer,* 91, and in the translated volume at 82. The original reads:

> la *sacratio* configura una doppia esclusione, tanto dallo *ius humanum* quanto dallo *ius divinum,* tanto dall'ambito religioso che da quello profano. La struttura topologica, che questa doppia eccezione disegna, è quella di una duplice esclusione e di una duplice cattura, che presenta più di una analogia con la struttura dell'eccezione sovrana. [...] Come, infatti, nell'eccezione sovrana, la legge si applica al caso eccezionale disapplicandosi, ritirandosi da esso, così l'*homo sacer* appartiene a Dio nella forma dell'insacrificabilità ed è incluso nella comunità nella

forma dell'uccidibilità. *La vita insacrificabile e, tuttavia, uccidibile, è la vita sacra.*

On the state of exception and its role in the French Revolution, with an argument not unrelated to Agamben's view that 'exception' is a covert application of law rather than its suspension, see Dan Edelstein, *The Terror of Natural Right: Republicanism, the Cult of Nature, and the French Revolution* (Chicago: University of Chicago Press, 2009).

40 Harrison, *Dominion*, 147–8, 144, 145. Philippe Ariès, *Essais sur l'histoire de la mort en Occident du Moyen-Age à nos jours* (Paris: Seuil, 1975), 223–37.

For Gorman, the purpose of the Unknown Soldier Memorial was to prolong grieving, so that the dead could explain the war to the living ('The Return of the Dead,' 72).

41 On visibility, see Ministero dei lavori pubblici, Direzione generale, edilizia: 'Sepoltura del Soldato Ignoto sull'Altare della Patria' (1921–26), divisione 5, busta 39, fascicolo 107, Archivio Centrale dello Stato, Rome (hereafter MLP, SdSI, 1921–26). On the men not chosen to be the Unknown Soldier, and on him as reminder of all those not even retrieved from the battlefield, see *Il secolo illustrato*, 30 Oct. 1921. The original reads, 'le donne di Aquileia sembravano ancor più invase di pietà di fronte ai dieci esclusi dall'apoteosi di Roma.'

42 *L'illustrazione Italiana*, 13 Nov. 1921. Overall, the article by Otello Cavara insists on how absence, of eloquent figures, of speeches, of music is an appropriate response to a suffering that nothing can fully express.

43 D'Annunzio, 'Agli uomini di pena,' 583. The original reads: 'eroi tuttavia imprigionati nel silenzio come quelli del Buonarroto nello scarpellato sasso.'

44 On the 'grande mole michelangiolesca,' see Miniero, *Da Versailles*, 212.

45 D'Annunzio, *Notturno*, 303. The original reads: '[il] disegno sublime [di] Michelangelo [...] Il Cristo titanico, avendo sforzato il pesante coperchio del sepolcro, tiene tuttora un piede nel sasso cavo. Ma con il capo levato, con levate le braccia, con l'impeto e con la rapina di tutta la sua passione, si scaglia verso il cielo.'

The model here is clearly Michelangelo's *non finito* and specifically his two Pietà. Of course, the *Pietà Rondanini* is unfinished, perhaps defaced by Michelangelo himself; but even more like the present of 'Resurgo' is the *Saint Peter Pietà*: for it bears Michelangelo's signature, but instead of the traditional past perfect 'me fecit' it says in the imperfect 'Michel Angelus Bonarrotus Florent faceb[at]' and indeed the end of the 'facebat' is missing, making it even more incomplete or imperfect. Also, this Pietà is explicitly about sacrifice and resurrection, since Mary is presented like an altar, clothed in a single piece of fabric like an altar cloth, upon which the Host is being sacrificed.

D'Annunzio's inscription, and this image, recalls a possibile inscription that was rejected by the British for their soldier: 'This is my body broken for you.' For Laqueur ('Names, Bodies') it was replaced with 'In Christ shall all be made alive' because its 'sacralization would have been perhaps too literal' (135). But since the Christian reference is just as explicit, I would surmise that it was rejected because it pauses too severely on the Passion as death, breaking, and impenetrable mystery rather than on resurrection and intelligibility.

On the *non finito,* see Jean-Pierre Barricelli, 'Michelangelo's Finito: In the Self, the Later Sonnets, and the Last Pietà,' *New Literary History* 24/3 (1993): 597–616. More broadly, on the difference between a future that is genuinely open and one whose contingency we can objectify, in effect, creating a double negative 'contingency of contingency,' see Hans Ulrich Gumbrecht, 'How Is Our Future Contingent? Reading Luhmann against Luhmann,' *Theory Culture Society* 18/1 (2001): 56–7.

46 Leed, *No Man's Land,* 21–2, 32–3.

47 Agamben, *Homo sacer,* 111, 69, 210, 211, and in the translation at 83, 188. The originals read: 'l'essere non è che il bando dell'ente,' 'vita che è la sua forma e resta inseparabile da essa.'

Introduction to Part II: The Undead Body, the Photographic Image, and the Religious Icon

1 For avoiding the humiliation of the mass grave, and practices of side-by-side burial, see Capdevila and Voldman, *Nos morts,* 176–7. A rare case of showing exhumations in all their gore, albeit years later, is in Henri Storck's 1932 Belgian film, *Histoire du poilu inconnu.*

2 Scarry, *The Body in Pain,* 162, 168, 166, 168, 180.

3 Harrison, *Dominion,* 92. Vovelle (*La mort et l'Occident,* 650) persuasively argues that both the creation of stone *caveaux* and other ways to preserve bodies, and the popularity of cremation, were part of the same strategy to deny decay in the nineteenth century.

Though the analyses of Ariès and Vovelle are methodologically different, in particular, as the former is more concerned with long-term trends and overall structures, whereas the latter focuses more on microhistory, they both agree on the general eclipsing of dead bodies from modern Western culture, and Vovelle in particular, reads late nineteenth-century fascination with decay as a sort of of 'return of the repressed.' For a comparison of their work see Roger Chartier, 'Histoire de la mort et histoire culturelle,' in Roger Chartier, ed., *La Mort aujourd'hui* (Marseille: Rivages, 1982), 111–24. For his comments on Vovelle's work, see Philippe Ariès, 'Du livre de Michel Vovelle "La Mort en Occident,"' in ibid., 157–68.

For the denial of decay as part of the emergence of new notions of sexuality and the body, in correlation with modern social and political discourse, see Catherine Gallagher and Thomas Walter Laqueur, *The Making of the Modern Body: Sexuality and Society in the Nineteenth Century* (Berkeley: University of California Press, 1987). The story of dual sexual opposition as culturally significant, hence, the link between the body, decay, mortality, and the female, begins with necrophilia in Thomas W. Laqueur, *Making Sex: Body and Gender from the Greeks to Freud* (Cambridge: Harvard University Press, 1990), 1–24.

4 Leed, *No Man's Land*, 18, 21–2.

5 'Distributive' is in Laqueur, 'Names, Bodies,' 126. For the experience of presence, positive or negative, as something abstraction and language cannot fully contain, see Hans Ulrich Gumbrecht, *The Production of Presence: What Meaning Cannot Convey* (Stanford: Stanford University Press, 2004).

6 On Comerio, see Carla Manenti, Nicolas Monti, and Giorgio Nicodemi, *Luca Comerio: Fotografo e cineasta* (Milan: Electa, 1979). Comerio's footage was not the mainstream fare for the public, which was treated rather to the reassuring populist vision of war found in films such as the 1916 *Maciste alpino*; see Mario Isnenghi, *Le guerre degli italiani: Parole, immagini, ricordi 1848–1945* (Bologna: Il Mulino, 2005), 144ff.

For mothers reacting to Comerio's footage, see Yervant Gianikian and Angela Ricci-Lucchi, *Futurism at 100: The Measure of a Centenary, 1909–2009*, presented at Harvard University, 17–18 Oct. 2009.

I do not mean to say that cinema and literature work the same way as regards the uncanny realness of the image, merely that as regards First World War literature, cinema highlights a more general break in previously normal patterns of representation. For the uniqueness of the photographic image, a classic discussion is André Bazin, 'The Ontology of the Photographic Image,' *Film Quarterly* 13/4 (1960): 4–9. Bazin sees in photography 'an hallucination that is also a fact' (9) – a very suggestive formulation as regards photography's ability to reawaken trauma. For photography in the First World War, and its role in inaugurating the beginning of censorship, see Capdevila and Voldman, *Nos morts*, 34–42.

7 Hans Ulrich Gumbrecht, *In Praise of Athletic Beauty* (Cambridge: Belknap Press of Harvard University Press, 2006), 52, 46–7, 47.

E.A. Grosz, *Volatile Bodies: Towards a Corporeal Feminism* (Bloomington: Indiana University Press, 1994). Grosz makes a similar contrast in comparing non-Western and Western understandings of tattoos and bodily markings: 'What differentiates savage from civilized systems of inscription is the sign-ladenness of the latter, the creation of bodies as sign systems, texts, narratives, rendered meaningful and integrated into forms capable of being read in terms of personality, psychology, or submerged subjectivity. Ours, we believe, is not a superficial identity but an enigma, a mystery to be uncovered, a secret to be explored through a reduction of the body to a symptom of the self' (141).

8 Gianikian and Ricci-Lucchi are mentioned in Martin Lefebvre, *Landscape and Film* (London and New York: Routledge, 2006), 250. They are interviewed on their film *From the Pole to the Equator* in Scott MacDonald, *A Critical Cinema* (Berkeley: University of California Press, 1998), 274ff. Their films' zoom into the most violent or abject details of Comerio's footage has a double valence reminiscent of the desire to 'go to the limit' in seeking embodied experiences that Hans Ulrich Gumbrecht, in *Making Sense in Life and Literature* (Minneapolis: University of Minnesota Press, 1992), finds in modern intellectuals, which he reads as a sign of 'how difficult (impossible) it has become for them to be in their bodies at all' (287).

9 The Veil of the Veronica has a complex history, which appears to begin in the Middle Ages; it is mentioned in Dante's *Paradiso*. It is associated with the Mandylion of the Eastern Church, and also known in Italy as 'il volto santo,' which is echoed in Douhet's newspaper's claim about Fiume as the location of the 'santo volto' (see discussion of Fiume in Chapter 5 for more details). Folk etymology traces the origin of the name Veronica in 'vera' 'eikon,' i.e., true image. See the following: Ewa Kuryluk, *Veronica and Her Cloth: History, Symbolism, and Structure of a 'True' Image* (Cambridge and Oxford: Blackwell, 1991); Frank Knight Lord, *Image Vision and Faith: Viewers' Responses to the Mandylion, Veronica's Veil and the Shroud of Turin* (PhD thesis, University of North Carolina and Chapel Hill, Art Department, 2003); Bissera V. Pentcheva, *Icons and Power: The Mother of God in Byzantium* (University Park: Pennsylvania State University Press, 2006).

On the veil at the end of Dante's *Vita Nuova*, where it announces a new mode of representation and prefigures the *Commedia*, see Robert Pogue Harrison, *The Body of Beatrice* (Baltimore: Johns Hopkins University Press, 1988), 127ff.

Michel Tournier, 'Les suaires de Véronique,' in *Le Coq de bruyère* (Paris: Gallimard, 1978), 170, 171, 166, 160, 158, 159. The *écorché* is an anatomical drawing in which the skin has been peeled away to reveal the musculature.

At the opposite extreme, aesthetically, we have the *Veiled Christ* scultpure by Giuseppe Sanmartino in the Sansevero Chapel in Naples: a deposition in which drapery accentuates the fragility and preciousness of the human body. On 'Les suaires' and the aura, see Susanne Sara Thomas, 'Petrarch, Tournier, Photography, and Fetishism: The Veil in the Rime Sparse,' *Romance Quarterly* 43 (1996): 131–42.

10 Tournier's analysis of consciousness is also a response to Sartre. Michel Tournier, *Vendredi ou les limbes du Pacifique* (Paris: Gallimard, 1967), 224, 220, 96, 97. The originals read: 'qui n'existe que de façon intermittente'; 'connaissance des choses *par autrui*'; 'objets phosophorescents par eux-mêmes, sans rien d'extérieur qui les éclaire'; 'la conscience que j'ai d'un objet est cet objet même, l'objet est connu, senti, etc., sans personne qui connaisse, sente, etc.'

On mystical consciousness and Buddhist awareness, see James, *Varieties of Religious Experience*, and Eleanor Rosch, 'How to Catch James' Mystic Germ: Religious Experience, Buddhist Meditation, and Psychology,' *Journal of Consciousness Studies* 9/9–10 (2002): 37–56.

Francisco J. Varela, Evan Thompson, and Eleanor Rosch, *The Embodied Mind: Cognitive Science and Human Experience* (Cambridge: MIT Press, 1991). It has been demonstrated that 'the actual source of nihilism is objectivism. We have already discussed how the basis of objectivism is to be found in our habitual tendency to grasp after regularities that are stable but ungrounded. In fact, nihilism too arises from this grasping mind [...] the absence of a solid ground is immediately reified into the objectivist abyss' (240).

Jean-Pierre Dupuy, *The Mechanization of the Mind: On the Origins of Cognitive Science* (Princeton: Princeton University Press, 2000). Dupuy warns against such nihilism when he concludes that cybernetics failed to capitalize on the following insight, implicitly giving in to a reification of the empty subject: 'The attributes of subjectivity are emergent effects produced by the spontaneous, self-organized functioning of a complex organization in the form of a network' (160). See also Gumbrecht, *Athletic Beauty*, 169, 171. However, in *The Production of Presence*, Gumbrecht also warns against excessively facile claims to non-Western, and in particular, Zen wisdom (149).

11 Tournier, *Vendredi*, 97. The original reads: 'un déclic se produit. Le sujet s'arrache à l'objet [...] quelque chose a craqué dans le monde [...] [qui] se meurt sous mon regard sceptique. [...] Un objet a brusquement été dégradé en sujet.' In *The Production of Presence*, at 142 and 148, Gumbrecht similarly questions the '*exclusive* dominance of the Cartesian worldview' and, like Tournier, refuses to equate his emphasis on 'presence' (an experience he describes in terms not unlike the 'radiant' objects imagined by Tournier)

with the '*theological*,' arguing that reference to it is not necessary to understanding 'Being' as 'the things of the world devoid of any conceptual grid.'

12 On the ritual use of icons, see Pentcheva, *Icons and Power*. Tournier, 'Les suaires,' 171. The original reads: 'impression [...] étouffante.'

In *Sexes and Genealogies*, at 36, Irigaray uses the image of the 'veil' in a way that is close to Tournier's return to veiling rather than unveiling when she writes that the return to the maternal womb (and thus also the placental veil) cannot be grasped or reified, but that the veil somehow is reconstituted in 'the sympathy between two bodies capable of mututally decoding one another,' in the encounter between two selves that it at once unites and separates, for which it is a medium. This sort of non-specular relation to the other, which is of confusion, fluidity, but not complete absorption, is also described in Irigaray's essay 'La Mystérique,' in *Speculum of the Other Woman*, at 97, where she associates it with a feminimized Christ; this is similar to the eperience evoked by the Veil of the Veronica, especially given the confusion between the woman (the Veronica) and her veil (which bears the face of Christ). Robert Harrison, in *The Body of Beatrice*, at 127ff., points to a similar feminization of Christ at the end of Dante's *Commedia*, when the Veil of the Veronica is also invoked.

13 For an understanding of death seen as a journey, see Stanislav Grof, *The Ultimate Journey: Consciousness and the Mystery of Death* (San Francisco: MAPS, 2006). On healing veterans' trauma by helping them with this journey, specifially through storytelling, see Tick (*War and the Soul*) and Paulson (*Haunted by Combat*).

For the notion that any representation of a traumatic past is more 'true' to its subject when it incorporates the blockages and displacements of memory, see Shoshana Felman's introduction of Claude Lanzmann's comments on his film, *Shoah*, in Lanzmann and Felman, 'The Obscenity of Understanding,' 204. For the inevitable doubleness that this occasions in any representation of trauma, see Cathy Caruth, *Unclaimed Experience: Trauma, Narrative, and History* (Baltimore: Johns Hopkins University Press, 1996), 7. From the point of view of cultural history, from a global perspective, this also implies that notions of 'distinction' and 'cultural capital' are now part of multiple, intertwined stories, as argued by Palumbo-Liu in his introduction to David Palumbo-Liu and Hans Ulrich Gumbrecht, *Streams of Cultural Capital: Transnational Cultural Studies* (Stanford: Stanford University Press, 1997).

14 On memory bridge, see Saunders, *Matters of Conflict*, 15.

15 On the marginalization of those who are associated with suffering and death, Amy Hollywood writes:

> We are left, then, with the problem of how to acknowledge trauma and loss and allow for mourning and recognition of its bodily effects without forcing women and other oppressed people to bear the weight of this work through their symbolic association with the mortal body, and without succumbing to a valorization of trauma as the sole site of the real. [...] Although our experiences of loss, limitation, and death are always embedded in the social, and hence in the ethical and the political, they are not reducible to these realms. [...] we may also need sites for more fully embodied practices. (*Sensible Ecstasy*, 278)

16 On the mourning in 2003, see Maria Zegarelli, 'Duecentomila per i caduti di Nassiriya: Sfilano, piangono, si chiedono perché,' *L'Unità*, 18 Nov. 2003; 'Il caso di monsignor Nogaro,' *Repubblica*, 18 Nov. 2003. The originals read: 'li chiamiamo eroi ma sono martiri'; 'non esaltare il culto dei martiri e degli eroi.' Tick, *War and the Soul*, 61, 259, 211, 223, 244. Hollywood, *Sensible Ecstasy*, 278. Kertzer (*Ritual, Politics, and Power*), concludes on a similar question when he considers John Kenneth Galbraith's view of the disappearance of ritual in modern political life, pointing out as well that while ritual allowed Fascism to rise and rule in Italy, it was also ritual that preserved 'an alternative basis of social solidarity' (183). Vattimo, 'Ritrovare le parole,' *L'Unità*, 20 Nov. 2003; Red, 'Un lungo, interminabile applause,' *L'Unità*, 18 Nov. 2003; Red, 'Nogaro: 'Dico che il terrorismo,' *L'Unità*, 18 Nov. 2003; 'Il caso di monsignor Nogaro,' *L'Unità*, 18 Nov. 2003. For related views on the 'return' of the religious, see Richard Rorty, Gianni Vattimo, and Santiago Zabala, *The Future of Religion* (New York: Columbia University Press, 2005).

The originals read: 'smettere il culto dei martiri'; 'Frontegeremo gli assassini. Ma senza odio'; 'il terrorismo non si combatte con le armi.' Tick, *War and the Soul*, 261.

4 Embodiment and *Imbestiamento*

1 D'Annunzio, *Notturno*, 88–9. The original reads:

> Ecco che la battaglia lontana della Mosa entra nel mio incendio. I battaglioni ubriachi di etere vengono innanzi come quelle zone di pineta ardenti, chiamate 'controfuoco' nella mia Landa d'esilio, quasi mandre di fiamme animali [...]. Li vedo attraverso i pali e gli spini dei reticolati. [...] I carnai diventano roghi. Non si consumano, non s'inceneriscono. Bruciano a lungo senza fiammeggiare, come la torba.
>
> Rimango tutta la notte disteso contro un reticolato che sbarra la collina. Conto i cadaveri.

S'impigliano nei roveti di ferro, si serrano negli attorcigliamenti di fili rotti, penzolano tra palo e palo come i ladroni male inchiodati alle croci, si torcono come le bestie incappate nei lacci.

Non hanno palpebre, non hanno labbra. Vedi gli occhi fissi e nudi; vedi i denti fissi e nudi.

Vedo il sangue colare giù pel legno e pel ferro, aggrumarsi, annerarsi, viscoso come la pania che impiastra le verghe.

Non c'è più rugiada, non c'è più alba sul mondo.

2 D'Annunzio, *Notturno*, 66–7, 98–9. The original reads:

> Entriamo nella bara oscura. [...]
>
> D'improvviso, ho paura. Qualcuno è con me. [...]
>
> Mentre scrivo nel buio, il pensiero mi si rompe e la mano si arresta. [...]
>
> Ho un brivido di spavento. E rimango immobile, con tutto il corpo rigido, non osando più tracciare un segno nelle tenebre.
>
> [...] Poi sole esistono le mie ossa, solo esiste il mio scheletro fasciato di carne. E nello scheletro è come una coagulazione improvvisa della vita.
>
> La vita s'aggruma, s'accaglia come il sangue che non scorre più. E' un orribile peso. [...]
>
> Ogni moto della vita interna è abolito. Ho interamente perduta la forza di muovere e di rimuovere le grandi masse incoerenti di sostanza lirica ond'è formata la mia malinconia. [...]
>
> Il mio patimento è vile e senza potere. [...] Prima di sprofondare nell'orrore delle trasformazioni, sento la mia bocca divenire di metallo nel respiro rallentato del sonno.

3 D'Annunzio, *Notturno*, 27. The original reads: 'due 'dentro ad un fuoco'; ma il fuoco non è diviso.' This is a quote from Dante's *Inferno* XXVI, where Ulysses and Diomedes are said to be as one flame divided into two.

4 Leys, *Trauma*, 3–4, 18–19, 83–119.

5 Laura S. Brown, 'Not Outside the Range: One Feminist Perspective on Psychic Trauma,' in Cathy Caruth, ed., *Trauma: Explorations in Memory* (Baltimore: Johns Hopkins University Press, 1995), 100–12. Brown's argument that something 'outside the range' is dangerously also beyond the law, and thus deprived of rights that can be debated in a democracy is reminiscent of Dan Edelstein's comments on the Terror making an exception to the law, and rendering 'death inevitable,' for the individual who is 'an enemy of the human race.' 'War and Terror: The Law of Nations from Grotius to the French Revolution,' *French Historical Studies* 31/2 (2007): 259. *DSM-IV-TR*,

retrieved from http://www.ncptsd.va.gov/ncmain/ncdocs/fact_shts/fs_dsm_iv_tr.html. *DSM-III-TR*, Available: http://www.cirp.org/library/psych/ptsd/.

Kirkwood is in Paulson, *Haunted by Combat*, xv–xxi.

6 I borrow the term *imbestiamento* from Bonadeo, *Mark of the Beast*. In this book he documents how Italian and British writers who fought in the First World War overwhelmingly come back to images of *imbestiamento*, which means 'bestialization' but also, as I use the term, 'animality' (arguably less monstrous, less dualistic).

Robert Pogue Harrison, *Forests: The Shadow of Civilization* (Chicago: University of Chicago, 1992). Harrison submits that 'the forest remains a margin of exteriority with respect to civilization. [...] Because we exist first and foremost *outside ourselves*, forests become something like an ancient and enduring correlate of our transcendence' (201). For Harrison, dwelling is the relationship between this outside and the human. In the case of First World War *imbestiamento*, what we see is a breakdown of this relationship, as humans dwell in a new, human-made, yet still 'outside' location, the trenches; the journey home of the Unknown Soldier can thus be seen as an attempt to re-establish dwelling in the sense defined by Harrison, of properly separating the outside and the inside once again.

7 Brown, 'Not Outside the Range,' 109.

8 On modernity, human violence, and the capacity for self-destruction, see Michel Serres, *Hominescence: Essais* (Paris: Pommier, 2001).

9 See Ozouf, 'Le Panthéon,' where she notes, at 174: 'Alors que le culte des grands hommes, si conformisant, est antinomique à toute rupture, le Panthéon, conçu pour la mise en scène quasi religieuse du rassemblement national, est le lieu même de la rupture entre les Français: sur lui ne parvient pas à s'effacer la marque originelle de la révolution Française' (Even as the cult of great men, so demanding of conformity, is antithetical to all rupture, the Pantheon, created for the almost religious mise en scène of national unity, is the place of rupture among the French: upon it the originary mark of the French Revolution cannot be erased).

On the awkward association of Gambetta with the Unknown Soldier, see Inglis, 'Entombing Unknown Soldiers,' 24.

Strenski observes, in *Contesting Sacrifice*, at 96, that Dreyfus himself was divisive in a particularly dangerous way, for he could represent opposite notions of sacrifice, being seen by different parties as innocent victim or traitorous agent.

On the Italian Pantheon as too 'partisan' because of associations with the monarchy, and the refusal to bury Garibaldi there, see Labita, 'Il Milite

Ignoto,' 129. Miniero (*Da Versailles,* 124) adds that the nationalists and the Fascists liked the idea of the Pantheon. This was presumably because they associated it with revolutionary patriotism, i.e., they saw the Pantheon transformng the Unknown Soldier into a hero, in contrast with Douhet, who saw the Unknown Soldier transforming the Pantheon into a more democratic monument.

For the link between Garibaldi and the Unknown Soldier, as both are 'revolutionary,' though not necessarily in the same way later intended by Fascism, see Claudio Fogu, 'Fascism and Historic Representation: The 1932 Garibaldian Celebrations,' *Journal of Contemporary History* 31/2 (1996): 328. This link, important to Douhet, was ultimately downplayed in the implementation and inauguration of the Memorial.

10 On the contested history of Sainte Geneviève and the Pantheon, see Richard Wittman, *Architecture, Print Culture, and the Public Sphere in Eighteenth-Century France* (New York: Routledge, 2007). Worry over disruptions is from MdI, OaSI, 1921. On the arditi and the Arditi del popolo, see the following: Staid, *Gli Arditi del popolo*; Francescangeli, *Gli Arditi del popolo*; Giorgio Rochat, *Gli arditi della Grande guerra: Origini, battaglie e miti* (Milan: Feltrinelli economica, 1981); Ferdinando Cordova, *Arditi e legionari dannunziani* ([Padua]: Marsilio, 1969).

For d'Annunzio as an inventor of a 'style' of contestation that both left and right borrowed, see Ledeen, *D'Annunzio,* 202. For an evaluation of the notion of 'fascist left' in the work of Giuseppe Parlato, see Danilo Breschi, 'Fascist Historiography,' *Telos* (2006): 32–6.

11 On debates about the Parisian Pantheon, and final burial on 28 Jan. 1921 under the Arc de Triomphe, see Jagielski, *Le soldat inconnu,* 69–83, 115–20.

12 The original reads: 'Ritorna Garibaldi, ricorda [...] e ammonisce.' For d'Annunzio as a 'new Garibaldi,' especially in his takeover of Fiume, see Pamela Ballinger, 'La Pentecoste italiana: Accogliere il verbo della fede italica; Lo sviluppo del rituale nazionalistico a Fiume, 1919–1921,' *Qualestoria,* no. 2 (2001): 127.

The new geography is described in Tobia, *L'altare della patria,* 66. After the Lateran Pacts the via della Conciliazione would connect the Vatican to the city once again, but as a satellite of the central Vittoriano. On its political implications, see Catherine Brice, *Il Vittoriano.* Also on the Vittoriano and Fascism, see David Atkinson and Denis Cosgrove, 'Urban Rhetoric and Embodied Identities: City, Nation, and Empire at the Vittorio Emanuele II Monument in Rome, 1870–1945,' *Annals of the Association of American Geographers* 88/1 (1998): 28–49.

On the refusal to bury Garibaldi in the Pantheon, see Labita, 'Il Milite

Ignoto,' 129. On associating Garibaldi and the Unknown Soldier, see Fogu, 'Fascism and Historic Representation,' 328. See also 'Garibaldi,' in Mario Isnenghi, *Il luoghi della memoria: Personaggi e date dell'Italia unità* (Bari: Laterza, 1997), 25–46.

David Kertzer, *Prisoner of the Vatican: The Popes' Secret Plot to Capure Rome from the New Italian State* (Boston: Houghton Mifflin, 2004), 290–5. According to Kertzer, Italian unification was undermined by an anticlericalism that popular Catholicism could not quite come to terms with; the antagonism represented by Garibaldi's statue marching from the Gianiculum towards the Vatican had to be undone by the Lateran Pacts, and it is eclipsed in contemporary Italy's return to the Church, and the right, as exemplified in John Paul II's visit to the Italian House and Senate with Prime Minister Berlusconi on 20 Sept. 2002.

On the law mentioning the 'emblemi delle diverse fasi del Risorgimento nazionale' along with 'vedova' and 'madre,' see AP, Senato, 1921, 10 Aug. 1921. On Sacconi's proposal for sculptures of 'great men,' and dislike of the equestrian statue, and on the subsequent shift in emphasis from 're galantuomo' to 'Patria' as the statue of Rome became more important, see Brice, *Il Vittoriano*, 277, 275.

On the existence of a 'civic religion' in liberal era Italy, different from the later religion of the nation that evolves into Fascism, see Catherine Brice, 'La religione civile in Italia: Piccoli e grandi rituali,' in Maurizio Ridolfi, ed., *Rituali civili* (Rome: Gangemi, 2006), 97–115. The association between Garibaldi and the Unknown Soldier, as an essentially subversive one, underplayed by the state, is taken up again in an ironic vein in the novel discussed in Chapter 6, Testa, *Io sono il milite ignoto.*

Notably, the location of the Unknown Soldier Memorial was imposed on the Artistic Committee of the Vittoriano, and its members initially were offended at not having been consulted; it brought a certain completion, then, but not to everyone's satisfaction. See AP, Senato, 1921, 10 Aug. 1921; MLP, SdSI, 1921–26.

13 The maquette is in MLP, SdSI, 1921–26.

14 The Timavo monument proposal is in *Il secolo illustrato*, vol. 4 (1919), 133. For the Trench of the Bayonets, see Winter, *Sites of Memory, Sites of Mourning*, 99–101. Another manifestation of a widespread rejection of abstraction is in the refusal of cremation, mentioned in Laqueur, 'Names, Bodies,' 135. For the notion that our longing for a primordial nature, while appealing, may be a denial of our own 'domesticatedness,' see Robert Pogue Harrison, 'Not Ideas about the Thing but the Thing Itself,' *New Literary History* 30/3 (1999): 670–1.

15 Laqueur ('Names, Bodies,' 131) notes that this focus on the landscape, seen in Churchill's proposal, but also in many others, reflected a fear of erasure.

For the d'Annunzian memorial, see *L'Ardente*, July 1921, 10, 12, 5. This text is based on his visit to a ruined church used as a camp hospital, with the main altar covered with disinfectants and the wounded soldiers like living depictions of the Stations of the Cross; this is published in d'Annunzio, *Taccuini*, 871. After the 1921 version in *L'Ardente*, it is republished as 'La preghiera di Doberdò,' in 1922, in *Il libro ascetico della giovane Italia*, 636–41; see d'Annunzio, 'Tre preghiere.'

For the comparison of the Carso to Golgotha, see d'Annunzio, 'Comento meditato,' 521. For blood quenching the Carso's thirst, see d'Annunzio, *Notturno*, 229.

For d'Annunzio's visit to Doberdò very soon after he left his convalescence in Venice, see Martinelli, *La guerra di d'Annunzio*, 137. A similar image occurs in d'Annunzio's 'Cantano i morti con la terra in bocca e le carene valicano i monti' (*Prose*, 1922, *Il libro ascetico*, at 781), a text that shocked some because it evoked the execution of infantrymen who had mutinied at Santa Maria la Longa in 1917. Bonadeo, in *D'Annunzio and the Great War*, at 77, sees in the latter a prime example of d'Annunzio's fascination with death as 'inexorable fate' but does not note the surprising fact that d'Annunzio puts traitors and patriots together in the same image of *pena umana* (human suffering).

The originals read: 'chiesa sense preghiere'; 'sembrano corpi formati di terra con in sommo un viso di carne che duole'; 'San Francesco lacero e logoro'; 'ero un tronco sanguinoso che abbeverava di sé qualche sasso del Carso.'

16 D'Annunzio's rejection of monuments is in the 1919 'L'erma bifronte,' in his 'Il sudore di sangue,' 932. The original reads: 'oggi i fanti, contadini e non contadini, gridano a una voce, di sopra la terra e di sotterra: "Non vogliamo monumenti."' On remorse, *Avanti!* 25 Oct. 1921, also in Labita, 'Il Milite Ignoto,' 138. The original reads: 'Forse tra le rocce delle Alpi sarebbero rimaste mute e senza eloquio quelle ossa: a Roma esse parlano. A noi sono un monito: ai cultori dell'altare della "patria" un rimorso.'

'Tribolata carne' (suffering flesh) is in *L'illustrazione Italiana*, 6 Nov. 1921, 530. The original reads: 'una realtà travagliata, sudata, insanguinata, un blocco di angoscia, di torture, di strazi.'

17 D'Annunzio, *Notturno*, 55, 58, 59, 70. The original reads:

> E intravedo, a un guizzo della fiamella, una macchia scura sotto il lettuccio, rabbrividendo.

La bandiera è tolta dal capezzale; la garza è tolta dal viso irriconoscibile. [...] La spoglia si scompone. [...]

C'è una specie di orrore impietrato, intorno a me. [...]
rivedo quella macchia oscura che avevo scoperta sotto il lettuccio.

E sangue, è sangue e siero colati a traverso il materasso ... Mi pare che il cuore mi cada giù, sotto il calcagno. [...]

Quattro marinai sollevano la cassa [...] metto le mani sotto il fondo e sostengo il peso. [...]

Ho il petto pieno di grido e non odo la mia voce.

Il letto oscilla, sbanda, e poi precipita. Un deserto di sasso, sgretolato e forato, viene precipitosamente incontro al mio occhio che non si chiude. [...] Ecco il Carso, pallido.

d'Annunzio, 'Agli uomini di pena,' 573.

18 Ungaretti, *Vita d'un uomo: Tutte le poesie,* 41, and *A Major Selection,* 69. The original reads:

Sono una creatura
Valloncello di Cima Quattro il 5 agosto 1916

Come questa pietra
del S. Michele
così fredda
così dura
così prosciugata
così refrattaria
così totalmente
disanimata

Come questa pietra
è il mio pianto
che non si vede

La morte
si sconta
vivendo

19 'Perché?' Carsia Giulia (1919), in Ungaretti, *Vita d'un uomo: Tutte le poesie,* 56, and *A Major Selection,* 89. The original reads:

Reggo il mio cuore
che s'incaverna
e schianta e rintrona

come un proiettile
nella pianura
ma non mi lascia
neanche un segno di volo

Il mio povero cuore
sbigottito
di non sapere

20 On trauma as 'liminal,' the difficulty of remembrance, and loss of control over the past, see Paulson, *Haunted by Combat*, 35, 137.

21 D'Annunzio, *Notturno*, 71, 73, 75. The original reads:

> Il fante ritorna a me con un fascio di pellicce macchiate di bruno, chiazzate di sangue risecco. [...] Sono le spoglie di Alfredo Barbieri [...]
> L'aria s'è fatta come di cristallo gelidissimo. Ha la medesima qualità di quello spirituale masso di ghiaccio che serra la testa del cadavere nella prima ora. La mia corsa diritta la fende come il diamante riga il vetro. E lo stridore mi divide il cervello. [...]
> La testa gloriosa aveva due fóri. Dall'uno il sangue aspergeva i compagni, dall'altro sfavillava nel vento mattutino… [...]
> Non v'è parte che non sia aspersa. Il sangue è ormai fisso; eppure gronda sul mio capo quando mi curvo tra le ruote del carrello, non osando inginocchiarmi davanti ai testimoni estranei. Le stille a miriadi si riscaldano e s'avvivano e si rinvermigliano, come la reliquia bruna che di subito rifiammeggia nell'ampolla. [...]

22 D'Annunzio, *Notturno*, 57, 61. The original reads:

> Il cadavere è ormai separato da me, è chiuso, è solo, è già nella tomba. [...]
> Mi pareva ancor mio, dianzi, se bene disfatto, se bene difformato. Ora è in prigione. [...]
> Non si potrebbe levare, neppure se Cristo lo chiamasse. [...]
> E là, lo vedo a traverso la coltre e il legno. Iersera era infoscato, fumoso, gonfio. Un'altra notte è passata. E il terzo giorno. Lo sfacelo continua.
> Ho nelle ossa un freddo terribile. Toccare la morte, imprimersi nella morte, avendo un cuore vivo!
> Eppure siamo anche una volta soli, noi due, soli come nella carlinga in volo.

23 D'Annunzio, *Notturno*, 258. The original reads:

Il cadavere di Roberto Prunas, gonfio e bianchiccio, s'è arrestato contro uno sbarramento di torpedini. Sta sospeso e oscilla, macerandosi nel suo gabbano come in un sacco slegato.

Nessuno scriverà su quella tomba marina il suo nome.

24 D'Annunzio, *Notturno,* 20, 27. The original reads:

Il mio compagno è morto, è sepolto, è disciolto.

Io sono vivo, ma esattamente collocato nel mio buio com'egli nel suo.

[...]

V'è un luogo dell'anima, là dove il nero fiume e il fiume chiaro confluiscono.

Beati i due compagni eroi le cui ossa irriconoscibili sono mescolate nella barella come due tizzoni fumanti!

25 The first two lines are from 'A riposo,' Ungaretti, *Vita d'un uomo: Tutte le poesie,* 26. The second is from 'In dormiveglia,' Valloncello di Cima Quattro il 6 agosto 1916, Ungaretti, *Vita d'un uomo: Tutte le poesie,* 42. Respectively, Ungaretti, *Selected Poems,* 21, Ungaretti, *A Major Selection,* 71. The originals read:

E piombo in me
E m'oscuro in un mio nido.

Assisto la notte violentata
L'aria e crivellata

For the connection between access to death via the corpse and the transformation of land into place, see Harrison, *Dominion,* 17–36.

26 Detailed plans and timetables for the transport from Aquileia to Rome, as well as reports of unwanted honours such as the one from the Arditi del popolo, are in MdI, OaSI, 1921. The originals read: '*sono vietati i discorsi.* Da tutti sarà osservato un religioso silenzio'; 'non false pompe aurate, non mercato di morti gloriosi, ma fiamme di fede, ma auspicio di elevate speranze offrono con semplice anima proletaria al loro fratello ignoto gli arditi del popolo.'

Accounts of popular rebellion against state impositions of silence are numerous. For Cadeddu's overall view that 'A stento il servizio d'ordine [...] riuscì a trattenere la folla' (The forces of order were barely able [...] to restrain the crowd), see Miniero, *Da Versailles,* 193.

On crowds reaching out to touch the coffin, and other spontaneous expressions of grief, such as kneeeling crowds, thousands of hands reaching out, people hanging on to the coffin and having to be torn off (especially mothers), as well as light shows, and finally the preference of some newspapers for 'povere figure isolate' (poor isolated figures) over 'parate ufficiali' (official parades), see Miniero, *Da Versailles*, 195–8.

On the contrast between day and night, see Labita, 'Il Milite Ignoto,' 147–8. On dualism overcome versus goliardic joy, *Corriere della Sera*, 5 Nov. 1921, and *Avanti!* 2 Nov. 1921. The originals read: 'Il dualismo fra la tendenza al raccoglimento e la sensibilità all'entusiasmo sta per essere superato. Come attraverso le campagne durante il pellegrinaggio da Aquilea a Roma gli episodi furono soprattutto sentimentali, l'imperiale sfondo di Roma esalta e induce il popolo alla glorificazione'; 'pazza gioia: canti, musiche, battimani, inni, 'alte uniformi,' coreografia.'

For Miniero (*Da Versailles*, 231–3), it was the reverse, for the day, in Rome, was the moment for a more sombre inhumation, whereas ceremonies continued into the evening with 'electric lights,' which were at the time associated with a more celebratory, 'fairlike' atmosphere. From this perspective it is interesting that in 1923 the Ministry of Public Works refused an offer of free permanent electrical illumination of the Unknown Soldier's Tomb, considering it too frivolous; later the eternal flame was added instead. See MLP, SdSI, 1921–26.

27 'Comprehension,' as Kirkwood intends it, thus implies a certain slowness in reading, a not hastening to conclusions, as advocated in Paolo Valesio, 'Il metodo come strada accidentata,' *Lettore di Provincia* 30/104 (1999): 47–53.

28 Giono, *Le grand troupeau*, 595, and *To the Slaughterhouse*, 61. The originals read: 'veiller le corps absent'; 'pauvre, que tu es mort dans la terre comme une bête, tout seul!'

On Giono's *Le grand troupeau*, see the following: W.D. Redfern, 'Against Nature: Jean Giono and Le Grand Troupeau,' in Holger Klein, ed., *The First World War in Fiction: A Collection of Critical Essays* (London: Macmillan, 1976), 73–83; Luc Rasson, '"On fait la vie avec le sang": A propos du Grand troupeau,' in Christian Berg, Walter Geerts, Paul Pelckmans, and Bruno Tritsmans, eds., *Retours du mythe* (Amsterdam: Rodopi, 1996), 183–94.

29 Giono, *Le grand troupeau*, 545, 546, 547, 548, 549, 550, and *To the Slaughterhouse*, 11, 12, 15, 16. The originals read: 'Elle mangera vos béliers, vos brebis et vos moissons'; 'las à mort'; 'une grande bête toute noire qui avait du sang sous le ventre'; 'avait du sang caillé sur ses dents et dans ses babines'; 'dehors, le grand troupeau coule'; 'les bêtes maintenant étaient malades. On n'en pouvait plus de cette longueur de troupeau, de tout ce mal, de toute cette vie qu'on usait sur la route'; 'Le bélier était toujours là par terre, les jambes écartées. Le sang s'était mis à couler de lui.'

30 Giono, *Le grand troupeau*, 551, 564, 547, 560, and *To the Slaughterhouse*, 18,

14, 29. The originals read: 'Et le bélier vient de mourir. Il a relevé d'un seul coup sa lourde tête branchue, comme sur un ordre; il a regardé le ciel d'entre les branches de ses cornes: un long regard interminable. Le cou tendu, il a eu un petit gémissement d'agneau; il a écarté les cuisses, étiré les jambes; il a lâché un paquet de sang noir et tripes avec un bruit de ballon qui se crève.'

And: 'quand les temps auront passé, si je suis encore en vie, je reviendrai le chercher.'

31 On the apocalyptic tone on Giono, and on apocalyspse as a preserving of life outside of time, see Winter, *Sites*, 197–9. Giono's chapter titles sound like the Old Testament but are not direct quotes.

32 Giono, *Le grand troupeau*, 616, 611, 612, and *To the Slaughterhouse*, 81. The originals read: 'Et il n'y aura point de pitié' 'Il [Olivier] se retourna: c'était, sur l'autre bord du trou, un homme couché et qui avait la figure toute noire; sa cervelle coulait par une large blessure en coin. Il ne regardait pas; [...] l'oeil pourri et plein de boue.' 'Il sentit dans lui comme une grande envie de vomir.'

33 Giono, *Le grand troupeau*, 620–1, and *To the Slaughterhouse*, 91–2. The original reads:

> Les morts avaient la figure dans la boue, ou bien ils émergeaient des trous, paisibles, les mains posées sur le rebord, la tête couchée sur les bras. Les rats venaient les renifler. Ils sautaient d'un mort à l'autre. Il choisissaient d'abord les jeunes sans barbe sur les joues. Ils reniflaient la joue puis ils se mettaient en boule et ils commençaient à manger cette chair d'entre le nez et la bouche [...] pour les yeux, ils les sortaient à petits coups de griffes, et ils léchaient le trou des paupières [...] les corbeaux arrivaient à larges coup d'ailes tranquilles. Ils cherchaient le long des pistes et des chemins les gros chevaux renversés. A côté de ces chevaux, aux ventres éclatés comme des fleurs de câprier, des voitures et des canons culbutés mêlaient la ferraille et le pain, la viande de ravitaillement encore entortillée dans son pansement de gaze et les baguettes jaunes de la poudre à canon. [...] Là, toutes les corvées de la nuit laissaient des hommes. Ils étaient étendus, le seau de la soupe renversé dans leur jambes, dans un mortier de sang et de vin. Le pain même qu'ils portaient était crevé des déchirures du fer et des balles. [...] cette partie du cou où est le partage de la barbe et du poil de la poitrine. C'était là tendre et tout frais, le sang rouge y faisait encore la petite boule. Ils se mettaient à becqueter là, tout de suite, à arracher cette peau [...]. Les morts bougeaient. Les nerfs se tendaient dans la rainure des chairs pourries et un bras

se levait lentement dans l'aube. Il restait là, dessant vers le ciel sa main noire toute épanouie; les ventres trop gonflés éclataient et l'homme se tordait dans la terre [...] Il reprenait une parcelle de vie.

34 Bonadeo, *Mark of the Beast*, 145–6. He cites Redfern, 'Against Nature.'

35 On latency, see Gumbrecht, 'How (If at All).' On the disappearance of agriculture, see Serres, *Hominescence.*

36 Giono, *Le grand troupeau*, 700, 723–4, and *To the Slaughterhouse*, 186, 208, 214–15. The originals read: 'deux [...] jambes mortes'; and

C'est bon signe quand les bêtes sont là pour les naissances. Une bête de plus sur la terre. [...]
Il souffle sur la bouche du petit.
'Le vert de l'herbe,' il dit.
Il souffle sur l'oreille droite du petit.
'Les bruits du monde,' il dit.
Il souffle sur les yeux du petit.
'Le soleil.
Bélier, viens ici. Souffle sur ce petit homme pour qu'il soit, comme toi, un qui mène, un qui va devant, non pas un qui suit.'
[...]

And: 'Si Dieu m'écoute, il te sera donné d'aimer lentement, lentement dans tous tes amours, comme un qui tient les bras de la charrue et qui va un peu plus profond chaque jour.'

37 Giono, *Le grand troupeau*, 718, 719, 664, and *To the Slaughterhouse*, 208, 209, 142. The originals read: 'porter la marque'; 'J'ai trop de sang'; 'l'agneau.'

38 In my discussion of Marinetti, I will use the Freudian term 'disavowal' rather than 'denial' to indicate the coexistence and/or oscillation of denial and its suspension. On disavowal and its relation to the Nietzschean death of God in *Mafarka*, see Laura Wittman, '"Cuore-proiettile": From Marinetti's 'Mafarka to World War One and Ungaretti,' *L'anello che non tiene* (2011).

'La carne congelata' is in Filippo Tommaso Marinetti, *Gli amori futuristi: Programmi di vita con varianti a scelta* (Cremona [Piacenza?]: Casa Editrice Ghelfi, 1922). It is republished in Filippo Tommaso Marinetti, *Scatole d'amore in conserva* (Rome: Edizioni d'arte Fauno, 1927). It is again republished, with the title, 'Come si nutriva l'ardito,' in Filippo Tommaso Marinetti, *Novelle colle labbra tinte: Simultaneità e programmi di vita con varianti a scelta* (Milan: Mondadori, 1930).

'Come amava Guzzo l'Ardito,' ms. in the Filippo Tommaso Marinetti Papers, General Collection, Beinecke Rare Book and Manuscript Library, Yale University, Gen Mss 130, (hereafter Marinetti Papers), Box 19, no.

1226. This manuscript contains minor variations; the major change appears at the end, but is made between 1927 and 1930 apparently. 'Beasts' is from Marinetti, *Novelle*, 398. The original reads: 'belve indomabili ma generosissime.'

39 For encounters between Marinetti and d'Annunzio on the battlefield, see Filippo Tommaso Marinetti, *Taccuini* (Bologna: Il Mulino, 1987), 110. In his 2 June 1917 entry, Marinetti writes in particular about the recent death of 'Randazzo' (Randaccio), and d'Annunzio's funerary oration for him, about to be published in the *Corriere della Sera.*

On the 'ambivalent' attitude towards violence typical of Marinetti, and often of Futurism, and its importance for Italian Modernism, see Paolo Valesio: 'Foreword: After *The Conquest of the Stars*,' in Somigli and Moroni, eds., *Italian Modernism*, and 'La "macchina morbida" di Marinetti,' *Annali d'Italianistica* 27 (2009): 243–62.

For Marinetti and decadent aspects of his work, see the following: Marjorie Perloff, *The Futurist Moment: Avant-Guarde, Avant-Guerre, and the Language of Rupture* (Chicago: University of Chicago Press, 1986); Alice Yaeger Kaplan, *Reproductions of Banality: Fascism, Literature, and French Intellectual Life* (Minneapolis: University of Minnesota Press, 1986); Giusi Baldissone, *Filippo Tommaso Marinetti* (Milan: Mursia, 1986); Barbara Spackman: 'Mafarka and Son: Marinetti's Homophobic Economics,' *Modernism/Modernity* 1/3 (1994): 89–107, and *Fascist Virilities: Rhetoric, Ideology, and Social Fantasy in Italy* (Minneapolis and London: University of Minnesota Press, 1996); Cinzia Sartini Blum, *The Other Modernism: F.T. Marinetti's Futurist Fiction of Power* (Berkeley: University of California Press, 1996); Paolo Valesio, 'The Most Enduring and Honored Name,' *Selected Poems and Related Prose [of F.T. Marinetti]* (New Haven: Yale University Press, 2002), i–xix; Luca Somigli: *Legitimizing the Artist: Manifesto Writing and European Modernism, 1885–1915* (Toronto: University of Toronto Press, 2003), and 'The Mirror of Modernity: Marinetti's Early Criticism between Decadence and "Renaissance latine,"' *Romanic Review* 97/3–4 (2006): 331–52; Christine Poggi: *In Defiance of Painting: Cubism, Futurism, and the Invention of Collage* (New Haven: Yale University Press, 1992), and *Inventing Futurism: The Art and Politics of Artificial Optimism* (Princeton: Princeton University Press, 2009); Tatiana Cescutti, *Les origines mythiques du futurisme: F.T. Marinetti, poète symboliste français (1902–1908)* (Paris: Presses de l'Université Paris-Sorbonne, 2009).

On liquidity in Marinetti, and its violence, see Enrico Cesaretti, 'Dyspepsia as Dystopia? F.T. Marinetti's Le Roi Bombance,' *Romanic Review* 97/3–4 (2006): 353–69. On the mystical aspect of Marinetti's famous car accident in the Founding Manifesto, see Antonino Musumeci, 'Marinetti: A Mystical

Experience on the Way to Futurism,' *Romance Languages Annual* 3 (1991): 263–6. On its relationship to speed, see Jeffrey T. Schnapp, 'Politics and Poetics in Marinetti's Zang Tumb Tuuum,' *Stanford Italian Review* 5/1 (1985): 75–92, and 'Propeller Talk,' *Modernism/Modernity* 1/3 (1994): 153–78.

On Marinetti and war, see Lucia Re, 'Valentine de Saint-Point, Ricciotto Canudo, F.T. Marinetti: Eroticism, Violence and Feminism from Prewar Paris to Colonial Cairo,' *Quaderni d'Italianistica* 24/2 (2003): 37–69, and 'Futurism, Seduction, and the Strange Sublimity of War,' *Italian Studies* 59 (2004): 83–111.

40 Marinetti, *Novelle*, 397–8, 398, 399. For 'molecular rage,' see Gilles Deleuze, 'Coldness and Cruelty,' in Leopold von Sacher-Masoch and Gilles Deleuze, *Masochism: Coldness and Cruelty Followed by Venus in Furs* (New York: Zone Books, 1989), 119.

The originals read: 'non esagerare la tua angoscia [...] questa volontà deve irrigidire i nostri nervi contro ogni viltà sentimentale'; 'eccoci di nuovo lanciati in una nuova conflagrazione'; 'ceffi spavaldi selvaggi tracotanti e giocondi che hanno già superato in ferocia tutti i Marat e in eroismo tutti gli arditi grigio-verdi dell'ultima guerra'; 'pre[si] a pugni'; 'ricompensati'; 'scatena[ti] contro il nemico'; 'vero Saraceno magro agile scattante, grandi occhi neri dolcissimi, ma la bocca cattiva e deformata da una rasoiata alla guancia sinistra.'

41 Marinetti, *Novelle*, 400, 401, 402. This notion of suspension is related to the 'deferral' that Andrew Hewitt sees as the main connection between Fascism and Marinetti's libidinal economy. *Fascist Modernism: Aesthetics, Politics, and the Avant-Garde* (Stanford: Stanford University Press, 1993), 155–6.

The originals read: 'si guasta a mostarla'; 'ammaestrare'; 'ringhiano'; 'Nasce la luna con gioielli rubati all'acqua bollente di risate. Corrono dei riflessi furbissimi'; 'si incrociano, svegliando miracolosamente fogliami e fusti fantastici d'argento lieve, case color smeraldo, spiagge di madreperla, forme spettrali, isolotti di ghiaccio, paesaggi d'oro liquido. Guzzo fruga sempre nel suo zaino, mentre la forbice tagliente formata de due proiettori si apre con lento mistero. Sono due scope lunghe d'argento pulverulento. L'acqua s'infiamma di mille estasi [...] velluti carnali con pallori umani che vibrano, implorano.'

42 Marinetti, *Teoria e invenzione futurista*, 971, 1012. The originals read: 'lago di untuoso chiaro di luna stemperato'; 'morbida musica di affetti'; 'sovvrumana frescalata Distrazione dell'Arte [che] operava la metamorfosi.' Blum, *The Other Modernism*, 127–8, 129, 129–30.

43 In Marinetti's dream, the untamables are guarded by day by violent 'black men' and by night by 'musical' people, who represent 'spiritual trends.' A

brief beginning of the dream is in the published diaries (Marinetti, *Taccuini*, 503–4). Most of the dream is unpublished, and can be found in Sept. 1920, Diary LII, Marinetti Papers. For the quote, see 31 March 1917, Diary IIA, ibid.; also excerpted in Filippo Tommaso Marinetti, 'Excerpts from the Unpublished Diaries,' *Modernism/Modernity* 1/3 (1994): 4. See also Laura Wittman, 'Introduction to Part Three: Stars-in-Freedom and the Dark Night of Futurism,' in Lawrence Rainey, Christine Poggi, and Laura Wittman, eds., *Futurism: An Anthology* (New Haven: Yale University Press, 2009), 409–18.

Marinetti's diaries were published as a document of the war experience and the unusual editorial decision was made to cut all passages that appeared to be notes or drafts for literary works, notably those preceded by the heading 'per il romanzo' (for the novel), which referred to various works in progress, including *Gli indomabili*, *Amori futuristi*, and *L'Alcova d'acciaio*. My examination of all omitted passages – which run to about 150 typescript pages – revealed much fascinating reworking of the war into both decadent and more Futurist-sounding fantasmagoria.

Gli indomabili has been recently republished, edited and with an introduction by Luigi Ballerini; see Filippo Tommaso Marinetti, *Gli indomabili: Con un'antologia di scritti futuristi sull'arte meccanica e d'avanguardia* (Milan: Mondadori, 2000).

The original reads: 'Sempre sempre, nella mia vita si alternarono questi due stati d'animo torturanti: Nel letto più voluttuoso e felice mi sono ad un tratto strappato alle braccia d'una amante deliziosa e buona, e seduto fremente ho teso l'orecchio ad un lontano immaginario bombardamento. Nostalgia desiderio d'eroismo e di violenza Nella trincea fango broda delle bombarde mostruosa edera rampicante di corpi nudi di donne dalle mie tempie fino alla luna. Nostalgia torrida di lussuria. Perchè? Perchè?'

44 Marinetti, *Novelle*, 402–5. The original reads:

> Guzzo in ginocchio, tutto acceso di splendori, estrae dallo zaino qualcosa che splende più di lui. Avvicinati! Non ti sente, è assorto. [...] Ha veramente fra le mani un pezzo di carne. Un pezzo di donna nuda decapitata senza braccia e senza gambe! Il tronco grazioso di una piccola donna! Sembra cesellato e imbrillantato di sale prezioso che luce. Le mammelline tonde soavi, tremano, forse parleranno tanto più che la testa perduta lontana non parla più! Il ventre umile, timido, ingenuo, mansueto s'incurva verso il lieve sognante piccolo giardino fra le cosce sensuali troncate a metà. I due tagli delle cosce sono chiusi da cuffie di seta nera aderente. Così pure il collo.
>
> 'Cara, sono qui con te, con te. Sono caduto giù dall'alto dei miei

eroismi, giù! La morte ci dà quest'ultima ora di festa. Vedi? Tutto è bello intorno a noi. [...]'

Ora tu apri le braccia che non hai più e mi tendi le labbra che non hai più! [...] Sono capitombolato giù dal cielo disperato nero, squartato e rullante della mia vita morta, giù giù su te, in te, nel tuo calore rovente umido succhiante! Paradiso inferno iddio mio, mio mio. Bevo l'infinito in te, piccola, immensa divina [...]'

Piange, e con voce flebile di donna ripete a memoria un brano di lettera: 'Portami con te in guerra. Porta con te la mia carne e baciami giorno e notte, baciami, mangiami la faccia di baci, divorami, sono tua! tua!'

Poi Guzzo cambiando voce: 'Ho fatto ciò che volevi! [...] tutto ho mangiato con milioni di baci! Ed ora sei me stesso! Nelle mie vene il mio sangue ti culla, piccola, con la tenerezza che le madri morte hanno in Paradiso!'

45 Blum, *The Other Modernism*, 95, 97. Jessica Benjamin, quoted in Blum, ibid., 63.

46 Filippo Tommaso Marinetti, 'La Ville d'émeraude,' Marinetti Papers, Gen Mss 130, Box 39, Folder 1681, 262r–265r. The original reads:

> Je porte en moi, dans ma poitrine une bête
> noire, visqueuse, gluante avec des petites pat-
> tes elastiques et munie de ventouses [...] ma petite âme de lux-
> ure [...]
> Et grand Dieu
> quel mâle lui faudra-t-il? quel mâle vien-
> dra jamais s'accoupler avec une aussi sale,
> gluante, et monstrueuse bête? Car il n'est
> pas d'animaux qui lui ressemblent. D'ailleurs
> tout est fermé dans ma poitrine close.

47 Marinetti, *Novelle*, 406–7. The original reads:

> Gli arditi si ammucchiano poi davanti nel guado che sognando travolge succhia mastica raggi di stelle. [...] Guzzo si caccerà con un balzo contro una palla che lo cercava scherzosamente pur volendolo scansare. Colpito cadrà nell'acqua.
>
> Tu allora non esitare, stacca lo zaino e caricatelo sulle spalle, poi svincola il cadavere di Guzzo dalle erbacce e spingilo nella corrente che lo porterà al mare. Poi avanti!
>
> Mentre camminerai fra gli spudorati splendori scandalosi dei proiettori che ti cercano, sentirai piangere vicino a te, su di te, in te! Piangerà il corpo della bella di Guzzo, già adultera, tutta tremante d'amore per te! E il suo pianto colerà sulla tua schiena a gocce lente, voluttuose.

48 Céline, *Voyage*, 12, 14. Louis-Ferdinand Céline, *Journey to the End of the Night*, trans. Ralph Manheim (New York: New Directions, 2006), 4, 3, 5. The originals read: 'race française'; 'la Patrie no. 1'; 'la Patrie no. 2.'

On Céline, see the following: Henri Mondor, 'Voyage au bout de la colère,' *Nouvelles Littéraires* (1962): 3; John Leeds, James Flynn, and C.K. Mertz, '"Possibilities for Compassion": An Alternate Reading,' *Understanding Céline* (Seattle: Genitron, 1984), 125–32; Isabelle Blondiaux, 'La Représentation de la pathologie psychique de guerre dans Voyage au bout de la nuit,' *Roman 20–50* 17 (1994): 105–15; Fredric Jameson, 'Céline and Innocence,' *South Atlantic Quarterly* 93/2 (1994): 311–19; Tom Quinn and Henri Godard, *The Traumatic Memory of the Great War, 1914–1918, in Louis-Ferdinand Céline's Voyage au bout de la nuit* (Lewiston: Mellen, 2005); Valerio Magrelli, '"Avant que ça soye la nuit partout": Il Baudelaire di Céline fra parodia e riscrittura,' *Confronto Letterario* 23/45 (2006): 147–59.

49 Céline, *Voyage*, 124, 253, 257, 169, 18, 22, 26, and *Journey*, 105, 287, 206, 156, 8–9, 12, 15. The originals read: 'les riches n'ont pas besoin de tuer eux-mêmes pour bouffer'; 'c'est par les odeurs que finissent les êtres, les pays et les choses'; 'soldats inconnus'; 'mille morts'; 'j'étais devenu du feu et du bruit moi-même. Et puis non, le feu parti, le bruit est resté longtemps dans ma tête, et puis les bras et les jambes qui tremblaient comme si quelqu'un vous les secouait par derrière. Ils avaient l'air de me quitter et puis ils sont restés quand même, mes membres'; 'il y en avait pour des kilos et des kilos de tripes étalées, de gras en flocons jaunes et pâles, de moutons éventrés avec leurs organes en pagaïe, suintant en ruisselets ingénieux dans la verdure d'alentour [...] Et puis du sang encore et partout, à travers l'herbe, en flaques molles et confluentes qui cherchaient la bonne pente [...] j'ai dû céder à une envie de vomir, et pas qu'un peu, jusqu'à l'évanouissement.'

50 Céline, *Voyage*, 56, 60, 60, 124, 41, 66, 66, 67, 71, 71, and *Journey*, 39, 43, 64, 124, 27, 48, 48, 49, 52, 52. On Hemingway's claim that inability to suspend the imagination is what makes the soldier a coward, see Bonadeo, *Mark of the Beast*, 6. The originals read: 'il n'y a que la bravoure au fond qui est louche. Être brave avec son corps? Demandez alors à l'asticot d'être brave, il est rose et pâle et mou, tout comme nous'; 'l'âme, c'est la vanité et le plaisir du corps tant qu'il est bien portant, mais c'est aussi l'envie d'en sortir du corps dès qu'il est malade ou que les choses tournent mal. On prend des deux poses celle qui vous sert le plus agréablement dans le moment et voilà tout! Mais moi, je ne pouvais plus choisir, mon jeu était fait.' And: 'monstre'; 'courage'; 'carnages'; 'dans ce métier d'être tué, faut pas être difficile, faut faire comme si la vie continuait, c'est ça le plus dur, ce mensonge'; 'Stand des Nations'; 'Sur moi aussi qu'on tire Lola!'; 'bêtises'; 'c'est même à partir de ce moment-là, je crois, que ma tête est devenue si difficile à tran-

quilliser, avec ses idées dedans'; 'ça vous donne un autre vous-même. On est deux'; 'sale attrait mystique.'

51 Céline, *Voyage*, 127, 160, 193, 214–15, 241, 242–3, 247, 245, 245, 246,and *Journey*, 98, 98, 125–6, 151, 171–3, 193, 195, 195, 195, 197, 197, 197. The originals read: 'le Nord au moins ça vous conserve les viandes'; 'suce' 'bouffe' 'fend' 'colle' 'roustille'; 'on va vite à pourrir, dans les verdures'; 'le Blanc lui s'empoisonne, cloisonné qu'il est entre son jus acide et sa chemise en cellular. Aussi malheur à qui l'approche.' And:

> Passage du métro aérien. Il bondissait en face, entre deux rues, comme un obus, rempli de viandes tremblotantes et hachées, saccadait à travers la ville lunatique de quartier en quartier. [...] on voit bien qu'ils ne cherchent pas à comprendre, eux, le pourquoi qu'on est. [...] Ils ont toujours la conscience tranquille. J'en avais trop vu moi des choses pas claires pour être content. J'en savais trop et j'en savais pas assez. [...] On n'a plus beaucoup de musique en soi pour faire danser la vie, voilà. Et où aller dehors, je vous le demande, dès qu'on n'a plus en soi la quantité suffisante de délire? La vérité, c'est une agonie qui n'en finit pas. La vérité de ce monde c'est la mort. Il faut choisir, mourir ou mentir. Je n'ai jamais pu me tuer moi.

Then: 'nous n'avons pas besoin d'imaginatifs dans notre usine'; 'un autre Ferdinand'; 'c'était un vrai corps que je voulais toucher, un corps rose en vraie vie silencieuse et molle'; 'me refaire une âme'; 'un coeur infini'; 'il était trop tard pour me refaire une jeunesse'; 'cette envie de m'enfuir partout, à la recherche de je ne sais quoi.'

52 Céline, *Voyage*, 307, 290, and *Journey*, 248–9, 234. The original reads: 'depuis le retour de Robinson, j'étais devenu bien étrange dans ma tête et mon corps et les cris de ce petit innocent me firent une impression abominable. Quels cris, mon Dieu! Quels cris. Je n'en pouvais plus.'

53 Céline, *Voyage*, 523–4, 526, 528, and *Journey*, 425, 427, 428–9. The originals read: 'Mais si! [...] Que j'en ai du courage! et sûrement autant que toi! ... Seulement moi si tu veux tout savoir … Tout absolument … Eh bien, c'est tout, qui me répugne et qui me dégoûte à présent! Pas seulement toi! ... Tout! ... [...] Tu fais la sentimentale pendant que t'es une brute comme pas une … Tu veux en bouffer de la viande pourrie? Avec ta sauce à tendresse? ... Ça passe alors? ... Pas à moi!' And: 'une espèce de calme'; 'C'était pas comme un malade ordinaire, on ne savait pas comment se tenir devant lui. C'était comme s'il essayait de nous aider à vivre à présent nous autres. Comme s'il nous avait cherché à nous des plaisirs pour rester. Il nous tenait par la main. Chacun une.'

54 On increasing the Catholic presence, Atkinson and Cosgrove ('Urban Rhetoric and Embodied Identities,' 42–4); 'real presence' (ibid., 40).

My own research shows that the casket had to be moved in 1923 because of work that still needed to be done on the statue of Rome and friezes around it: temporary plaster versions were to be replaced by stone ones. Initial estimates claimed the casket needed only to be moved for one month, but this was later increased to nine, and by then, in 1924, as the work dragged on, proposals to move it permanently into the crypt began, as detailed below. MLP, SdSI, 1921–26, letter of 7 April 1923.

On Sacconi's more secular design, see Brice, *Il Vittoriano,* 274. Paul Baxa (*Roads and Ruins,* 92–3, 91) tells part of this story, asserting that for the Fascist regime the Unknown Soldier was too closely associated with 'the image of liberal Italy.'

55 Acciaresi's comments are from an article in *Il Messaggero,* 26 Feb. 1924. A clipping of it is in MLP, SdSI, 1921–26. The original reads: 'in tutti i tempi le divinità o gli eroi a cui si decretò il culto, o gli illustri uomini ai quali si eressero mausolei, ebbero l'ara propiziatrice o la cella funeraria nella parte più recondita e più nobile del tempio.' See also Primo Acciaresi, *Giuseppe Sacconi e l'opera sua massima: Cronaca dei lavori del Monumento Nazionale a Vittorio Emanuele II illustrata da 330 incisioni* (Rome: Tipografia dell'Unione Editrice, 1911).

56 For the response of the Artistic Committee, and the printed leaflet making the move official, see MLP, SdSI, 1921–26. The originals read: 'simboli civili e patriottici'; 'Per la Cripta Sacra e le Onoranze Perpetue al Milite Ignoto.'

57 Mussolini's letter, and other letters referring to the replacement of the plaster versions with stone ones, are in MLP, SdSI, 1921–26. The originals read: 'pietra tombale dell'Ignoto Milite'; 'ignominiosamente negletta'; 'Occorre cioè trovare una sistemazione che assicuri la perfetta conservazione della salma (che nella parete ove era stata murata sembra sia stata esposta ad infiltrazioni di acqua che a lungo andare la avrebbero seriamente danneggiata)'; 'esigenze della pietà, del patriottismo, della buona conservazione e dell'estetica.'

58 On the Artistic Committee's objections, as well as clippings from debates in *La Tribuna,* see MLP, SdSI, 1921–26. For the Unknown Soldier's final return 'from the silent battlefield' (*la silente trincea*) to bask in 'the sun of Italy' (*il sole d'Italia*), see the speech by the Minister of War Rodinò to the Parliamentary Commission, AP, 1921–23, no. 202 SdSI. The originals read: 'Il milite ignoto deve tornare nella sua tomba'; 'il buio della cripta dove il Milite Ignoto dovrebbe essere sepellito.'

59 Fogu, 'Fascism and Historic Representation,' 339. The original reads: 'L'idea di onorare il soldato ignoto [...] parve, in un primo tempo, poco felice e molto pericolosa.' Gessi continued, saying that by then, 2 Nov. 1921, the danger had been excluded; this is noted in Miniero, *Da Versailles,* 193n145. On a literal level this was true, as unrest turned out to be minimal; however, on an imaginative level the Unknown Soldier Memorial did turn out to be a subversive idea, as I hope to show in this book.

5 Mutilation and Spectacle

1 'Urgentissima,' 1 Nov. 1921, from Questore di Roma to Prefetto, MdI, OaSI, 1921. The original reads: 'I mutilati e invalidi di guerra [...] [saranno collocati] anzichè nel recinto riservato agli ex combattenti (Piazza Cinquecento) sul marciapiedi del fabbricato della Stazione lato arrivi a conveniente distanza dalla Saletta Reale, [...]. Questa concessione Autorità Militari hanno dovuto fare per calmare viva agitazione mutilati non soddisfatti del posto loro precedentemente assegnato.'

2 3 Nov. [1921], 'riservata urgente' from ibid. The original reads: 'si vanno affiggendo per le cantonate manifesti intestati Associazione Nazionale Combattenti e firmati Comitato provvisorio coi quali si incita a rompere i cordoni di truppa e di impossessarsi della salma del Milite Ignoto per protesta contro la rigida ufficialità delle manifestazioni di esaltazione del milite stesso.'

3 On the relationship between the Associazione Nazionale Combattenti (ANC) and its growth from the Associazione Nazionale Mutilati e Invalidi di Guerra (ANMIG), see Marco Unia, *Il soldato torna in Patria: E adesso pover'uomo?* (book review of Hobsbawm, Sabbatucci, Gentile, Isnenghi, Gibelli, Leed, and Winter), retrieved from http://www.storiain.net/arret/num74/artic3.asp.

The presence of mutilated soldiers is reported, among other places, in *Corriere della Sera,* 5 Nov. 1921.

4 The maps are mentioned in Coglitore G. Garufi, *Il Milite Ignoto,* 2008, retrieved from http://www.alfamodel.it/modules/smartsection/item.php?itemid=180&keywords=eroi. One of them is reproduced in Talamo and Marco, eds., *Dalla vittoria al milite ignoto,* 53.

For the *stacconate* constructed to contain the public, as well the extra 16,200 troops brought in to reinforce the police, see MdI, OaSI, 1921. The archive also contains reports from numerous cities in Italy on their plans to 'contain' the public; notable is one from Perugia, which has a letter from a local Socialist who claims 'you won't lay the dead to rest with these choreographic speculations' (con simili speculazioni coreografiche non si placano i morti). Donna Haraway, *Modest_Witness@Second_Millennium. FemaleMan©Meets_OncoMouse™: Feminism and Technoscience* (New York: Routledge, 1997), 135–6. Mary Louise Pratt, *Imperial Eyes: Travel Writing and Transculturation* (London; New York: Routledge, 1992). Vetri Nathan connects Haraway and Pratt in his 2009 dissertation, *Bodies in Migration: The Ambivalent Locations of Post-Coloniality in Italy.*

5 For the placement law, or law on 'collocamento dei mutilati e invalidi di guerra,' see AP, Senato, 10 Aug. 1921. For all the figures, see MdI, OaSI, 1921.

6 On women, see Tobia, *L'altare della patria*, 78. Canal, 'La retorica della morte,' 660–1. The mother who regenerates the dead son recalls Spackman's description (in *Fascist Virilities*) of the 'virile woman' in the d'Annunzian and Fascist imagination. Caroll (*Madonnas that Maim*, 68–9) shows that there are many Virgins in Italian popular Catholicism, related to each other, yet with distinct personalities and roles. He also shows that the Madonna can be a dangerous figure, as well, who must be placated, lest she bring destruction; I believe this vengeful Madonna is connected to the 'hysterical' female mourners that the state saw as dangerous to social order. Grayzel, *Women's Identities at War*, 239, 237.

7 On the French inauguration, see Le Naour, *Le Soldat Inconnu*, 13, 12–13, 60; Soudagne, *L'histoire incroyable du soldat inconnu*, 102, 112. On the British one, see Hanson, *Unknown Soldiers*, 293, 294, 295, 290. On the one in the United States, see *Tomb of the Unknowns: 'Here Rests In Honored Glory An American Solider Known But To God,'* retrieved from http://www.military.com/veterans-day/tomb-of-the-unknowns.htm. One witness emphasizes the great pomp, order, and glory of the U.S. event, noting the presence of political and military leaders far more than that of soldiers of any sort (ibid.).

8 Amy Hollywood (*Sensible Ecstasy*, 260–5) cites Judith Butler and comments. For the notion of the Other as defined by a suffering that always exceeds any capacity to alleviate it, see Edith Wyschogrod, *Saints and Postmodernism: Revisioning Moral Philosophy, Religion and Postmodernism* (Chicago: University of Chicago Press, 1990), 254–7. Wyschogrod argues, in particular, that 'heedfulness to what is abyss-like [or also cataclysmic, linked to total war, among other things] in history need not produce an eros for it but can elicit a postmodern ethic that is sensitive to the postmodern cry for difference' (xxi).

Freud's contrast between mourning and melancholy, written in 1915, clearly in part at least in reaction to the First World War, is in 'Mourning and Melancholia,' *The Standard Edition of the Complete Psychological Works of Sigmund Freud*, vol. 14, trans. James Strachey (London: Hogarth Press and the Institute of Psycho-Analysis, 1995). For the connection between spectacle and the mourning of political values, in the passage from the Middle

Ages to Absolutism, when the body of the nation usurps actual physicality, but the theatre is where it can still exist and we can feel its loss, see Jean-Marie Apostolidès, *Le prince sacrifié: Théâtre et politique au temps de Louis XIV* (Paris: Éditions de Minuit, 1985).

9 Vincenzo Nardelli, *L'Arcangelo* (Rome: Stock, 1931). Andreoli, *Il vivere inimitabile.* 'Libro riparatore per i mutilati del corpo ed i mutilati dell'anima' is in *L'Ardente,* Oct.-Nov.-Dec. 1921. *L'illustrazone Italiana,* 27 Nov. 1921, describes *Notturno* as 'luce dalle tenebre' (light from the darkness) and praises d'Annunzio's evocation of his illness as allowing us to relive 'la nostra Guerra' (our war).

10 On Cendrars, including quotes from *La main coupée,* see Claude Leroy's introduction in Cendrars, *La vie et la mort,* 18–19, 22. For an analysis of Cendrars' pseudonym in relation to the loss of his hand and to his works as a whole, see Leroy, *La main de Cendrars.*

The originals read: 'si elle a été ensevelie et non pas jetée aux ordures dans un charnier avec d'autres restes ou dépouilles ou abattis'; 'une grande fleur épanouie, un lys rouge, un bras humain tout ruisselant de sang, un bras droit sectionné au-dessus du coude et dont la main encore vivante fouillait le sol des doigts comme pour y prendre racine.'

The hand is described as having fallen from the sky in an area far away from any battle: Blaise Cendrars, *La Main coupée,* in *Oeuvres complètes,* ed. Raymond Dumay and Nino Frank (Paris: Club français du livre, 1968), vol. 10, 276–7. See also: Elisa Bricco, 'La Main coupée di Cendrars: La guerra come gioco,' *Quaderni del Dipartimento di Lingue e Letterature Straniere Moderne* 9 (1997): 255–67; Elisabetta Carta, 'Grande guerra e memoria del corpo: Stigmate del tempo e dell'esperienza,' *Compar(a)ison* 1–2 (2005): 185–91. On Cendrars and Gance, see Abel Gance, 'Blaise Cendrars et le cinema,' *Mercure de France,* no. 345 (1962): 170–1.

11 Kirkwood, in Paulson, *Haunted by Combat,* xv; on initiation, see Chapter 9 and esp. pp. 92–101. Tick, *War and the Soul,* chapter 3.

12 Caruth, *Unclaimed Experience,* 2–3, 5, 2, 9.

13 Leys, *Trauma,* 9, 6; she cites Young, *The Harmony of Illusions,* 5. Young's point, at 9–10, is not that trauma or PTSD are not 'real,' but that we cannot separate them from their historical constructedness, which has cultural, social, and political implications. His title, 'the harmony of illusions,' refers to the way the researcher is never passive in the creation of new cognitive schemas, so that a new idea, such as that of trauma and later PTSD emerges from the interaction of the researcher and his subject, which corresponds to Leys' claim that mimetic and anti-mimetic views of trauma are not separable, since trauma emerges both as a 'creation' and a 'discovery.'

14 Leys, *Trauma*, 3–4.

15 'The phantom [limb] is an expression of nostalgia for the unity and wholeness of the body, its completion. It is a memorial to the missing limb' (Grosz, *Volatile Bodies*, 73).

Ariel Glucklich, *Sacred Pain: Hurting the Body for the Sake of the Soul* (New York: Oxford University Press, 2001). For Glucklich, phantom pain is a hallucination in the sense that it is generated by 'the absence of incoming neuronal signals'; however, insofar as it is experienced as pain, and tends to undo the cohesion of our body image, it is also a type of sensory overload or 'reverse hallucination,' which results in states of dissociation and, at the extreme, of ego loss (57–9).

D'Annunzio and Cendrars modulate their experience of loss via self-mythologizing, anticipating the role of storytelling in healing PTSD, as seen, e.g., in the use of the 'Warrior' archtype in Tick, *War and the Soul.* For a contrast between lack and 'a lapse and "loss" that are "structurally" rather than contingently irrecoverable by memory' and the relation of the latter to the saint's relation to the other, see Wyschogrod, *Saints and Postmodernism*, 108.

16 Sigmund Freud, *Beyond the Pleasure Principle*, trans. James Strachey (New York: Norton, 1990), 14–16.

17 Ibid., 16.

18 For Irigaray, in the essay 'Belief Itself' (*Sexes and Genealogies*), the omission of Freud's daughter's death allows the fiction of a possible return to symbiosis with the maternal to persist; she asserts, in contrast, 'there can be no return to that first dwelling place' (33). This would imply that the oceanic is always an illusion. However, in other essays, including 'Divine Women' (62ff., in the same book), Irigaray suggests that while no 'return' to the maternal is possible, another sort of symbiotic, interpenetrating, fluid relation between subjects is possible, and related to the divine, in particular, the female divine that eschews conceptual mirroring. For a thorough discussion of Irigaray's changing views of mystical experience and thus of the oceanic, see Hollywood, *Sensible Ecstasy*, 187–235.

Jacques Derrida's discussion is in his *L'écriture et la différence* (Paris: Seuil, 1967), translated by Alan Bass, as *Writing and Difference* (Chicago: University of Chicago Press, 1978). For how Derrida 'invites psychoanalysis to mourn its own past' and to be open to ethical and political reinterpretations of mourning, see Alessia Ricciardi, *The Ends of Mourning: Psychoanalysis, Literature, Film* (Stanford: Stanford University Press, 2003), 67. Both Irigaray and Derrida are discussed in Dennis King Keenan, *The Question of Sacrifice* (Bloomington: Indiana University Press, 2005). See Sigmund Freud, *Civi-*

lization and Its Discontents, trans. James Strachey (1930; New York: Norton, 1961). Freud's discussion of the oceanic is part of a complex dialogue with Romain Rolland: see William Barclay Parsons, *The Enigma of the Oceanic Feeling: Revisioning the Psychoanalytic Theory of Mysticism* (New York: Oxford University Press, 1999).

19 Freud, *Civilization and Its Discontents*, 20–1.

20 C.G. Jung, *Memories, Dreams, Reflections*, trans. Clara Winston and Richard Winston (New York: Vintage, 1989), 180ff.

21 Parsons analyses the correspondence of Freud and Romain Rolland regarding the oceanic, making two points especially relevant to our discussion. First, what Rolland, along with many contemporaries such as William James, calls the 'oceanic' is a specifically modern mystical tradition, related to ancient and non-Western ones, but not identical to them, in particular because it is a 'mystical psychoanalysis': this creates a link between the modern discourse of mysticism and that of trauma that is historically situated around the latter part of the nineteenth century and finds a specific instantiation in First World War experiences of simultaneous 'horror' and 'eternity,' such as those of Ungaretti. Second, unlike Freud, Jung, and other contemporaries such as Heinz Kohut proposed models for non-pathological forms of mysticism, and Parsons rightly asserts that their heritage has yet to be fully claimed: Jung's response to Freud suggests that examining the role of violence in mystical experience and in its interpretation might be a relevant distinction to explore. See Parsons, *The Enigma of the Oceanic Feeling*, 145, 167, 163.

22 For the 'sainted face' (*santo volto*), a clear reference to the Banner of Randaccio, see *Il Dovere*, 5 Nov. 1919. On d'Annunzio's charisma, see Ledeen, *D'Annunzio*, 202, 909. The importance of Fiume rituals to the creation of the Unknown Soldier Memorial is attested to by the 6 Nov. 1921 issue of *L'illustrazione Italiana*, which describes at length the Memorial's inauguration, but also includes an article on 'a Fiuman ceremony' celebrated by d'Annunzio recently at his then new home on Lake Garda. On Fiume see the following: Claudia Salaris, *Alla festa della rivoluzione: Artisti e libertari con D'Annunzio a Fiume* (Bologna: Il mulino, 2002); Gentile, *Il culto del littorio*; Gumbrecht, Kittler, and Siegert, *Der Dichter als Kommandant*. On Fiume and d'Annunzio see: Valesio, *The Dark Flame*; Ballinger, 'La Pentecoste italiana.' On d'Annunzio's ritualization of violence, see Spackman, 'Il verbo (e) sangue.'

Testimonies of some participants are published in Elena Ledda, *Fiume e D'Annunzio: Pagine di storia* (Chieti: M. Solfanelli, 1988).

For one of the earliest historical examinations of the relation between d'Annunzio, Fiume, and Fascism, see Renzo De Felice, 'D'Annunzio e la vita politica italiana dal 1918 al 1936,' in *D'Annunzio politico, 1918–1938.* In De Felice's view ('D'Annunzio nella vita politica italiana,' 471), d'Annunzio's entourage should be divided between those who were 'legalitarian' and those who sought 'innovation' and 'liberation,' 'many of whom [i.e., the latter] became anti-Fascists.'

Gumbrecht ('I rendentori della vittoria,' 262–3), makes the important connection between redemption theology, with its notion that the redeemed state is never quite present, and d'Annunzio's willingness while at Fiume to procrastinate in terms of action, yet also to stage new dramas of sacrifice and redemption.

An important testimony on Fiume is Giuseppe Moscati, *Le cinque giornate di Fiume* (Milan: Carnaro, 1931). The Unknown Soldier Memorial was also described as a 'potlach,' in Canal, 'La retorica della morte,' 661–2. On declining interest in trauma after the First World War, see Leys, *Trauma,* 5.

23 Ledeen, *D'Annunzio,* 23, 25, 27, 29, 30, 33–4. De Felice, cited in Ledeen, ibid., 45. Ledeen, ibid., 40, 50, 61, 66ff.

24 Ibid., 83, 121, 147. On Keller's Yoga, see Salaris, *Alla festa della rivoluzione,* 47, 73. See Guglielmo Negri, 'La Carta del Carnaro,' *Il Veltro,* 40/5–6 (1996): 495–505.

Jeffrey Meyers, *A Fever at the Core: The Idealist in Politics* (London: London Magazine Editions, 1976). For Meyers, d'Annunzio was 'the master of patriotic eroticism,' having 'the ability to convert sexual energy into political action' (89) and knowing that 'politics is linked with theatre' (93); he is condemned as a 'political chameleon' (110). In contrast, Bertazzoli ('Retorica fiumana,' 361, 357), emphasizes the 'style' invented by d'Annunzio at Fiume, defining it as one of 'antithesis' and quoting d'Annunzio: 'Voglio essere un capo sense partigiani e un condottiero senza seguaci' (I want to be a head without an entourage and a leader without followers).

The originals read: 'unione di spiriti liberi tendenti alla perfezione' and:

> La YOGA vi invita ad una nuova danza – luci profumi e musiche – gli elementi eccitatori più idonei a spirituali suggestioni di bellezza [...]
> La danza negli abissi di marine profondità.
> La danza nelle foreste africane.
> La danza al di là del bene e del male.

25 Ledeen (*D'Annunzio,* 164ff., 184–5), shows that the Socialists refused an alliance with d'Annunzio, following Soviet instructions, and that a notable turn to the right at the end of the Fiuman enterprise was a desperate attempt to find allies. Also, he points out that the anti-Fascist Arditi del popolo appear

after the Fiuman adventure. De Felice ('D'Annunzio nella vita politica italiana,' 474) quotes Giulietti's notion of Fiume as 'la Svizzera di tutti i sowersivi' (the Swtizerland of all subversives).

26 'Oceanic' is also in Ledeen, *D'Annunzio*, 201.

27 For fire, and the horse, see ibid., 72, 158.

28 See Martinelli, *La guerra di d'Annunzio*, on the ninth Isonzo (140–1). D'Annunzio's own diaries show him further back at this specific moment, though Randaccio and other witnesses testified to his courage in this battle, and perhaps more unusual, to his humility in assisting the soldiers with first aid (147–9). On the two flags and on the tenth Isonzo (149–59). Martinelli reports the apocryphal story from Giovanni Artieri's *Il Re, i soldati, e il generale che vinse* (Bologna: Cappelli, 1951), 148.

29 'Dalla ringhiera del Campidoglio,' on 'terre irredente,' is in d'Annunzio, 'Il sudore di sangue,' 883–4. It was also published in 1919 in his *Contro uno e contro tutti* (Rome: Presso La Fionda, 1919). See d'Annunzio, *Siamo spiriti azzurri e stelle*, 37.

The originals read:

> Questa, Romani, questa, Italiani, questa, compagni, è la bandiera di quest'ora.
>
> L'imagine sublime del fante, che vi poggiò la testa, v'è rimasta effigiata.
>
> Ed è l'imagine di tutti i morti; ché tutti quelli che sono morti per la Patria e nella Patria si somigliano.
>
> E' il sudario del sacrifizio.
>
> [...] voglio abbrunare la mia bandiera finché Fiume non sia nostra, finché la Dalmazia non sia nostra.
>
> Ogni buon cittadino, in silenzio, abbruni la sua bandiera, finché Fiume non sia nostra, finché la Dalmazia non sia nostra.

'Sangue e sanie' appears in d'Annunzio, *Siamo spiriti azzurri e stelle*, 158, and recurs in the 3 Jan. 1921 funeral oration for the dead of Fiume (Moscati, *Le cinque giornate di Fiume*, 173).

30 The assertion 'ma in te, o Fiume nostra, santità del Quarnaro, in te ha loco il Santo Volto' (but in you, o our Fiume, saintliness of Quarnaro, in you is kept the sainted face) is in *Il Dovere*, 11 May 1919. On how famous this gesture became, see Ledeen, *D'Annunzio*, 59. D'Annunzio vowed to take the flag back from Rome to Fiume and the battlefields of the First World War, in an itinerary that was, *ante litteram*, the reverse of the Unknown Soldier's. On *Hiroshima mon amour*, see Caruth, *Unclaimed Experience*, 37ff.

In his introduction to d'Annunzio's *Siamo spiriti azzurri e stelle*, at xix, Pietro Gibellini notes that an appropriate epigraph for d'Annunzio's wartime and postwar writings, and in particular for the 'Comento meditato,' in which his fall and the Banner of Randaccio come together, would be the phrase from the *Libro Segreto*, 'L'anima è misteriosa ma anche il corpo lo è' (The soul is mysterious but so is the body).

31 For uses of the Banner of Randaccio, see Ledeen, *D'Annunzio*, 69, 95, 126, 136. For its use at the end of the Fiuman enterprise, see Moscati, *Le cinque giornate*, 132, 161. D'Annunzio connects the flag to the Unknown Soldier in *Notturno*, 333. The original reads: 'E perché il feretro del Grande Offeso [il Soldato Ignoto] non era ammantato dalla bandiera del Timavo [...]?'

32 The documented history of the Shroud of Turin goes back to the fourteenth century, but there is textual history for one or more similar objects back to the year 300. Though scientists have questioned its age, they remain puzzled by how blood and the other present substances might have been applied to the linen to produce an image that is three dimensional and 'not painted.' See Stefano Rizzelli, *Dalla sindone a Padre Pio: I segreti del sacro, Viaggio nell'Italia dei miracoli e dei misteri* (Rome: Newton Compton, 1999), 119–25, 148–54.

Etymologically 'labaro' (banner) refers to a depiction of Christ's face; hence, it connects Randaccio to the Shroud.

The quote is from d'Annunzio, *La riscossa*, 162–3, 165. The original reads:

> Il vostro [eroismo] è come le vostre ossa, è dentro di voi come il vostro scheletro; è l'armatura interna, è sempre là; regge la vostra carne misera e la tiene esposta di continuo alla distruzione più orrenda. La vostra vita è come il drappo della bandiera e il vostro coraggio è come l'asta. Ci sono bandiere che il vento lacera e rapisce. L'asta rimane. [...] Eroi; e pure non sembravate neppure uomini ma cose, ma povere cose come le pietre tritate, come i sacchi sventrati, come gli elmetti sformati, come le scatole vuote, come le bottiglie rotte [...].
>
> Si dice che c'è una vita e si dice che c'è una morte. Si vive nella vita e nella morte si muore. Sembra vero. Ma per il fante c'è qualche cosa che non è la vita e qualche cosa che non è la morte; c'è un elemento nuovo, una specie di limite sospeso, una specie di orlo misterioso e irrespirabile, dove egli pure respira e spesso ride e spesso canta, e non perisce; perché io dico che quell'elemento è l'infima immortalità dell'uomo.

The English translation is from Gabriele d'Annunzio, *The Rally*, trans. Magda Sindici (Milan: Casa Editrice d'Arte Bestetti & Tuminelli, 1918), 153–6. I have amended the last phrase, which reads 'I maintain that that element is man's lesser immortality': this translation implies that there is another, greater immortality, an implication that is not present in the original. Valesio, *The Dark Flame*, 118–19.

33 The diary entry for d'Annunzio's visit to Doberdò is in *Taccuini*, 870–1. The originals read: 'gli elmetti amaccati, scrostati, forati, l'un su l'altro, grigi come la cenere, col cuoio dentro macero di sudore, intriso di sangue'; 'Il cristo – Le scarpe gli elmetti spoglie dei morti – Il Cristo che porta la croce – la VI Stazione sull'altar maggiore Appoggiate al muro le barelle macchiate di sangue.'

34 For the wreath, see *L'illustrazione Italiana*, 13 Nov. 1921, 584. The quote is from d'Annunzio, 'Comento meditato,' 543. The original reads:

> Quando Lazaro solleva il ginocchio fuor dal sepolcro, egli si esprime più che se gettasse il grido immortale della sua risurrezione. [...] Io vi dicevo dianzi: 'Voglio aiutare i morti a esprimersi.'
>
> Questo ginocchio ferito, quest'osso di fùsolo messo a nudo, questo malleolo schiacciato, tutto questo dolore che si scarnisce e s'impoverisce, ecco, io lo sollevo con la mia volontà.
>
> Non v'è cagione d'inorridire. Così mi esprimo. Questa è la forza, questo è il rilievo della mia espressione. [...] Ora taccio. Non parlo più.

On reopening Randaccio's tomb, see D'Annunzio, 'La preghiera di Aquileia,' Ognissanti 1918, in d'Annunzio, 'Tre preghiere.' The original reads: 'Domani laggiù i nostri morti primi, i nostri martiri primi, sorgeranno e scoperchieranno la tomba di Giovanni Randaccio.'

On Miraglia's corpse, see d'Annunzio, *Notturno*, 57, 61. The original reads:

> Non si potrebbe levare, neppure se Cristo lo chiamasse. [...]
>
> E il terzo giorno. Lo sfacelo continua.
>
> Ho nelle ossa un freddo terribile. Toccare la morte, imprimersi nella morte, avendo un cuore vivo!

35 Lazarus' encounter with Jesus is described by Gabriele d'Annunzio, *Le faville del maglio e altri studi del vivere inimitabile* (Milan: Treves, 1924), 227. The original reads: 'le bende funebri intrise d'unguento radunate in mucchio rendevano odore.'

Lazarus reappears various times in nineteenth- and twentieth-century Italian and French literature, and d'Annunzio's sources include Charles Baudelaire, 'Le peintre de la vie moderne,' *Oeuvres complètes* (Paris: Seuil, 1968); Maurice Barrès, *Le Culte du moi*, vol. 3, *Le Jardin de Bérénice* (Paris: Perrin, 1891).

36 D'Annunzio, *Notturno*, 106–8. The original reads:

> Voi mi bendate la fronte, mi fasciate le palpebre, mi lasciate nell'oscurità.
> E io vedo, vedo, sempre vedo. E di giorno e di notte, sempre vedo. [...]
> Guardo la borsa dell'acqua nell'occhio gonfio, il mio viso consunto e smorto, la mia bocca livida e piegata dalla tristezza, i nuovi fili bianchi nella mia barba negletta, il mio collo scarnito: una immagine di miserabile accoramento, che si fissa nella retina e vi rimane. [...] E incomincia la trasformazione.
> La tristezza umana è divenuta una materia plastica. Non so qual pollice misterioso la modelli incessantemente. [...]
> Non sento più i confini del mio scheletro.

Martinelli, *La guerra di d'Annunzio*, 146. The original reads:

> C'è un braccio là. Non lo pestate. E rivolto contro il nemico. Ha il pugno quasi chiuso. Vedete. Ha le dita rattrappate, tranne l'indice. Vedete. [...] Vi segna la nostra linea [...] Questo morto rivuole il suo fucile. Non lo pestate. Si disotterra da sé. Stanotte fà da guida alla brigata ... e quel braccio di martire, tutto un seccume forte di cartilagini e di tendini e di ossi, pareva ogni volta tremare come se anch'esso mi fosse dentro.

37 'I fanti mordevano l'azzurro' is in d'Annunzio, *La riscossa*, 151, 161, and in *The Rally*, 143–4; it reappears in 'Il Vittoriale,' 16 Aug. 1919, in d'Annunzio, 'Sette documenti,' 655. On the gold medals and pension, see Ridolfi, *Le feste nazionali*, 156. That this was an issue of social class (the 'Fante' had to be lowly, yet elevated) is confirmed by two somewhat comical imaginations of the Unknown's 'return,' in *L'Ardente*, 1 July 1921 and Aug.–Sept. 1921, respectively: in the first, an elegant happening is announced in which 'Capitano Landi' will play the 'Fante' to which a noblewoman will offer a laurel wreath; the second reports an imaginary dialogue between the Unknown Soldier and Dante, Cino [da Pistoia], and Guido [Cavalcanti], who instruct him on love. On the livid wound, see d'Annunzio, 'Comento meditato,' 550, 553. The original reads: 'I fanti mordevano l'azzurro. Ma l'azzurro mi rosseggiava. Mi pareva che tutti avessero il mio cuore per insegna vermiglia [...] Vi lascerò entrare nel mio petto attraverso questa lividura che sembra un varco azzurro.'

38 Michel Serres, *Les cinq sens* (Paris: Grasset, 1985), 13–17.

39 For Ungaretti's 'Sempre notte,' and 'Un'altra notte,' the reference to the 'notte violentata' in 'In dormiveglia,' 'Malinconia,' and 'Dolina notturna,' see Ungaretti, *Vita d'un uomo: Tutte le poesie*, 71, 72, 42, 63. Translations are mine, in the first case, and in the others are respectively drawn from Ungaretti, *Selected Poems*, 57, and *A Major Selection*, 67, 99.

The originals read:

Sempre notte
Vallone il 18 aprile 1917

La mia squallida
vita si estende
più spaventata di sé

In un
infinito
che mi calca e mi
preme col suo
fievole tatto

Un'altra notte
Vallone il 20 aprile 1917

In quest'oscuro
colle mani
gelate
distinguo
il mio viso

Mi vedo
abbandonato nell'infinito

Malinconia
Quota Centoquarantuno il 10 maggio 1916

Calante malinconia lungo il corpo avvinto
al suo destino

Calante notturno abbandono
di corpi a pien'anima presi
nel silenzio vasto
che gli occhi non guardano
ma un'apprensione

Abbandono dolce di corpi
pesanti d'amaro
labbra rapprese
in tornitura di labbra lontane
voluttà crudele di corpi estinti
in voglie inappagabili [...]

40 On the 'Risorgimento relic,' see Miniero, *Da Versailles*, 232. On use of the banner with the Lupi, see Martinelli, *La guerra di d'Annunzio*, 137. On lacerated flags, see Miniero, *Da Versailles*, 150, 159.

41 The article on blind soldiers at the inauguration is cited in Miniero, *Da Versailles*, 208.

42 Pozzi, 'Il Duce e il Milite ignoto,' 341, 342, 343, 347, 348, 355. Fogu, in *The Historic Imaginary*, at 76, makes a similar point about Mussolini's relationship to Garibaldi, noting that the latter could not be celebrated too directly by the Duce lest he eclipse him; Mussolini's solution was to honour Anita Garibaldi. See also Claudio Fogu, 'Il Duce taumaturgo: Modernist Rhetorics in Fascist Representations of History,' *Representations* 57 (1997): 24–51.

On Mussolini's body, see also Luisa Passerini: *Mussolini* (Bari: Laterza, 1997), and *Mussolini immaginario: Storia di una biografia 1915–1939* (Rome: Laterza, 1991). For 1922 and onward, see Ridolfi, *Le feste nazionali*, 159. Carlo Delcroix is mentioned in Dogliani, 'Redipuglia,' 385.

For a definition of *bonifica* as far more than the literal draining of the Pontine marshes, and as a program of modernization and national regeneration, see Ben-Ghiat, *Fascist Modernities*, 4–15.

The originals read: 'il Duce è la forma vivente del Milite Ignoto, la sua reincarnazione dinamica'; 'cadavere polimorfo'; 'schermo bianco'; 'corpo universale'; 'in nessuna delle molte affabulazioni sullo stato fisico del Milite ignoto viene detto che potrebbe essere mutilato,'; 'ripristin[o]'; 'Vittoria mutilata'; 'Mussolini bonifica in se stesso il male sociale che ha accolto in se stesso.'

43 Ridolfi, *Le feste nazionali*, 159. Brice, 'Le Vittoriano,' 351–6, and *Il Vittoriano*, 327ff. This sort of advance absolution is satirized in Alberto Moravia's *The Conformist*, in which the main character, Marcello, who is wondering whether he is capable of murdering a former professor and friend in order to ingratiate himself with the Fascist regime, goes to confession and demands to be absolved of the sins he is about to commit; he receives absolution upon reassuring the priest that he will do nothing that does not safeguard the nation. This in turn refers to Dante, *Inferno* XXVII.

44 Dogliani, 'Redipuglia,' 378–9. For 'ripulire,' see 382; on 'comandanti,' 384; on the Duca d'Aosta, 385; on the second 'sacrario' and on 18 Sept. 1938, with anti-Jewish speech, 386.

The originals read: 'ripulire, riordinare, discipinare anche i luoghi di culto, smantellando gli orginari cimiteri sorti spontaneamente durante e subito dopo la guerra'; 'affermare e ideologizzare anche visivamente lo schiacciante rapporto di superiorità dello Stato sul cittadino'; 'un insieme omogeneo di marmo bianco'; 'tra massa e capo'; 'comandanti che non morirono, come i soldati, sui campi di battaglia o sucessivamente per ferite di guerra, bensì di morte naturale anni dopo il conflitto'; 'corpo sacro dei Savoia.'

On the great contrast between the original cemetery and the Fascist clean-up, see Pamela Ballinger, *History in Exile: Memory and Identity at the Borders of the Balkans* (Princeton: Princeton University Press, 2003), 53–4. On the notion that homogeneity and blankness regarding death is a way for Fascism to accumulate a sort of emotional capital by *not* attributing any specific meaning to dead bodies but instead endlessly projecting that meaning into the future, see Vander, *L'estetizzazione della politica.*

45 Gentile, *Il culto del littorio,* 67–9, 53–4. The originals read: 'l'appello divenne il *rito fascista* per eccellenza, la testimonianza più alta della loro religiosità'; 'trasfigura[re] i riti di morte in *riti di vita.*'

46 On fire and ardent heterogeneity, and the fire and iron image in Boehme, see Evelyn Underhill, *Mysticism* (1911; New York: Doubleday, 1990), 284. 'Teste di ferro' is how d'Annunzio named his followers; see Ledeen, *D'Annunzio,* 82. 'Blaze,' 'Calvary,' and 'meaning of my enterprise' are respectively in Ledeen, ibid., 72, 120, 141. On the imagery of the flame at Fiume but also in the earlier play, *La Gloria,* whose hero is named Ruggiero Flamma, see Choate, 'D'Annunzio's Political Dramas,' 375–6.

47 On Keller's mission, and the Dadaists, see Ledeen, *D'Annunzio,* 142, 144. Kochnitzky, in ibid., 151.

48 Salomé in Huysmans, *A rebours,* 95. D'Annunzio and Saint Sebastian, in Ledeen, *D'Annunzio,* 149–50. Orpheus over Christ, in ibid., 148.

49 Ledeen, *D'Annunzio,* 148–9.

50 In ibid., 158.

51 Cendrars, *La vie et la mort,* 41–2. The original reads:

> Mais l'ancêtre de tous les chevaux, des légendaires et des domestiques, c'est l'etalon qui porta Prométhée jusqu'au sommet du Caucase à la conquête du feu [...].
>
> Et la voix chevrotante des vieux bardes se raffermissait pour dévoiler les arcanes du poème [...] l'obscure généalogie pour acclamer dans un crescendo triomphal les hauts faits du fondateur de la race, du premier membre de la tribu, du père du clan, Uûr, l'étalon fantastique [...].

Cendrars, ibid., 58–60. The original reads:

> Derrière un tertre, il tombe sur un cheval crevé. Il ne saurait dire comment il est arrivé là. Mais il n'est pas le premier. [...] Ils ont hâte d'en finir car leur tournée sera longue cette nuit. Ils arrachent les fers du

cadavre, fendent la peau le long du ventre, le long des jambes, dépiautent fébrilement la bête; puis ils s'en vont, le barbu emportant la peau roulée et toute dégoulinante [...] Maintenant tout le monde crie comme s'il s'agissait pour chacun de couvrir la voix les autres. Les hommes ont sorti leur couteaux de la gaine et chacun s'attaque à son morceau de prédilection. [...] Deux hommes s'emparent du crâne, un autre fouille par terre comme un chien. [...] Ceux qui vont manger en commun auront à échanger des morceaux de cheval et deviendront ainsi frères pour la vie, frères de sang. [...] Ils boivent d'abord le jus de la viande [...] quand ils s'en sont bien régalés, chacun attrape son morceau de viande, chacun le frotte contre ses joues, puis, avant de le porter à sa bouche, chacun fait la part du feu [...] alors chacun se met à mastiquer [...] Ne dirait-on pas qu'ils sont ivres? Mais, non, ils sont tout simplement absents, loin. Chacun d'eux est visité et chaque homme de s'entretenir avec Iron [descendant of Uûr].

Though *faire la part du feu* can mean 'to cut a firebreak,' i.e., to create a 'counterfire' that stops an existing one, in this case 'la part du feu' or 'fire's share' refers to the practice of not consuming all of the flesh of the sacrificed animal, leaving some for 'the fire,' i.e., for the gods.

52 Leroy in Cendrars, ibid., 127–8. On horse sacrifice, see Eliade, *Shamanism*, 197, 192, 193, 198, 199. On visions, see Cendrars, *La vie et la mort*, 61. The original reads: 'visions funèbres qui le faisaient claquer des dents, [où il] voyait son fils tomber d'un train, son bonnet blanc se ouater de sang, des serpents couler de ses yeux, une grosse mouche se poser sur sa bouche.'

53 Trance, the Captain, and 'double anonymity,' in Cendrars, *La vie et la mort*, 60, 62–3, 156, 146. The originals read: 'dans un état très voisin de la transe qui s'empare des moullahs et des dervishes tourneurs'; 'vaillants montagnards'; 'L'un d'eux, qui se nommerait paraît-il Ostapyth et qui est d'après les dires des blessés le chef-chanteur de son escadron, arrêta à lui seul un train militaire bondé de soldats allemands en fuite et le fit rétrograder en gare, où tout le monde fut fait prisonnier'; 'Le Tsar fit établir une enquête qui établit que le cosaque n'existait pas'; 'double anonymat.'

54 For El-Nar, see d'Annunzio, *Notturno*, 239–42. For wounds transformed into roses at Doberdò, see d'Annunzio, 'Tre preghiere,' 638.

55 On 'rumination' as a metaphor for the meeting of the body and the spirit, as relevant to mourning, in Saint Augustine and others, see Edward Sieveking Emery, 'In Light of Mourning: Spiritual Transformations between Trauma and Presence,' *Psychoanalytic Review* 95/4 (2008): 625–54.

D'Annunzio, *Notturno*, 101–2. The original reads:

> quei soldati ciechi [...] [m]i sono fratelli. [...] la povera carne messa fuori
> combattimento, la bocca inquieta di chi non vede. [...] Ero nudo d'ogni privilegio, senza singolarità, senza rilievo, senz'altra gloria che il mio umile sacrifizio.

Ibid., 116. The original reads:

> La lacrimazione dell'occhio e il sudore delle tempie mi colano fin sul labbro. E lambisco le gocciole salse. E mi sembra di lambirle con la bocca di mia madre, con quella bocca deformata che pesa in me, che soffre in me contraffatto. [...]
>
> Non è un imagine immobile. Si muove, si muta. [...] Apre la bocca, e non può parlare la parola umana; non può se non masticare l'anima, biascicare la desolazione.

In the 'Comento meditato,' at 561, where d'Annunzio speaks with the voice of the Unknown Soldier, he also imagines his mother covering his sick body with Randaccio's flag, in a gesture reminiscent of that of Veronica.

D'Annunzio, *Notturno*, 119. The original reads:

> Credo di udire dentro di me le grida di mia madre che, quando nacqui, non penetrarono le mie orecchie sigillate. [...]
>
> E' la sesta stazione: il sudario della Veronica. [...]
>
> Guardo quel viso. [...] Orribile e sublime, veramente, con uno sguardo che non mi vede, che non mi riconosce, oscurato e fisso, dove l'amore non è se non tristezza senza nome, tristezza sino alla morte e di là dalla morte.
>
> Mia madre!
>
> Una povera creatura avvilita, percossa, sfigurata.

The word 'sudario' instead of the normal 'velo' (as the *Veronica* is far too small to be a shroud and only bears a facial imprint) confirms d'Annunzio's imaginative confusion of the Randaccio flag, the Shroud of Turin, and the Veil of the Veronica. Luisa de Benedictis, d'Annunzio's mother, died on 27 Jan. 1917.

56 Apocryphal tradition has at times identified Veronica with the 'bleeding woman' cured by Christ, as though to confirm the association of the veil with an excessive boundary-less 'circulation' of blood. See Luke 8:43–8. 'Throwing his lot in with humanity' is from Eamon Duffy, *Madonnas That Maim? Christian Maturity and the Cult of the Virgin*, retrieved from http://www.bfpubs.demon.co.uk/duffy.htm. For the connection between an art of 'dis-

figuring' and the 'death of God,' see Taylor, *Disfiguring.* For the argument that the structure of torture and of war is 'the inverted form' or another sort of disfiguring of 'the relation between physical pain and imagining,' and a reflection on the relationship between 'unmaking' and 'making,' see Scarry, *The Body in Pain.*

57 For the *theotokos,* see Duffy, *Madonnas that Maim?* For d'Annunzio, Michelangelo's art was clearly the model for the commemoration of wartime death, and specifically what one might call Michelangelo's 'nocturnal' art, the *non finito* and the incomplete, defaced works. See, e.g.: 'Ci sono eroi tuttavia imprigionati nel silenzio come quelli del Buonarroto *appresi* nello scalpellato sasso' (There are heroes still imprisoned in their silence like those of Michelangelo that *catch fire* in the chiselled stone) and 'L'anima non fugge ma è tuttora *appresa* alla ferita come alla face lo splendore che nella raffica si spicca e si rappicca, cessa e si riattiva, si piega e si risolleva, non tenuto se non da un legame invisibile che la volontà di ardere rende più forte della tempesta' (The soul does not flee but *catches fire* in the wound as does from the embers the splendour that in the fire shoots forth and returns, stops and is reactivated, bends and rises, held only by an invisible will to burn made stronger by the storm) (emphasis added in both cases), respectively in d'Annunzio, 'Agli uomini di pena,' 583, and *Notturno* 15.

On the purity of the mothers, see *Corriere della Sera,* 6 Nov. 1921, also in Labita, 'Il Milite Ignoto,' 151. 'Sedute, le mani in grembo, niente le distrae.' Labita concludes on this, even as he mentions other less noble episodes. Tobia (*L'altare della patria,* 78) sees the mourning mothers as being in perfect syntony with the celebratory yet sombre tone of the ceremony. In contast, for a view of how in France feminists sought to turn mourning mothers to their cause and to pacifism, see Siegel, 'To the Unknown Mother.' For the more overwrought mothers, see *Gazettino,* 30 Oct. 1921, and *Corriere della Sera,* 2 Nov. 1921; they are mentoned also in Labita, 'Il Milite Ignoto,' 142, and Tobia, *L'altare della patria,* 79.

58 Hollywood, *Sensible Ecstasy,* 268ff. For the related notion of 'carnal generality' that 'precludes hypostasized values,' see Wyschogrod, *Saints and Postmodernism,* 50. On corporeality in d'Annunzio, and the claim that in his work the word becomes 'esigenza del corpo' (demand for a body) so that 'il "mistero" è invece per d'Annunzio realtà corporea, biologica, da indagare fisicamente con una coscienza estesa dei sensi all'*anima profonda*' (the 'mysterious' [or also numinous] is for d'Annunzio a corporeal realilty, to be investigated by a consciousness expanded by the senses to the *deeper soul,*' see Niva Lorenzini, *Il segno del corpo: Saggio su d'Annunzio* (Rome: Bulzoni,

1984), 35, 55; see also Niva Lorenzini, 'D'Annunzio, l'irrazionale, il linguaggio del corpo,' *Il Verri*, nos. 5–6 (1985): 136–52.

6 Mourning Transcendence and Re-enchanting the Flesh

1 Jeremy Rifkin, *The Empathic Civilization: The Race to Global Consciousness in a World of Crisis* (New York: Tarcher/Penguin, 2009), 8.

2 Olivier Barbarant, *Douze lettres d'amour au soldat inconnu* (Seyssel: Champ Vallon, 1993); Testa, *Io sono il milite ignoto.* For 2003 mourning and suicides, see the following: 'Chiuso il Vittoriano, 1 milione di persone in 24 ore,' *Corriere della Sera,* 18 Nov. 2003; Lazzaro and Macrì, 'Marinaio si uccide all' Altare della Patria,' *Corriere della Sera,* 19 Jan. 2000; Spadaccino, 'Giù dal Vittoriano, la fine terribile scelta da Francesca,' *Corriere della Sera,* 2 June 2005.

3 On Pasolini, Paolo Valesio, 'Pasolini come sintomo,' *Italian Quarterly* 21–2/82–3 (1980): 31–42. Luce Irigaray, *This Sex Which Is Not One* (Ithaca: Cornell University Press, 1985), 78. On the Virgin in South America, see Marcella Althaus-Reid, *Indecent Theology: Theological Perversions in Sex, Gender, and Politics* (London: Routledge, 2000).

4 Barbarant, *Douze lettres,* 14, 30, 93; Arnold van Gennep, *The Rites of Passage* (Chicago: University of Chicago Press, 1960), 149. The originals read:

> et, davantage que la dissmulation de l'horreur me scandalise [...] cette poésie à deux sous [...] à Rome ou à Madrid [...] le corps engoncé dans la gloire, l'impossibilité de représenter autrement qu'en un masque épais le visage, la peau, le battement véritable des chairs.
>
> Je voudrais, même, que vous soyez enterré sous une dalle de verre, qu'on puisse assister à votre décomposition. One serait forcé de plonger les yeux dans le vôtres, au lieu de s'y incliner sans péril [...].

Ariès (*L'homme devant la mort*) discusses these changes and comments on *The Death of Ivan Ilyich,* citing '"quelque chose de terrible se passait en lui [...] et il était le seul à le savoir"' (559). Vovelle goes further (*La mort et l'Occident*), showing that the doctor had a new religious function (as seen in his appearance on ex-votos), in that his role was not only to help with physical pain, but to ensure a 'good death' which, increasingly, meant keeping both the 'patient' and his family away from 'hysteria' (528). As the association of death with folly, despair, alcoholism, suicide, and violence increased both in the media and in art, in the late nineteenth and early twentieth centuries, Vovelle finds that the body became the 'controversial' locus where conflict was played out (585, 608, 650).

5 Barbarant, *Douze lettres,* 20. The original reads: 'C'est cet archange de chair

pourrie et de boue qui m'ordonne de vous écrire. Ce sont ses lèvres de presque humus que j'embrasse, pour vous saluer.'

One of the merits of Santanu Das' work (*Touch and Intimacy in First World War Literature*) on touch in connection with the First World War is to recognize that the existential or humanizing component of touch is far more important than its sexual one.

6 Barbarant, *Douze lettres*, 16. D'Annunzio, *Notturno*, 119. Barbarant, *Douze letters*, 18, 81, 81. The originals read:

> Sachez donc que je n'avais moi non plus guère de visage, avant de me dédier à vous parler. Et la solitude ne fait pas que vous défigurer: elle vous anéantit, comme tout à l'heure, quand j'avais l'impression de ne porter que des pas de brouillard, une idée d'être vers vous.
>
> 'Rassurez-vous,' 'je m'efforcerai de ne pas vous prêter mes traits.'
>
> Et puis sachiez que je traîne un peu trop de cadavres le long de mes chemins. Tout à l'heure c'était un autre, de nouveau dans un lit d'hôpital, qui venait de fermer les yeux. [...] Que je n'ai plus droit depuis bientôt dix ans à la légèreté, à la grâce, et qu'il faudrait apprendre à dépasser leur perte, ce que je croyais tenter en vous réinventant.
>
> C'est un échec, évident ce soir, donc j'aurais mieux fait de pas vous parler. Loin de vous en extraire, je me confonds à votre boue.

7 Céline, *Voyage*, 14. Barbarant, *Douze lettres*, 90. The original reads: 'Schizophrène [...] un médecin racontait [...] s'excusant d'écrire [...] sa phrase à cet endroit louvoyait, se perdait dans les détails du crépuscule, reculant d'autant l'intolérable souvenir: le récipient débordait d'yeux humains, de minces billes brillantes encore, comme du blanc d'oeuf, suivie chacune d'un ressort rouge et noir.'

8 Barbarant, *Douze lettres*, 16, 20, 24, 23–4, 31, 71. For the curtain of the temple torn in two, see Mark 15:38. The originals read: 'vous prêt[er] une impossible présence'; 'Nous étions fort loin du "poilu" du folklore, auquel d'ailleurs je n'ai jamais cru, et la jeunesse de cette portion de visage déjà presque repris par la terre faisait violence à qui s'y attardait'; 'epiphanies minuscules.' And:

> Sans doute est-ce ma seule faiblesse qui m'incite ici à vous inventer une présence. Il est trop facile vraiment d'affirmer que les morts se maintiennent, persistent ou nous font signe dans ces beautés que le monde distribue au hasard [...] il n'y a rien derrière cette lueur pesante et chaude où je baigne en aveugle [...] Il me faut cependant chaque fois m'en convaincre [...] Il m'est permis de croire que vous avez pu connaître

> comme moi certains étés éblouissants de blés [...] le scandale aussitôt se manifeste: on accepte de perdre des êtres, mais, aussi criminel que cela paraisse, on n'accepte pas la perte de ce qui fut leurs états.
>
> Moins de vingt ans après l'établissement de votre culte, une femme, jeune encore, avant moi l'écrivait: 'Et si l'Evangile omettait toute mention de la Résurrection, la foi me serait plus facile. La Croix seule me suffit.' Imbécile qui ne verrait ici que le goût du sacrifice, quand il ne s'agissait que de remettre à l'endroit la pureté que peut contenir quelquefois de croire, montrer que la reconnaissance du malheur ferait l'Humanité moins brutale [...].
>
> Donnez-moi une part, même infime, de votre ténèbre, et je vous prêterai la mienne, et ce sera comme la légende du manteau qu'on déchire en deux, qui ne suffit pas, ne protège pas, mais déjà commence à ouvrir un ciel.

9 Barbarant, *Douze lettres,* 99. The original reads:

> Mon ami, qui dira un jour ce qui brûle dans la plus banale des approches, comment la peau simplement de plus près dans les voiles de la lumière hivernale révèle un secret porté après toute la vie, que j'ai subi et pris par force pour une clé du monde, le résumé de tout ce qu'on appelle réalité. [...] la grâce soudain d'un corps qui vous écartèle, où il ne s'agit pas, ou pas seulement, de désir, mais du temps comme un gant retourné [...] il n'y a pas moyen d'aller si l'on pense de vous-même à la vie publique, de cette révolution en soi ordonnée à ce qui dépasse le nombre deux. Cependant j'insiste, et c'est votre nuque, vos épaules, le bras doré dans le carré de jour [...] qui s'opposent à la barbarie.

See also: 'comme s'il était affreux de vous trouver beau, ou confus à l'excès de prendre ensemble le coeur et les tranchées, les draps de lit et la politique, la relation humaine et celle des Nations [...] Comme si l'amour de la chair n'était pas le commencement de la morale' (ibid., 46).

For the claim that the Unknown Soldier Memorial disrupts our common notion of time as linear, as well as our tendency to dual rather than trinal logic, see Ingersoll and Nickell, 'The Most Important Monument,' 200, 206–7. For the connection between radical openness to the other and a sense of time that 'unfolds in errancy and dissemination' (based on Plotinus), see Wyschogrod, *Saints and Postmodernism,* 121, 162. For the related idea of death as 'the limit at which subjective consciousness cannot return to itself or reappropriate itself in its self-identity,' see Thomas A. Carlson, *Indiscretion: Finitude and the Naming of God* (Chicago: University of Chicago

Press, 1999), 192. This emphasis on time resonates with recent reflections on trauma and PTSD as 'a disease of time' (Young, *The Harmony of Illusions*, 7).

10 Running away is in Testa, *Io sono il milite ignoto*, 7. This refers to the problem of traitors discussed in Chapter 3, as is confirmed by a later reference to Caporetto (ibid., 146). Testa is also author of *Il terrorismo: La strategia che viene dall'alto* (Bologna: Thyrus, 1986). On the powerful, Testa, *Io sono il milite ignoto*, 10; on spirituality, mysticism, yellow, and sacrality, see ibid., 15, 18, 20, 26.

The originals read: 'kamikaze giapponesi'; 'giovani dell'Islam estremo'; 'ammaestrati dai signori della brutalità, dai potenti sempre capaci di usare delle vite appena fiorite per raggiungere fini spesso inconfessabili'; 'ma dov'è la spiritualità che si addice a un morto?' And:

> è inutile affidarsi a improbabili favole, a miti e a leggende per rendere palese che voi stessi siete la vostra coscienza, voi siete Dio, materia che si fà spirito, spirito che si fà materia, il resto è propaganda di chi, per interesse personale, per la propria gloria, vende ogni giorno prodotti divini.
>
> [...] Forse avranno esagerato quei mistici che avevano visioni, eremiti erotizzati dall'astinenza sessuale, forse avranno ecceduto in retorica i poeti e i filosofi, ma in tutti loro c'era almeno un profondo desiderio di conoscenza trascendentale [...]

And: 'nelle montagne dell'Irpinia'; 'il paese dei lupi'; 'i miei occhi, allora, erano saturi di giallo'; 'una sua sacralità.'

11 Testa, *Io sono il milite ignoto*, 51, 43, 107, 112. Mattia Pascal eventually finds that his new state is no better than his old one. See Thomas Harrison, 'Regicide, Parricide, and Tyrannicide in Il fu Mattia Pascal: Stealing from the Father to Give to the Son,' in Gian-Paolo Biasin and Manuela Gieri, eds, *Luigi Pirandello: Contemporary Perspectives* (Toronto: University of Toronto Press, 1999), 189–213. Also Thomas Harrison, *Essayism: Conrad, Musil, and Pirandello* (Baltimore: Johns Hopkins University Press, 1992).

On Pasolini's sense of himself as 'surviving' after his own death, see, among many other examples, his poem 'Le ceneri di Gramsci,' and his description of his 'ex-vita' in 'L'usignolo della Chiesa Cattolica.' The originals read: 'non ci si può parlare'; 'essere nessuno.'

12 Testa, *Io sono il milite ignoto*, 115. The originals read: 'non certo sul monte Subasio'; 'amore' and 'regola.'

13 Testa, ibid., 35, 122, 120, 121, 78, 131, 79, 124, 133, 131, 131, 132, 145–55, 159. For how the dignity of the body is not present, see Testa's screed against the body in Dante's *Paradiso*, XXXIII, contrasted with his defence

of Paolo and Francesca (58). The originals read: 'estasi mistiche'; 'signori potenti'; 'sto facendo la rivoluzione'; 'mio padre, così stronzo'; 'lasciai in sospeso [...] la questione-Dio.' And:

> mi avvidi che il silenzio non è soltanto una categoria dell'udito ma è un fenomeno fisico palpabile, mi sembrava di poterlo toccare, il silenzio di quel momento era fatto di mille suoni e mille colori diversi, il fruscìo delle foglie sopra di me leggero leggero.
> fui preso prigioniero nel reticolato, i chiodi di ferro penetrarono nella divisa e mi stracciarono la pelle [...] le braccia larghe come nei crocifissi, la testa penzoloni. [...] vidi a terra il mio sangue confondersi con il fango. Dopo pochi istanti l'affanno cessò [...] Non salvatemi l'anima, pregavo mentre morivo, non voglio che mi salviate l'anima, ridatemi il mio corpo con il quale io possa vivere, amare, farmi amare, l'anima non esiste senza il corpo [...] Signore mio, se esisti [...] restituiscimi la dignità del mio corpo [...]. Divenni leggero come una piuma, ormai sembravo davvero una povera bandiera svolazzante a qualsiasi vento. [...] Ero un leggero Cristo in croce.

L'Ardente, Aug.–Sept. 1921, published a rather surreal dialogue between Cino, Dante, Guido, and a young man who died in the Great War, to whom love is being explained; though there is a d'Annunzian nostalgia to this piece, love is invoked, not entirely successfully, to counter ideological appropriations, as in Barbarant and Testa.

14 On the need to touch, see 'Un silenzioso abbraccio,' *L'Unità*, 17 Nov. 2003; 'Solenni funerali di Stato,' *Repubblica*, 18 Nov. 2003. On 'carezze alle bare,' see Giulio Anselmi, 'L'identità nel dolore,' *Repubblica*, 19 Nov. 2003. On silence on TV, see 'Giornata di lutto nazionale i funerali in diretta tv,' *Repubblica*, 17 Nov. 2003. On no advertisements, see 'L'Italia si ferma per i funerali,' *Repubblica*, 18 Nov. 2003. On the need to *provvedere* for one's family as motivation, see d'Avanzo, 'Il funerale infinito delle famiglie, il dolore di chi ha perso figli, papà e mariti dietro quelle divise,' *Repubblica*, 18 Nov. 2003. On grief, see fainting women in Zegarelli, 'Duecentomila per i caduti di Nassiriya,' *L'Unità*, 18 Nov. 2003. Tearful Berlusconi, photo, is in *Repubblica*, 17 Nov. 2003, echoed in the article by Concita de Gregorio, 'Le lacrime del presidente,' *Repubblica*, 18 Nov. 2003.

The outraged father is mentioned in the following: 'Pochi posti per i familiari; Alla fine fine seduti davanti i politici,' *Repubblica*, 18 Nov. 2003; 'Militari caduti in Iraq, il giorno dell'addio,' *Corriere della Sera*, 18 Nov. 2003; Pardi, 'Morti in guerra o in missione di pace?' *L'Unità*, 17 Nov. 2003; Vattimo, 'Ritrovare le parole,' *L'Unità*, 20 Nov. 2003. Vattimo is reacting to Cardinal Ruini's homily in San Paolo and expanding on the claim that the nineteen 'are not heroes but martyrs,' made by Zegarelli ('Duecentomila per i caduti di Nassiriya,' *L'Unità,* 18 Nov. 2003). Zucconi, 'Il Vietnam italiano,' *Repubblica,* 18 Nov. 2003. Berselli, 'Il cordoglio e il consenso,' *Repubblica,* 20 Nov. 2003. On Farfaglia, see 'Aperte tre inchieste,' *Repubblica,* 18 Jan. 2000. Lazzaro and Macrì, 'Marinaio si uccide all' Altare della Patria,' *Corriere della Sera,* 19 Jan. 2000

15 On the Unknown Soldier's mother, see Siegel, 'To the Unknown Mother.' The poem is in Alda Merini, *Santo Padre Divina Poesia* (Milan: Giuliano Grittini, Incisione Arte, 2009). The original reads:

> [...] voglio
> bene al mio corpo che mi porta
> all'altare della patria, come un soldato morto.

16 'Chi sei,' and 'La presenza di Orfeo' in Alda Merini, *La presenza di Orfeo; La terra santa,* vol. 1, ed. Grazia Carpena (Milan: Scheiwiller, 2005), 41, 18. Alda Merini, *La carne degli angeli* (Milan: Frassinelli, 2003). For autobiographical prose on her time in the asylum, see Alda Merini, *Delirio amoroso* (Genoa: Il Melangolo, 1989). The originals read:

> Sei Colui che ha due volti: uno di luce
> pascolo delle anime beate,
> ed uno fosco
> indefinito, dove son sommerse
> la gran parte dell'anime, cozzanti
> contro la persistente
> ombra nemica: e vanno, in quelle tenebre,
> protendendo le mani come ciechi ...
>
> giacerò nell'informe
> fusa io stessa, sciolta dentro il buio,
> per quanto possa, elaborata e viva,
> ridivenire caos ...

17 Merini, *La presenza di Orfeo; La terra santa,* vol. 2, (Milan: Scheiwiller, 2005), 42. Alda Merini, *Francesco: Canto di una creatura* ([Milan]: Frassinelli, 2007), 54. Merini, *La carne degli angeli,* 100. The originals read:

> ho mangiato anch'io la mela
> della tua onnipresenza
> e ne sono riuscita
> vuota di ogni sapienza,
> [...]

e il calice della tua vita
sfiorava tutte le rose.
Ora ti sei confusa
con gli oscuri argomenti della lira
[...]
[...] io verrò a cercarti
scaverò il tuo fermento,
madre [...]

'anch'io devo partorire il mio vero Dio';

Ho solo bisogno del tuo grembo
per conoscere il mio
e del mio grembo
per conoscere il tuo [...]

18 Merini, *Delirio amoroso,* 30. Alda Merini, *L'anima innamorata* (Milan: Frassinelli, 2000), 8. Merini, *Santo Padre Divina Poesia.* The originals read: 'violentat[e] dalla santità di Gesù'; 'l'uomo è stato deflorato da Dio al momento stesso della creazione.' And:

Qui con le ali convesse gli angeli
della terra che siamo noi rattristati da tempo
da una maternità non risolta
non cantiamo più alla pendice della
tua croce, e guardiamo sventolare
nel ventre la tua carne come ferita, che
pare strappata da mille demoni

19 'Inferocite' is in Franca Pellegrini, *La tempesta originale: La vita di Alda Merini in poesia* (Florence: Cesati, 2006), 134. Merini, *Delirio amoroso,* 17. Merini, *Francesco,* 100. Merini, *La presenza di Orfeo; La terra santa,* 108. The originals read: 'inferocite dai fiori'; 'a volte avverti un fruscio come di qualcosa che ti passa vicino, e stranamente esulti, e torna l'uomo amato, e la rosa finalmente si apre'; 'che si apre'; 'blocco / d'incertezza e di paura'; 'aspra,' 'accesa,' and 'castigata,' 'pausa di Dio.'

20 Alda Merini, *Corpo d'amore: Un incontro con Gesù* (Milan: Frassinelli, 2001), 33, 78. Alda Merini, *Magnificat: Un incontro con Maria,* ed. Ugo Nespolo (Milan: Frassinelli, 2002), 31. Merini, *Corpo d'amore,* 38. Alda Merini, *Superba è la notte* (Turin: Einaudi, 2000), 4. Merini, *L'anima inamorata,* 27. Merini, *La presenza di Orfeo; La terra santa,* 124, 15, 17.

The originals read: 'di tutti coloro che sono stati martirizzati per nulla'; 'variazione del sangue ... colore rosso del martirio'; 'una terra che diventa un fiore / e un fiore che diventa terra,' 'vergine ... madre' and 'madre ... vergine.' And:

perché tu, nuovamente colpito
da un amore umano,
possa dare sangue per noi.

Also 'frutto di dannata certezza'; '[il] sangue [...] si colora di Dio'; 'impronta di Dio e delle cose mutabili' (emphasis added); 'non so quale fantasma che mi calchi'; 'apparizione che dilegua ... tempo che intercorre fra due tappe'; 'divergenza di dolore.'

21 Merini, *Delirio amoroso*, 66. Alda Merini, *La terra santa*, ed. Maria Corti (Milan: All'insegna del pesce d'oro, 1984), 33. Alda Merini, *L'altra verità: Diario di una diversa* (Milan: Rizzoli, 2006), 78. Merini, *Santo Padre Divina Poesia*. The originals read: 'un altra figlia non nata'; '[una] bambina che non ha[i] religione [...] [una bambina] mai partorita'; 'adagiati sulle rose e sulle spine'; 'concepito una genesi.' And:

Avevi stanze per ogni dove,
Un vicolo per il dolore
in quei vicoli
tu andavi a morire
giorno dopo giorno
  ti ho amato perché anche io
ho dei vicoli aperti entro cui aspetto
che Dio mi preghi perché risolga la vita,
e al mattino quando mi alzo mi dico,
il mio corpo
ormai vecchio stanco ascolta
ancora la sua voce e allora voglio
bene al mio corpo che mi porta
all'altare della patria, come un soldato morto.

22 Merini, *Delirio amoroso*, 62. Althaus-Reid, *Indecent Theology*, 110. The originals read: 'le dolgie doloranti del parto'; 'penso che Gesù mi fosse inferiore per grandezza e per tolleranza.'

See also Merini's comment on the relationship between 'obscenity' and the contestation of traditional Christian theology in contrast with a more embodied sense of the sacred (*Delirio amoroso*, 54–5):

> Quando andai a Bergamo per ritirare un premio di poesia, mi regalarono una bella rosa di stoffa. Da allora me la porto appuntata sulla giacca, solenne, irrisoria e provocatoria. [...]

> Sì, mi offro al miglior offerente, soprattutto mi offro al panico, ma la rosa da allora me la porto sul petto; e si chiama Michele, si chiama Alda, si chiama matrimonio sconsacrato. Si chiama vergogna. Si chiama tutto fuorché 'fiore.' E una rosa stupida, inutile come la sofferenza umana quando l'uomo deve soffrire per colpa d'altri e non per sua propria volontà terrena.

> When I went to Bergamo to collect a poetry prize, they gave me a lovely fabric rose. From then on I have worn it pinned to my jacket, solemn, derisive, and provocative. [...]
>
> Sure, I give myself to the best buyer, most of all I give myself to panic, but the rose has been on my chest since then; and it is called Michele [Michele Pierri, Merini's second husband], it is called Alda, it is called deconsecrated marriage. It is called shame. It is called anything but 'flower.' It's a stupid rose, useless as human suffering is when man suffers because of others and not of his own terrestrial will.

For the argument that 'post-modern comedy is de-sublimised *Unheimlich*, or simply grotesque Kitsch,' with the implication that decadent and postmodern attitudes towards mortality and decay mirror each other, see Massimo Riva, '1888–1988: Some Remarks on Nihilism and Secularization,' *History of European Ideas* 11 (1989): 979–88. The extradordinarily massive scale of the Unknown Soldier Memorial and related monuments in Iraq are criticized by al-Khalil, in *The Monument*, at 28, 50, and 59, for their vulgarity, showing once again the ironic connection between kitsch and a critique of modern Western normalization of death and dying.

23 The name Ernesto Picciafuoco can loosely be translated as 'earnest catching fire' and clearly refers to this Bellocchio's quest for what constitutes genuine political and ethical rebellion. The interview with Bellocchio is included on the DVD release of *L'ora di religione*. The originals are: 'rifare i conti della sua vita'; 'ritorno della memoria.'

24 The originals are: 'sorridi, sorridi, sorridi come una ... una santa'; 'un protettore, un padrino, un patrono, un titolo' (*padrino* could mean 'godfather,' but given the duel that takes place in the film, 'second' might be better); 'di chi vuole sempre sfottere il mondo'; 'quella parte sua [...] che si ritrova, che vuole strapparsi di dosso.'

25 For the English translation, see Arsenii Tarkovsky, *Poems from Films*, translated by Alex Nemser and Nariman Skakov, retrieved from http://people.ucalgary.ca/~tstronds/nostalghia.com/TheTopics/Tarkovsky_Poems.html. For the Italian, cited in the film, see Arsenij Tarkovskij, *Poesie Scelte*, trans. Gario Zappi (Milan: Scheiwiller, 1989).

E' fuggita l'estate,
più nulla rimane.
Si sta bene al sole.
Eppur questo non basta.
Quel che poteva essere
una foglia dalle cinque punte
mi si è posata sulla mano.
Eppur questo non basta.
Ne' il bene ne' il male
sono passati invano,
tutto era chiaro e luminoso.
Eppur questo non basta.
La vita mi prendeva,
sotto l'ala mi proteggeva,
mi salvava, ero davvero fortunato.
Eppur questo non basta.
Non sono bruciate le foglie,
non si sono spezzati i rami...
Il giorno è terso come cristallo.
Eppur questo non basta.

26 Bellocchio is quoted in the web page on Tarkovsky, retrieved from http://www.russianecho.net/contributi/tarkovskij.asp. The original is: 'La poesia che viene recitata da uno dei personaggi, scritta dal padre di Tarkovskij è un momento chiave della vicenda, perché ci dovrebbe essere sempre, in ciascuno di noi, questa insoddisfazione assoluta e permanente, un cercare sempre qualcosa di nuovo, di meglio.'

For the novel, and Freud's analysis, see: Wilhelm Jensen, *Gradiva: A Pompeiian Fancy*, trans. Helen M. Downey (German original, 1903; New York: Moffat, Yard, 1918); Sigmund Freud, *Delusion and Dream in Jensen's Gradiva*, trans. Helen M. Downey (German original, 1907; New York: Moffatt, Yard, 1917).

27 Fabio Vighi, 'On the Real Limits of Self-Consciousness: Gazing Back at the Subversive Subject with Marco Bellocchio,' *Culture, Theory and Critique* 46/2 (2005): 161, 160. He refers to Slavoj Žižek, *The Sublime Object of Ideology* (London: Verso, 1989).

28 See Carlo Dossi, *I Mattoidi al primo Concorso pel Monumento in Roma a Vittorio Emanuele II* (Rome: Sommaruga, 1884). For demolition proposals, which interestingly, called for rebuilding the Unknown Soldier Memorial, not eliminating it, see *Processo al monumento Vittorio Emmanuele II a Roma*, retrieved

from http://www.fondazionebrunozevi.it/19781988/frame/pagine5/processovittoriano.htm. For a recent reference to the monument as 'orrendo' (horrid), including the often-quoted description of it by Giovanni Papini as a 'grande piscatoio di lusso' (huge luxury latrine) see Laura Laurenzi, 'Apre le porte alla gente l'orrendo vittoriano,' *Repubblica,* 31 March 1988.

29 Robert Bird, 'The Stalker's Dream: Mediation and Violence in the Cinema of Andrei Tarkovsky,' paper presented at the Descent of Grace Colloquium, Stanford University, 15 May 2009.

Robert Harrison, *Gardens: An Essay on the Human Condition* (Chicago: University of Chicago Press, 2008). Harrison makes a similar point when he contrasts the 'creative reordering or serenification' that he sees in Epicurus' Garden School with Modernism's 'relentless attacks' against the spiritual ills of modernity, noting that 'destroying false idols is only a first step on the path to rehumanization' and one that has a dangerous 'excessive promiscuity with the nihilism against which it took its stand' (160).

30 In this sense, the 'iconic' way of seeing is the opposite of a representation that makes the other a mirror image of the self; rather, it asserts a connection between our 'throwness' in the world and our responsibility to the other as existential givens that are larger than ourselves. This is close to Žižek's notion of the 'neighbor'; see Slavoj Žižek, 'Neighbors and Other Monsters: A Plea for Ethical Violence,' in Slavoj Žižek, Eric L. Santner, and Kenneth Reinhard, eds., *The Neighbor: Three Inquiries in Political Theology* (Chicago: University of Chicago Press, 2005), 140.

31 Tournier, *Vendredi,* 185–7, 201–9.

32 Tick, *War and the Soul,* 286, 107, 254. By insisting that we cannot own our common mortality, I wish also to stress that it is not a oneness in which the many are reabsorbed. On how the problem of evil is tied to the question of the one and the many, and the absorption of one by the other, see Michel Serres, 'L'homme est un loup pour l'homme,' in Michel Deguy and Jean-Pierre Dupuy, eds., *René Girard et le problème du mal* (Paris: Grasset, 1982), 301–9.

33 Neil Hanson, 'Flying the Flag for the Padre Once More: The flag that covered the Unknown Warrior – a potent symbol literally steeped in blood – must be restored to its rightful place, says Neil Hanson,' *Daily Telegraph,* 10 Nov. 2006; Cassandra Jardine and Richard Savill, 'Wootton Bassett: A Very British Way of Mourning,' *Telegraph,* 7 July 2009.

34 Susan Sontag, 'Regarding the Torture of Others,' *New York Times,* 23 May 2004. Edith Wyschogrod (*An Ethics of Remembering*) makes a similar point, drawing on Barthes: 'For Barthes, a photograph is not merely an image

that merges with its neural configuration, a percept, but rather a cultural artifact that is subject to critique that emanates from a point *within* the picture itself. [...] the real power of photographs is not lodged in reconciling the viewer and image but rather in destabilizing the viewer through the picture's punctum, a detail in the photograph that functions as a partial object that wounds or punctures the viewer' (77). See Roland Barthes, *Camera Lucida: Reflections on Photography* (New York: Hill and Wang, 1981). The 'punctum' can be associated with the iconic element, which in the case of the Unknown Soldier as with the Abu Ghraib photographs has to do with an uncanny sense of common embodiment, of physical continuity, which is distinct from visual identification. As Wyschogrod points out, this element is not a given but depends on our active relation to the photograph.

Turning away from the fascination of the photograph and towards a more iconic way of seeing can also be related to Gumbrecht's sense of aesthetic experience as oscillating between meaning and presence, such that ultimately the 'presence effect' does not lead to a hypostatization, hence, to a reification of presence (*The Production of Presence*, 104ff.).

Bibliography

Note: Newspaper articles, both from the First World War era and on more recent uses of the Unknown Soldier Memorial, have been omitted from the Bibliography but are cited in the notes.

Acciaresi, Primo. *Giuseppe Sacconi e l'opera sua massima: Cronaca dei lavori del Monumento Nazionale a Vittorio Emanuele II illustrata da 330 incisioni.* Rome: Tipografia dell'Unione Editrice, 1911.

Agamben, Giorgio. *Homo sacer: Il potere sovrano e la nuda vita.* Turin: Einaudi, 1995. Translated by Daniel Heller-Roazen as *Homo Sacer: Sovereign Power and Bare Life* (Stanford: Stanford University Press, 1998).

al-Khalil, Samir. *The Monument: Art, Vulgarity, and Responsibility in Iraq.* Berkeley: University of California Press, 1991.

Althaus-Reid, Marcella. *Indecent Theology: Theological Perversions in Sex, Gender, and Politics.* London: Routledge, 2000.

Anderson, Benedict R. *Imagined Communities: Reflections on the Origins and Spread of Nationalism.* London and New York: Verso, 1991.

Andreoli, Annamaria. *Il vivere inimitabile: Vita di Gabriele d'Annunzio.* Milan: Mondadori, 2000.

Apostolidès, Jean-Marie. *Le prince sacrifié: Théâtre et politique au temps de Louis XIV.* Paris: Éditions de Minuit, 1985.

– *Héroïsme et victimisation: Une histoire de la sensibilité.* Paris: Exils, 2003.

'Arc de triomphe de l'Étoile – Wikipedia.' Retrieved from http://fr.wikipedia.org/wiki/Arc_de_triomphe_de_l%27Étoile.

Ariès, Philippe. *Essais sur l'histoire de la mort en Occident du Moyen-Age à nos jours.* Paris: Éditions du Seuil, 1975.

– *L'homme devant la mort.* 2 vols. Paris: Éditions du Seuil, 1977.

– 'Du livre de Michel Vovelle "La Mort en Occident."' In Roger Chartier, ed., *La Mort aujourd'hui*, 157–68. Marseille: Rivages, 1982.

– *Images de l'homme devant la mort.* Paris: Éditions du Seuil, 1983. Translated by Janet Lloyd as *Images of Man and Death* (Cambridge, Mass.: Harvard University Press, 1985).

Artieri, Giovanni. *Il Re, i soldati, e il generale che vinse.* Bologna: Cappelli, 1951.

Ascoli, Albert Russell, and Krystyna Von Henneberg, eds. *Making and Remaking Italy: The Cultivation of National Identity around the Risorgimento.* Oxford and New York: Berg, 2001.

Atkinson, David, and Denis Cosgrove. 'Urban Rhetoric and Embodied Identities: City, Nation, and Empire at the Vittorio Emanuele II Monument in Rome, 1870–1945.' *Annals of the Association of American Geographers* 88/1 (1998): 28–49.

Atti Parlamentari, Camera dei Deputati, Legislatura XXVI, 1921–23, sessione 1921, documenti, disegni di legge e relazioni, nos. 202, no. 202-A, Sepoltura della salma di un soldato ignoto. Rome: Biblioteca Giovanni Spadolini.

Atti Parlamentari, Senato del Regno, Legislatura XXVI, 1921, discussioni, tornata del 10 agosto 1921. Rome: Biblioteca Giovanni Spadolini.

Atti Parlamentari, Senato del Regno, Legislatura XXVI, documenti del Senato, 1921, disegno di legge no. 163, 163A. Rome: Biblioteca Giovanni Spadolini.

Baldissone, Giusi. *Filippo Tommaso Marinetti.* Milan: Mursia, 1986.

Ballinger, Pamela. 'La Pentecoste italiana: Accogliere il verbo della fede italic; Lo sviluppo del rituale nazionalistico a Fiume, 1919–1921.' *Qualestoria*, no. 2 (2001): 123–42.

– *History in Exile: Memory and Identity at the Borders of the Balkans.* Princeton: Princeton University Press, 2003.

Banti, Alberto Mario. *La nazione del risorgimento: Parentela, santità e onore alle origini dell'Italia unita.* Turin: Einaudi, 2000.

Banti, Alberto Mario, and Roberto Bizzocchi. *Immagini della nazione nell'Italia del Risorgimento.* Rome: Carocci, 2002.

Barbarant, Olivier. *Douze lettres d'amour au soldat inconnu.* Seyssel: Champ Vallon, 1993.

Bàrberi Squarotti, Giorgio. 'D'annunzio scrittore "politico."' In *D'Annunzio politico atti del Convegno (Il Vittoriale, 9–10 ottobre 1985)*, 319–48. Milan: Garzanti Editore, 1987.

– *La scrittura verso il nulla: D'Annunzio.* Turin: Genesi, 1992.

Barrès, Maurice. *Le Culte du moi*, vol. 3, *Le Jardin de Bérénice.* Paris: Perrin et Cie, 1891.

Barricelli, Jean-Pierre. 'Michelangelo's Finito: In the Self, the Later Sonnets, and the Last Pietà.' *New Literary History* 24/3 (1993): 597–616.

Barthes, Roland. *Camera Lucida: Reflections on Photography*. New York: Hill and Wang, 1981.

Bassnett, Susan. 'A Passion for Dismemberment: Gabriele D'Annunzio's Portrayals of Women.' In Michael St John, ed., *Romancing Decay: Ideas of Decadence in European Culture*, 128–40. Aldershot: Ashgate, 1999.

Baudelaire, Charles. 'Le peintre de la vie moderne.' In *Oeuvres complètes*, 546–65. Paris: Éditions du Seuil, 1968.

Baxa, Paul. *Roads and Ruins: The Symbolic Landscape of Fascist Rome*. Toronto: University of Toronto Press, 2010.

Bazin, André. 'The Ontology of the Photographic Image.' *Film Quarterly* 13/4 (1960): 4–9.

Becker, Annette. 'Les Soldats Inconnus.' *Historiens et Geographes* 89/364 (1998): 135–9.

Becker, Ernest. *The Denial of Death*. New York: Free Press, 1973.

Ben-Ghiat, Ruth. *Fascist Modernities: Italy, 1922–1945*. Berkeley: University of California Press, 2001.

Berman, Russell A. *Cultural Studies of Modern Germany: History, Representation, and Nationhood*. Madison: University of Wisconsin Press, 1993.

– *Fiction Sets You Free: Literature, Liberty, and Western Culture*. Iowa City: University of Iowa Press, 2007.

Bernardi, Mario. 'Gabriele d'Annunzio e il Milite Ignoto.' *Quaderni dannunziani (Gardone, Fondazione 'Il Vittoriale degli Italiani')* 22–23 (1962): 1120–39.

Bertazzoli, Raffaella. 'Retorica fiumana e pensiero "notturno" nel messaggio dannunziano "Al legionario Alceste de Ambris."' *Quaderni di lingue e letterature (Padova)* 11 (1986): 353–66.

Bertoldi, Silvio. *Guerra: Italiani in trincea da Caporetto a Salò*. Milan: Rizzoli, 2003.

Bird, Robert. 'The Stalker's Dream: Mediation and Violence in the Cinema of Andrei Tarkovsky.' Paper presented at The Descent of Grace Colloquium, Stanford University, Stanford, California, 15 May 2009.

Blondel, Maurice. *Action: Essai d'une critique de la vie et d'une science de la pratique*. Paris: Alcan, 1893.

Blondiaux, Isabelle. 'La représentation de la pathologie psychique de guerre dans Voyage au bout de la nuit.' *Roman 20–50: Revue d'Étude du Roman du XXe Siècle* 17 (1994): 105–15.

Blum, Cinzia Sartini. *The Other Modernism: F.T. Marinetti's Futurist Fiction of Power*. Berkeley: University of California Press, 1996.

Boatti, Giorgio. *Preferirei di no le storie dei dodici professori che si opposero a Mussolini*. Turin: Einaudi, 2001.

Bobbio, Norberto. *Il problema della guerra e le vie della pace*. Bologna: Il Mulino, 1979.

– *Una guerra giusta? Sul conflitto del Golfo.* Venice: Marsilio, 1991.

Bobbio, Norberto, and Pietro Polito. *Il terzo assente: Saggi e discorsi sulla pace e la guerra.* Milan: Sonda, 1989.

Bonadeo, Alfredo. *Mark of the Beast: Death and Degradation in the Literature of the Great War.* Lexington: University Press of Kentucky, 1989.

– *D'Annunzio and the Great War.* Madison, London, and Cranbury: Fairleigh Dickinson University Press and Associated University Presses, 1995.

Bontempelli, Massimo. 'D'Annunzio o del martirio.' In *Pirandello Leopardi d'Annunzio,* 89–142. Milan: Bompiani, 1938.

Bosc, Olivier. *La foule criminelle.* Paris: Fayard, 2007.

Breschi, Danilo. 'Fascist Historiography.' *Telos: A Quarterly Journal of Radical Social Theory* 133 (2006): 15–44.

Bricco, Elisa. 'La Main coupée di Cendrars: La guerra come gioco.' *Quaderni del Dipartimento di Lingue e Letterature Straniere Moderne, Università di Genova* 9 (1997): 255–67.

Brice, Catherine. *Il Vittoriano: Monumentalità pubblica e politica a Roma.* Rome: Archivio Guido Izzi, 2005.

– 'La religione civile in Italia: Piccoli e Grandi Rituali.' In Maurizio Ridolfi, ed., *Rituali civili,* 97–115. Rome: Gangemi Editore, 2006.

– 'Building Nations, Transforming Landscape.' *Contemporary European History* 16/1 (2007): 109–19.

– 'Italia: Una allegoria debole? Sistema iconografico e identità nazionale nell'Italia della fine del XIX secolo.' *Memoria e Ricerca,* no. 25 (2007): 171–86.

– 'Riti della corona, riti del fascio.' In Emilio Gentile, ed., *Modernità totalitaria: Il fascismo italiano,* 171–90. Bari: Laterza, 2008.

– 'The Many Origins of the Risorgimento Mobilization in Italy: Problems of Interpretation (1790s–1860s); or, Radical Beginnings, the Resurrection of the Dead.' Paper presented at the California Interdisciplinary Consortium for Italian Studies Annual Meeting, Stanford University, Stanford, California, 6–7 March 2009.

Brose, Margaret. 'Metaphor and Simile in Giuseppe Ungaretti's L'Allegria.' *Lingua e Stile: Trimestrale di Linguistica e Critica Letteraria* 11 (1976): 43–73.

– 'The Politics of Mourning in Foscolo's Dei Sepolcri.' *European Romantic Review* 9/1 (1998): 1–34.

Brown, Laura S. 'Not Outside the Range: One Feminist Perspective on Psychic Trauma.' In Cathy Caruth, ed., *Trauma: Explorations in Memory,* 100–12. Baltimore: Johns Hopkins University Press, 1995.

Buonaiuti, Ernesto. *Il Programma dei modernisti.* Rome: Libreria Editrice Romana, 1907. Translated by Leslie Lilley as *The Programme of Modernism* (New York and London: G.P. Putnam's Sons, 1908).

Cadeddu, Lorenzo. *La leggenda del soldato sconosciuto all'altare della patria.* Udine: Gaspari, 2001.

Camus, Albert. *L'homme révolté.* Paris: Gallimard, 1952. Translated by Anthony Bower as *The Rebel: An Essay on Man in Revolt* (New York: Vintage Books, 1956).

Canal, Claudio. 'La retorica della morte: I monumenti ai caduti della Grande guerra.' *Rivista di Storia Contemporanea,* no. 4 (1982): 659–69.

Capdevila, Luc, and Danièle Voldman. *Nos morts: Les sociétés occidentales face aux tués de la guerre, XIXe–XXe siècles.* Paris: Payot, 2002.

Carlson, Thomas A. *Indiscretion: Finitude and the Naming of God.* Chicago: University of Chicago Press, 1999.

Carroll, Michael P. *Madonnas That Maim: Popular Catholicism in Italy since the Fifteenth Century.* Baltimore: Johns Hopkins University Press, 1992.

– *Veiled Threats: The Logic of Popular Catholicism in Italy.* Baltimore: Johns Hopkins University Press, 1996.

Carta, Elisabetta. 'Grande guerra e memoria del corpo: Stigmate del tempo e dell'esperienza.' *Compar(a)ison: An International Journal of Comparative Literature,* nos. 1–2 (2005): 185–91.

Caruth, Cathy. *Unclaimed Experience: Trauma, Narrative, and History.* Baltimore: Johns Hopkins University Press, 1996.

'Caso Bruneri-Canella; smemorato di Collegno – Wikipedia.' Retrieved from *http://it.wikipedia.org/wiki/Caso_Bruneri-Canella.*

Cavara, Otello. *Il milite ignoto.* Milan: Alpes, 1922.

Céline, Louis-Ferdinand. *Voyage au bout de la nuit.* 1932. Paris: Gallimard, 2006. Translated by Ralph Mannheim as *Journey to the End of the Night* (New York: New Directions, 2006).

Cendrars, Blaise. *La Main coupée.* In *Complete works,* vol. 10. Edited by Raymond Dumay and Nino Frank. Paris: Club français du livre, 1968.

– *La vie et la mort du soldat inconnu.* Cahiers Blaise Cendrars. Paris: H. Champion, 1995.

Cendrars, Miriam. *Blaise Cendrars: La vie, le verbe, l'écriture.* Paris: Denoël, 2006.

'Cenotaph – Wikipedia.' Retrieved from http://en.wikipedia.org/wiki/Cenotaph.

Cesaretti, Enrico. 'Dyspepsia as Dystopia? F.T. Marinetti's Le Roi Bombance.' *Romanic Review* 97/3–4 (2006): 353–69.

Cescutti, Tatiana. *Les origines mythiques du futurisme: F.T. Marinetti, poète symboliste français (1902–1908).* Paris: Presses de l'Université Paris-Sorbonne, 2009.

Chartier, Roger. 'Histoire de la mort et histoire culturelle.' In Roger Chartier, ed., *La Mort aujourd'hui,* 111–24. Marseille: Rivages, 1982.

Choate, Mark. 'D'Annunzio's Political Dramas and His Idea-State of Fiume.' *Forum Italicum* 31/2 (1997): 367–88.

Cohen, Margaret. *The Sentimental Education of the Novel.* Princeton: Princeton University Press, 1999.

Cohen, Margaret, and Christopher Prendergast. *Spectacles of Realism: Body, Gender, Genre.* Minneapolis: University of Minnesota Press, 1995.

Colvile, Georgiana M.M. *Blaise Cendrars, écrivain protéiforme.* Amsterdam and Atlanta: Rodopi, 1994.

Cordova, Ferdinando. *Arditi e legionari dannunziani.* Padua: Marsilio, 1969.

D'Annunzio, Gabriele. *Il Fuoco.* Milan: Treves, 1900.

– *La Leda senza cigno, seguita da una Licenza.* Milan: Treves, 1916.

– *La riscossa.* Milan: Casa Editrice d'Arte Bestetti & Tuminelli, 1918. Translated by Magda Sindici as *The Rally* (Milan: Casa Editrice d'Arte Bestetti & Tuminelli, 1918).

– *Contro uno e contro tutti.* Rome: Presso La Fionda, 1919.

– *Le faville del maglio e altri studi del vivere inimitabile.* Milan: Treves, 1924.

– *Cento e cento pagine dal libro segreto di Gabriele d'Annunzio tentato di morire.* Milan: Mondadori, 1935.

– *Le tre redazioni di un taccuino di guerra di Gabriele d'Annunzio.* Edited by Antonio Bruers. Milan: Mondadori, 1942.

– 'Agli uomini milanesi per l'Italia degli italiani.' *Prose di ricerca, di lotta, di comando, di conquista, di tormento, d'indovinamento, di rinnovamento, di celebrazione, di rivendicazione, di liberazione, di favole, di giochi, di baleni* [hereafter *Prose*]. 1922, *Il libro ascetico della giovane Italia* [hereafter *Il libro ascetico*], 564–71. Milan: Mondadori, 1947.

– 'Cantano i morti con la terra in bocca e le carene valicano i monti.' *Prose, Il libro ascetico,* 775–96. Milan: Mondadori, 1947.

– 'Comento meditato a un discorso improvviso.' *Prose, Il libro ascetico,* 495–563. Milan: Mondadori, 1947.

– 'Il sudore di sangue.' *Prose, La penultima ventura,* 797–1005. Milan: Mondadori, 1947.

– 'Messaggio del convalescente agli uomini di pena.' *Prose, Il libro ascetico,* 572–634. Milan: Mondadori, 1947.

– 'Sette documenti d'amore.' *Prose, Il libro ascetico,* 653–710. Milan: Mondadori, 1947.

– 'Tre preghiere dinanzi agli altari disfatti.' *Prose, Il libro ascetico,* 635–52. Milan: Mondadori, 1947.

– *Taccuini.* Tutte le Opere di Gabriele d'Annunzio. Edited by Egidio Bianchetti. Milan: Mondadori, 1965.

– *Di me a me stesso.* Milan: Mondadori, 1990.

– *Notturno.* 1921. Milan: Garzanti, 1995.
– *Siamo spiriti azzurri e stelle: Diario inedito (17–27 agosto 1922).* Florence: Giunti, 1995.
– *Il fastello della mirra.* Edited by Angelo Piero Cappello. Florence: Vallecchi, 2004.
D'Azeglio, Massimo. *Things I Remember (I miei ricordi).* London and New York: Oxford University Press, 1966.
Das, Santanu. *Touch and Intimacy in First World War Literature.* Cambridge and New York: Cambridge University Press, 2005.
De Felice, Renzo. 'D'Annunzio e la vita politica italiana dal 1918 al 1936.' *Quaderni dannunziani,* nos. 38–9 (1969): 84–102.
– 'D'Annunzio nella vita politica italiana.' *Il Veltro* 32/5–6 (1988): 469–78.
De Felice, Renzo, ed. *D'Annunzio politico, 1918–1938.* Rome and Bari: Laterza, 1978.
De Martino, Ernesto. *Morte e pianto rituale: Dal lamento funebre antico al pianto di Maria.* 1958. New ed. Turin: Bollati Boringhieri, 2000.
– *La terra del rimorso: Contributo a una storia religiosa del Sud.* 1961. New ed. Milan: Il saggiatore, 2008. Translated by Dorothy Louise Zinn as *The Land of Remorse: A Study of Southern Italian Tarantism* (London: Free Association, 2005).
Deguy, Michel, and Jean-Pierre Dupuy, eds. *René Girard et le problème du mal.* Paris: Grasset, 1982.
Demetz, Peter. *The Air Show at Brescia, 1909.* New York: Farrar, Straus, and Giroux, 2002.
Derrida, Jacques. *L'écriture et la différence.* Paris: Éditions du Seuil, 1967. Translated by Alan Bass as *Writing and Difference* (Chicago: University of Chicago Press, 1978).
Dogliani, Patrizia. 'Redipuglia.' In Mario Isnenghi and Ersilia Alessandrone Perona, eds., *I luoghi della memoria; Simboli e miti dell'Italia unita,* 377–89. Rome: Laterza, 1996.
Dossi, Carlo. *I Mattoidi al primo Concorso pel Monumento in Roma a Vittorio Emanuele II.* Rome: A. Sommaruga, 1884.
Douhet, Giulio. *Le Profezie di Cassandra: Raccolta di scritti.* Genoa: Lang e Padano, 1931.
Duffy, Eamon. *Madonnas That Maim? Christian Maturity and the Cult of the Virgin.* Retrieved from http://www.bfpubs.demon.co.uk/duffy.htm.
Dunkelman, Mark H. *Gettysburg's Unknown Soldier: The Life, Death, and Celebrity of Amos Humiston.* Westport: Praeger, 1999.
Dupuy, Jean-Pierre. 'Not to Know What One Knows: Some Paradoxes of Self-Deception.' *Diogenes* 43/169 (1995): 53–68.

– *Self-Deception and Paradoxes of Rationality*. Stanford: CSLI Publications, 1998.
– *The Mechanization of the Mind: On the Origins of Cognitive Science*. Princeton: Princeton University Press, 2000.
– *Pour un catastrophisme éclairé: Quand l'impossible est certain*. Paris: Éditions du Seuil, 2002.

Durkheim, Émile. *The Elementary Forms of the Religious Life*. [1912]. Translated by Carol Cosman and Mark Sydney Cladis. Oxford and New York: Oxford University Press, 2001.

Edelstein, Dan. 'War and Terror: The Law of Nations from Grotius to the French Revolution.' *French Historical Studies* 31/2 (2007), 229–62.
– *The Terror of Natural Right: Republicanism, the Cult of Nature, and the French Revolution*. Chicago: University of Chicago Press, 2009.

Eksteins, Modris. *Rites of Spring: The Great War and the Birth of the Modern Age*. Boston: Houghton Mifflin, 1989.

Eliade, Mircea. *Shamanism: Archaic Techniques of Ecstasy*. New York: Bollingen Foundation and Pantheon Books, 1964.

Emery, Edward Sieveking. 'In Light of Mourning: Spiritual Transformations between Trauma and Presence.' *Psychoanalytic Review* 95/4 (2008): 625–54.

Fabi, Lucio, ed. *1918. La guerra nella testa: Arte popolare, esperienza e memoria nel primo conflitto mondiale*. Trieste: Lint Editoriale, 1998.

Faust, Drew Gilpin. *This Republic of Suffering: Death and the American Civil War*. New York: Alfred A. Knopf, 2008.

Fawcett, Gabriel. 'The Wehrmacht Exhibition.' *History Today* 52/4 (2002): 2–3.

Flynn, Stewart. 'Soldiers.' *Modern Poetry in Translation* 18 (2001): 185.

Fogazzaro, Antonio. *Il Santo*. Milan: Mondadori, 1970.

Fogu, Claudio. 'Fascism and Historic Representation: The 1932 Garibaldian Celebrations.' *Journal of Contemporary History* 31/2 (1996): 317–45.
– 'Il Duce taumaturgo: Modernist Rhetorics in Fascist Representations of History.' *Representations* 57 (1997): 24–51.
– *The Historic Imaginary: Politics of History in Fascist Italy*. Toronto: University of Toronto Press, 2003.
– 'Fascism and Philosophy: The Case of Actualism.' *South Central Review: The Journal of the South Central Modern Language Association* 23/1 (2006): 4–22.

Fonck, Gérard. *Le soldat inconnu*, vol. 1, *Les démarches*. Reims: autoédition, 2004.

Foscolo, Ugo. *Le ultime lettere di Iacopo Ortis*. Milan: Mursia, 1985.

Foucault, Michel. *Madness and Civilization: A History of Insanity in the Age of Reason*. Translated by Richard Howard. New York: Pantheon Books, 1965.
– *The Essential Foucault: Selections from Essential Works of Foucault, 1954–1984*. Translated by Nikolas S. Rose. Edited by Paul Rabinow and Nikolas S. Rose. New York: New Press, 2003.

Francescangeli, Eros. *Gli Arditi del popolo.* Rome: Odradek, 2000.
Freccero, Carla. *Popular Culture: An Introduction.* New York: New York University Press, 1999.
Freud, Sigmund. *Delusion and Dream in Jensen's Gradiva* [1907]. Translated by Helen M. Downey. New York: Moffatt, Yard, 1917.
– 'Mourning and Melancholia' [1917]. In *The Standard Edition of the Complete Psychological Works of Sigmund Freud.* Vol. 14, 237–58. Translated by James Strachey. London: Hogarth Press and the Institute of Psycho-Analysis, 1995.
– *Beyond the Pleasure Principle.* [1920]. Translated by James Strachey. New York: W.W. Norton, 1990.
– *Civilization and Its Discontents.* [1930]. Translated by James Strachey. New York: W.W. Norton, 1961.
Fuller, Mia. *Moderns Abroad: Architecture, Cities and Italian Imperialism.* London and New York: Routledge, 2007.
Fussell, Paul. *The Great War and Modern Memory.* New York and London: Oxford University Press, 1975.
Gaillard, Jone. 'La Terza Italia e il suo ultimo "vate": Gabriele d'Annunzio.' *Romance Languages Quarterly* 7 (1995): 245–56.
Gallagher, Catherine, and Thomas Walter Laqueur. *The Making of the Modern Body: Sexuality and Society in the Nineteenth Century.* Berkeley: University of California Press, 1987.
Gance, Abel. 'Blaise Cendrars et le cinema.' *Mercure de France*, no. 345 (1962): 170–1.
Gellner, Ernest. *Nations and Nationalism.* 2nd ed. Malden: Blackwell, 2006.
Gemelli, Agostino. *Il nostro soldato, saggi di psicologia militare.* Milan: Treves, 1917.
Gentile, Emilio. *Le origini dell'ideologia fascista (1918–1925).* Rome and Bari: Laterza, 1975. Translated by Robert L. Miller as *The Origins of Fascist Ideology, 1918–1925* (New York: Enigma, 2005).
– *Il culto del littorio: La sacralizzazione della politica nell'Italia Fascista.* Bari: Laterza, 1994. Translated by Keith Botsford as *The Sacralization of Politics in Fascist Italy* (Cambridge, Mass.: Harvard University Press, 1996).
– *La grande Italia: Il mito della nazione nel XX secolo.* Rome: GLF editori Laterza, 2006.
Gentile, Giovanni. *Teoria generale dello spirito come atto puro.* Bari: Laterza, 1920.
Giono, Jean. *Le grand troupeau.* Edited by Robert Ricatte. Paris: Gallimard, 1971. Translated by Norman Glass as *To the Slaughterhouse* (London and Chester Springs: Peter Owen and Dufour Editions, 2004).
– *Le déserteur et autres récits.* Paris: Gallimard, 1973.
Girard, René. *La violence et le sacré.* Paris: Grasset, 1972. Translated by Patrick

Gregory as *Violence and the Sacred* (Baltimore: Johns Hopkins University Press, 1977).

– *La route antique des hommes pervers.* Paris: B. Grasset, 1985. Translated by Yvonne Freccero as *Job, the Victim of His People.* Stanford: Stanford University Press, 1987.

Glucklich, Ariel. *Sacred Pain: Hurting the Body for the Sake of the Soul.* New York: Oxford University Press, 2001.

Goebel, Stefan. 'Review Article: Beyond Discourse? Bodies and Memories of Two World Wars.' *Journal of Contemporary History* 42/2 (2007): 377–85.

Gorman, Keith Phelan. 'The Return of the Dead: Christian Images of Resurrection and the Postwar Cult of Remembrance.' *Proceedings of the Annual Meeting of the Western Society for French History* 22 (1995): 67–77.

Grayzel, Susan R. *Women's Identities at War: Gender, Motherhood, and Politics in Britain and France during the First World War.* Chapel Hill: University of North Carolina Press, 1999.

Greene, Roland. 'New World Studies and the Limits of National Literatures.' In Joan Retallack and Juliana Spahr, eds., *Poetry and Pedagogy: The Challenge of the Contemporary,* 43–67. New York: Palgrave Macmillan, 2006.

Grof, Stanislav. *The Ultimate Journey: Consciousness and the Mystery of Death.* San Francisco: MAPS, 2006.

Grosz, E.A. *Volatile Bodies: Towards a Corporeal Feminism.* Bloomington: Indiana University Press, 1994.

Guasco, Maurilio. *Modernismo: I fatti, le idee, i personaggi.* San Paolo: Cinisello Balsamo, 1995.

Gumbrecht, Hans Ulrich. *Making Sense in Life and Literature.* Minneapolis: University of Minnesota Press, 1992.

– 'I rendentori della vittoria: On Fiume's Place in the Genealogy of Fascism.' *Journal of Contemporary History* 31 (1996): 253–72.

– 'How Is Our Future Contingent? Reading Luhmann against Luhmann.' *Theory Culture Society* 18/1 (2001): 49–58.

– *The Production of Presence: What Meaning Cannot Convey.* Stanford: Stanford University Press, 2004.

– *In Praise of Athletic Beauty.* Cambridge, Mass.: Belknap Press of Harvard University Press, 2006.

– 'Reading for the Stimmung? About the Ontology of Literature Today.' *Boundary 2: An International Journal of Literature and Culture* 35/3 (2008): 213–21.

– 'How (If at All) Can We Encounter What Remains Latent in Texts?' *Partial Answers: Journal of Literature and the History of Ideas* 7/1 (2009): 87–96.

Gumbrecht, Hans Ulrich, Friedrich Kittler, and Bernard Siegert. *Der Dichter als Kommandant: D'Annunzio erobert Fiume.* Munich: Fink, 1995.

Hand, Vivienne. 'Ambiguous Joy: Contradictions and Tensions in Giuseppe Ungaretti's L'Allegria (1914–19).' *Italianist: Journal of the Department of Italian Studies, University of Reading* 16 (1996): 76–116.

– *Mirage and Camouflage: Hiding behind Hermeticism in Ungaretti's L'Allegria.* Leicester, UK: Troubadour Publishing, 2006.

Hanson, Neil. *Unknown Soldiers: The Story of the Missing of the First World War.* New York: Alfred A. Knopf, 2006.

Haraway, Donna. *Modest_Witness@Second_Millennium.FemaleMan©Meets_Onco-Mouse™: Feminism and Technoscience.* New York: Routledge, 1997.

Harrington, Anne. *The Cure Within: A History of Mind-Body Medicine.* New York: W.W. Norton, 2008.

Harrison, Robert Pogue. *The Body of Beatrice.* Baltimore: Johns Hopkins University Press, 1988.

– *Forests: The Shadow of Civilization.* Chicago: University of Chicago Press, 1992.

– 'Not Ideas about the Thing but the Thing Itself.' *New Literary History: A Journal of Theory and Interpretation* 30/3 (1999): 661–73.

– 'Tombstones.' In Steven P. Schneider, ed., *Complexities of Motion: New Essays on A.R. Ammons's Long Poems,* 167–80. Madison: Fairleigh Dickinson University Press, 1999.

– *The Dominion of the Dead.* Chicago: University of Chicago Press, 2003.

– *Gardens: An Essay on the Human Condition.* Chicago: University of Chicago Press, 2008.

Harrison, Thomas. 'D'Annunzio's Poetics: The Orphic Conceit.' *Annali d'Italianistica* 5 (1987): 60–73.

– *Essayism: Conrad, Musil, and Pirandello.* Baltimore: Johns Hopkins University Press, 1992.

– *1910: The Emancipation of Dissonance.* Berkeley: University of California Press, 1996.

– 'Regicide, Parricide, and Tyrannicide in *Il fu Mattia Pascal*: Stealing from the Father to Give to the Son.' In Gian-Paolo Biasin and Manuela Gieri, eds., *Luigi Pirandello: Contemporary Perspectives,* 189–213. Toronto: University of Toronto Press, 1999.

– 'Overcoming Aestheticism.' In Luca Somigli and Mario Moroni, eds., *Italian Modernism: Italian Culture between Decadentism and Avant-Garde,* 167–90. Toronto: University of Toronto Press, 2004.

Hewitt, Andrew. *Fascist Modernism: Aesthetics, Politics, and the Avant-Garde.* Stanford: Stanford University Press, 1993.

Hobsbawm, E.J. *Nations and Nationalism since 1780: Programme, Myth, Reality.* 2nd ed. Cambridge and New York: Cambridge University Press, 1992.

Hollywood, Amy M. *Sensible Ecstasy: Mysticism, Sexual Difference, and the Demands of History.* Chicago: University of Chicago Press, 2002.

Hubert, Henri, and Marcel Mauss. *Sacrifice: Its Nature and Its Function.* 1898. Chicago: University of Chicago Press, 1964.

Humble, Malcolm. 'The Unknown Soldier and the Return of the Fallen: The Political Dimension of Mourning in German Texts from the First World War to the Present.' *Modern Language Review* 93/4 (1998): 1034–44.

Huysmans, Joris-Karl. *A rebours.* 1884. Paris: Pocket, 1999.

Ingersoll, Daniel W., and James N. Nickell. 'The Most Important Monument: The Tomb of the Unknown Soldier.' In Gordon Bronitsky, ed., *Mirror and Metaphor: The Material and Social Constructions of Reality,* 199–225. Lanham: University Presses of America, 1987.

Inglis, Ken S. 'War Memorials: Ten Questions for Historians.' *Guerres Mondiales et Conflits Contemporains* 42/167 (1992): 5–21.

– 'Entombing Unknown Soldiers: From London and Paris to Baghdad.' *History and Memory: Studies in Representations of the Past* 5/2 (1993): 8–31.

Irigaray, Luce. *Speculum of the Other Woman.* Translated by Gillian C. Gill. Ithaca: Cornell University Press, 1985.

– *This Sex Which Is Not One.* Translated by Catherine Porter and Carolyn Burke. Ithaca: Cornell University Press, 1985.

– *Sexes and Genealogies.* Translated by Gillian C. Gill. New York: Columbia University Press, 1993.

Isnenghi, Mario. *I vinti di Caporetto nella letteratura di guerra.* [Padua]: Marsilio, 1967.

– *Il mito della grande guerra: Da Marinetti a Malaparte.* Rome and Bari: Laterza, 1973.

– *Operai e contadini nella Grande guerra.* Bologna: Cappelli, 1982.

– *Il luoghi della memoria: Personaggi e date dell'Italia unita.* Bari: Laterza, 1997.

– *Le guerre degli italiani: Parole, immagini, ricordi 1848–1945.* Bologna: Il mulino, 2005.

– *La Grande Guerra: Uomini e luoghi del '15–'18.* Turin: UTET, 2008.

Jagielski, Jean-François. *Le soldat inconnu: Invention et postérité d'un symbole.* Paris: Imago, 2005.

James, William. *The Varieties of Religious Experience: A Study in Human Nature.* New York: Modern Library, 1902.

Jameson, Fredric. 'Céline and Innocence.' *South Atlantic Quarterly* 93/2 (1994): 311–19.

Janz, Oliver. *Das symbolische Kapital der Trauer: Nation, Religion und Familie im italienischen Gefallenenkult des Ersten Weltkriegs.* Tubingen: Niemeyer, 2009.

Jensen, Wilhelm. *Gradiva: A Pompeiian Fancy.* [1903]. Translated by Helen M. Downey. New York: Moffat, Yard, 1918.

Jung, C.G. *Memories, Dreams, Reflections.* Translated by Clara Winston and Richard Winston. New York: Vintage Books, 1989.

Kantorowicz, Ernst Hartwig. 'Pro patria mori.' In *Selected Studies*, 308–24. Locust Valley, NY: J.J. Augustin, 1965.

Kaplan, Alice Yaeger. *Reproductions of Banality: Fascism, Literature, and French Intellectual Life.* Minneapolis: University of Minnesota Press, 1986.

Keenan, Dennis King. *The Question of Sacrifice.* Bloomington: Indiana University Press, 2005.

Kertzer, David I. *Ritual, Politics, and Power.* New Haven: Yale University Press, 1989.

– *Prisoner of the Vatican: The Popes' Secret Plot to Capure Rome from the New Italian State.* Boston: Houghton Mifflin, 2004.

Kirkwood, Jeffrey. 'Private Traumas, Personal Mythologies.' In Daryl S. Paulson and Stanley Krippner, eds., *Haunted by Combat: Understanding PTSD in War Veterans, Including Women, Reservists, and Those Coming Back from Iraq*, xv–xxi. Westport and London: Praeger Security International, 2007.

Kuryluk, Ewa. *Veronica and Her Cloth: History, Symbolism, and Structure of a 'True' Image.* Cambridge, Mass., and Oxford: Blackwell, 1991.

Labita, Vito. 'Il Milite Ignoto: Dalle trincee all'altare della patria.' In C. Grottanelli and S. Bertelli, eds., *Gli occhi di Alessandro: Potere sovrano e sacralità del corpo da Alessandro Magno a Ceaucescu*, 120–53. Florence: Ponte delle Grazie, 1990.

Lancellotti, Giancarlo. 'L'orazione dannunziana tra poetica e politica.' *Problemi* (Palermo), nos. 114–115 (1999): 259–66.

Landy, Joshua. 'Nietzsche, Proust, and Will-to-Ignorance.' *Philosophy and Literature* 26/1 (2002): 1–23.

– *Philosophy as Fiction: Self, Deception, and Knowledge in Proust.* New York: Oxford University Press, 2004.

Lanzmann, Claude, and Shoshana Felman. 'The Obscenity of Understanding: An Evening with Claude Lanzmann.' In Cathy Caruth, ed., *Trauma: Explorations in Memory*, 200–20. Baltimore: Johns Hopkins University Press, 1995.

Laqueur, Thomas W. *The Making of the Modern Body: Sexuality and Society in the Nineteenth Century.* Berkeley: University of California Press, 1987.

– *Making Sex: Body and Gender from the Greeks to Freud.* Cambridge, Mass.: Harvard University Press, 1990.

– 'Memory and Naming in the Great War.' In John R. Gillis, ed., *Commemora-*

tions: The Politics of National Identity, 150–67. Princeton: Princeton University Press, 1994.

– 'Names, Bodies, and the Anxiety of Erasure.' In Theodore R. Schatzki and Wolfgang Natter, eds., *The Social and Political Body*, 123–41. New York: Guilford Press, 1996.

Latour, Bruno. 'Why Has Critique Run Out of Steam? From Matters of Fact to Matters of Concern.' *Critical Inquiry* 30 (2004): 225–48.

Le Naour, Jean-Yves. *Le soldat inconnu vivant.* Paris: Hachette, 2002.

– 'Le dernier voyage du soldat inconnu.' *Histoire*, no. 336 (2008): 54–5.

– *Le Soldat Inconnu: La guerre, la mort, la mémoire.* Paris: Gallimard, 2008.

Leake, Elizabeth. *The Reinvention of Ignazio Silone.* Toronto: University of Toronto Press, 2003.

Ledda, Elena. *Fiume e D'Annunzio: Pagine di storia.* Chieti: Solfanelli, 1988.

Ledeen, Michael. *D'Annunzio, the First Duce.* 1977. New Brunswisk, NJ: Transaction Publishers, 2002.

Leed, Eric J. *No Man's Land: Combat and Identity in World War I.* Cambridge: Cambridge University Press, 1979.

Leeds, John, James Flynn, and C.K. Mertz. '"Possibilities for Compassion": An Alternate Reading.' *Understanding Céline*, 125–32. Seattle: Genitron, 1984.

Lefebvre, Martin, ed. *Landscape and Film.* London and New York: Routledge, 2006.

Leopardi, Giacomo. *Poesie e prose.* Edited by Rolando Damiani, Cesare Galimberti, and Mario Andrea Rigoni. 2 vols. Milan: Mondadori, 1987.

Leroy, Claude. *La main de Cendrars.* [Villeneuve d'Ascq]: Presses universitaires du Septentrion, 1996.

Lethaby, W.R. *Architecture, Mysticism and Myth.* [1891]. New York: George Braziller, 1975.

Lewis, Pericles. *Modernism, Nationalism, and the Novel.* Cambridge: Cambridge University Press, 2000.

Leys, Ruth. *Trauma: A Genealogy.* Chicago: University of Chicago Press, 2000.

Lyttelton, Adrian. 'Creating a National Past: History, Myth, and Image in the Risorgimento.' In Albert Russell Ascoli and Krystyna Von Henneberg, eds., *Making and Remaking Italy: the Cultivation of National Identity around the Risorgimento*, 27–74. Oxford and New York: Berg, 2001.

Livorni, Ernesto. 'Giuseppe Ungaretti e Dante: Dall' "effimero" all' "eterno."' *Canadian Journal of Italian Studies* 14/42–43 (1991): 122–31.

– *Avanguardia e Tradizione: Ezra Pound e Giuseppe Ungaretti.* Florence: Lettere, 1998.

– 'Caino, Abele ed il sacrificio: La poesia moderna ed il sacro.' *Semicerchio: Rivista di Poesia Comparata* 19/2 (1998): 6–15.

– '"In sé da simulacro a fiamma vera/errando": Ungaretti's Poetry from Bergson to Plato.' *Romanic Review* 97/3–4 (2006): 517–54.

Loisy, Alfred. *L'Evangile et l'église.* Paris: A. Picard, 1902.

Lombroso, Cesare. *Genio e follia.* Milan: Tip. e libreria Giuseppe Chiusi, 1864.

Lord, Frank Knight. *Image Vision and Faith: Viewers' Responses to the Mandylion, Veronica's Veil and the Shroud of Turin.* PhD. thesis, University of North Carolina at Chapel Hill, Art Department, 2003.

Lorenzini, Niva. *Il segno del corpo: Saggio su d'Annunzio.* Rome: Bulzoni, 1984.

– 'D'Annunzio, l'irrazionale, il linguaggio del corpo.' *Il Verri,* nos. 5–6 (1985): 136–52.

MacDonald, Scott. *A Critical Cinema.* Berkeley: University of California Press, 1998.

Magrelli, Valerio. '"Avant que ça soye la nuit partout": Il Baudelaire di Céline fra parodia e riscrittura.' *Confronto Letterario: Quaderni del Dipartimento di Lingue e Letterature Straniere Moderne dell'Università di Pavia* 23/45 (2006): 147–59.

Malaparte, Curzio. 'La rivolta dei santi maledetti.' [1921]. In Luigi Martellini and Giancarlo Vigorelli, eds., *Opere scelte.* 5–101. Milan: Mondadori, 1997.

Manenti, Carla, Nicolas Monti, and Giorgio Nicodemi. *Luca Comerio: Fotografo e cineasta.* Milan: Electa, 1979.

Marinetti, Filippo Tommaso. 'Come amava Guzzo l'Ardito.' Filippo Tommaso Marinetti Papers, General Collection, Beinecke Rare Book and Manuscript Library, Yale University, New Haven, Conn. [hereafter Marinetti Papers], Gen Mss 130, Box 19, Folder 1226.

– *La Ville d'émeraude.* Marinetti Papers, Gen Mss 130, Box 39, Folder 1681.

– *Gli amori futuristi: Programmi di vita con varianti a scelta.* Cremona [Piacenza?]: Casa Editrice Ghelfi, 1922.

– *Scatole d'amore in conserva.* Rome: Edizioni d'arte Fauno, 1927.

– *Novelle colle labbra tinte: Simultaneità e programmi di vita con varianti a scelta.* Milan: Mondadori, 1930.

– *Teoria e invenzione futurista.* Milan: Mondadori, 1983.

– *Taccuini.* Bologna: Il Mulino, 1987.

– 'Excerpts from the Unpublished Diaries.' *Modernism/Modernity* 1/3 (1994): 1–44.

– *Gli indomabili: Con un'antologia di scritti futuristi sull'arte meccanica e d'avanguardia.* Edited by Luigi Ballerini. Milan: Mondadori, 2000.

Martinelli, Vittorio. *La guerra di d'Annunzio.* Udine: Paolo Gaspari, 2001.

Mazza, Attilio. *D'Annunzio e l'occulto.* Rome: Edizioni Mediterranee, 1995.
– *D'Annunzio sciamano.* Milan: Edizioni Bietti, 2001.
Mazzoni, Cristina. *Saint Hysteria: Neurosis, Mysticism, and Gender in European Culture.* Ithaca: Cornell University Press, 1996.
– 'Parturition, Parting, and Paradox in Turn-of-the-Century Italian Literature (D'Annunzio, Aleramo, Neera).' *Forum Italicum* 31/2 (1997): 343–66.
– '"That in Giving Me Life, You Still Remain Alive": Fetal Beginnings and Maternal Endings at Two Centuries' Ends.' *Annali d'Italianistica* 18 (2000): 255–76.
– 'Mystical and Literary Texts: Meeting the Other, and Each Other, at the Borders of Language.' *Annali d'Italianistica* 25 (2007): 105–22.
Meilinger, P.S. 'Giulio Douhet and Modern War.' *Comparative Strategy* 12/3 (1993): 321–38.
Merini, Alda. *La terra santa.* Edited by Maria Corti. Milan: All'insegna del pesce d'oro, 1984.
– *Delirio amoroso.* Genoa: Il Melangolo, 1989.
– *L'anima innamorata.* Milan: Frassinelli, 2000.
– *Superba è la notte.* Turin: Einaudi, 2000.
– *Corpo d'amore: Un incontro con Gesù.* Milan: Frassinelli, 2001.
– *Magnificat: Un incontro con Maria.* Edited by Ugo Nespolo. Milan: Frassinelli, 2002.
– *La carne degli angeli.* Milan: Frassinelli, 2003.
– *La presenza di Orfeo; La terra santa.* 2 vols. Edited by Grazia Carpena. Milan: Libri Scheiwiller, 2005.
– *L'altra verità: Diario di una diversa.* Milan: Rizzoli, 2006.
– *Francesco: Canto di una creatura.* Milan: Frassinelli, 2007.
– *Santo Padre Divina Poesia.* Milan: Giuliano Grittini, Incisione Arte, 2009.
Meyers, Jeffrey. *A Fever at the Core: The Idealist in Politics.* London: London Magazine Editions, 1976.
Mignone, Mario B. 'L'Allegria ungarettiana come ricerca della parola sacrale.' In Florinda M. Iannace, ed., *Etica cristiana e scrittori del Novecento,* 193–202. Stony Brook: Forum Italicum, 1993.
Miniero, Alessandro. *Da Versailles al Milite ignoto: Rituali e retoriche della vittoria in Europa (1919–1921).* Rome: Gangemi, 2008.
Ministero dei lavori pubblici, Direzione generale, edilizia: 'Sepoltura del Soldato Ignoto sull'Altare della Patria' (1921–26). Archivio Centrale dello Stato, Rome, Divisione 5, busta 39, fascicolo 107.
Ministero dell'Interno, Direzione generale di pubblica sicurezza, anno 1921, F.D.1, Affari Generali: 'Onoranze al Soldato Ignoto.' Archivio Centrale dello Stato, Rome, C4, busta 80, 1921.

Mondor, Henri. 'Voyage au bout de la colère.' *Nouvelles Littéraires* (1962): 3.
Moretti, Franco. *Graphs, Maps, Trees: Abstract Models for a Literary History*. London: Verso, 2005.
Moroni, Mario. *Al limite: L'idea di margine e confine nel Novecento italiano*. Florence: Monnier Università, 2007.
Moscati, Giuseppe. *Le cinque giornate di Fiume*. 2nd ed. Milan: Carnaro, 1931.
Mosse, George L. *Fallen Soldiers: Reshaping the Memory of the World Wars*. New York: Oxford University Press, 1990.
– *The Image of Man: The Creation of Modern Masculinity*. New York: Oxford University Press, 1996.
Musumeci, Antonino. 'Marinetti: A Mystical Experience on the Way to Futurism.' *Romance Languages Annual* 3 (1991): 263–6.
Naas, Michael. 'History's Remains: Of Memory, Mourning, and the Event.' *Research in Phenomenology* 33 (2003): 75–96.
Nardelli, Vincenzo. *L'Arcangelo*. Rome: Stock, 1931.
Nathan, Vetri. *Bodies in Migration: The Ambivalent Locations of Post-Coloniality in Italy*. Doctoral dissertation, Stanford University, 2009.
'Nazario Sauro – Wikipedia.' Retrieved from http://en.wikipedia.org/wiki/Nazario_Sauro.
Negri, Guglielmo. 'La Carta del Carnaro.' *Il Veltro* 40/5–6 (1996): 495–505.
Osgood, Richard. *The Unknown Warrior: An Archaeology of the Common Soldier*. Stroud: Sutton, 2005.
Ozouf, Mona. 'Le Panthéon: L'école normale des morts.' In Pierre Nora, ed., *Les lieux de mémoire* (3 vols). Vol. 1, 155–78. Paris: Gallimard, 1997.
Palumbo-Liu, David, and Hans Ulrich Gumbrecht. *Streams of Cultural Capital: Transnational Cultural Studies*. Stanford: Stanford University Press, 1997.
Papa, Emilio. 'Discorrendo di d'Annunzio politico.' *Studi Novecenteschi* 26/58 (1999): 275–93.
Parsons, William Barclay. *The Enigma of the Oceanic Feeling: Revisioning the Psychoanalytic Theory of Mysticism*. New York: Oxford University Press, 1999.
Pasolini, Pier Paolo. *Le Ceneri di Gramsci*. Milan: Garzanti, 1957.
– *L'usignolo della Chiesa Cattolica*. Milan: Longanesi, 1958.
Passerini, Luisa. *Mussolini immaginario: Storia di una biografia 1915–1939*. 2nd ed. Rome: Laterza, 1991.
– *Mussolini*. Bari: Laterza, 1997.
Paulson, Daryl S., and Stanley Krippner, eds. *Haunted by Combat: Understanding PTSD in War Veterans, Including Women, Reservists, and Those Coming Back from Iraq*. Westport and London: Praeger Security International, 2007.
Pecci, Giuseppe. *D'Annunzio e il mistero*. Milan: Pan, 1969.

Pellegrini, Franca. *La tempesta originale: La vita di Alda Merini in poesia.* Florence: Cesati, 2006.

Pentcheva, Bissera V. *Icons and Power: The Mother of God in Byzantium.* University Park: Pennsylvania State University Press, 2006.

Perloff, Marjorie. *The Futurist Moment: Avant-Guarde, Avant-Guerre, and the Language of Rupture.* Chicago: University of Chicago Press, 1986.

Peterson, Thomas E. 'Schismogenesis and National Character: The D'Annunzio-Mussolini Correspondence.' *Italica* 81/1 (2004): 44–64.

Pieraccini, Giovanna. 'Il superuomo dannunziano come archetipo dei miti dell'ideologia di destra.' *Strumenti Critici* 10/3 (1995): 387–416.

Pireddu, Nicoletta. *Antropologi alla corte della bellezza: Decadenza ed economia simbolica nell'Europa fin de siècle.* Verona: Fiorini, 2002.

– 'Gabriele D'Annunzio: The Art of Squandering and the Economy of Sacrifice.' In Mark Osteen, ed., *The Question of the Gift: Essays across Disciplines,* 172–90. London: Routledge, 2002.

– 'Primitive Marks of Modernity: Cultural Reconfigurations in the Franco-Italian Fin de Siècle.' *Romanic Review* 97/3–4 (2006): 371–400.

Pius X. 'Pascendi dominici gregis: On the Doctrine of the Modernists.' 1907. Retrieved from http://www.papalencyclicals.net/Pius10/p10pasce.htm.

Pizzo, Marco. *Il Vittoriano: Guida storico-artistica.* Rome: Ministero per i beni e le attività culturali, 2002.

Poggi, Christine. *In Defiance of Painting: Cubism, Futurism, and the Invention of Collage.* New Haven: Yale University Press, 1992.

– *Inventing Futurism: The Art and Politics of Artificial Optimism.* Princeton: Princeton University Press, 2009.

Polanyi, Michel. *Personal Knowledge: Towards a Post-Critical Philosophy.* London: Routledge and Kegan Paul, 1958.

Pozzi, Enrico. 'Il Duce e il Milite ignoto: Dialettica di due corpi politici.' *Rassegna Italiana di Sociologia* 39/3 (1998): 332–57.

Pratt, Mary Louise. *Imperial Eyes: Travel Writing and Transculturation.* London and New York: Routledge, 1992.

Procacci, Giovanna. *Soldati e prigionieri italiani nella Grande guerra: Con una raccolta di lettere inedite.* Rome: Editori Riuniti, 1993.

Prost, Antoine. 'Le monuments aux morts: Culte républicain? Culte civique? Culte patriotique?' In Pierre Nora, ed., *Les lieux de mémoire* (3 vols). Vol. 1, 199–223. Paris: Gallimard, 1997.

Quinn, Tom. *The Traumatic Memory of the Great War, 1914–1918, in Louis-Ferdinand Céline's Voyage au bout de la nuit.* Lewiston, NY: Mellen, 2005.

Quintyn, Conrad, and Sarah Wagner. 'Dismantling a National Icon.' *Anthropology Today* 50/5 (2009): 7–9.

Railton, David. 'The Origins of the Unknown Warrior.' *Our Empire* 6/8 (1931): 34–6.
Rasson, Luc. '"On fait la vie avec le sang": A propos du Grand troupeau.' In Christian Berg, Walter Geerts, Paul Pelckmans, and Bruno Tritsmans, eds., *Retours du mythe*, 183–94. Amsterdam: Rodopi, 1996.
Re, Lucia. 'Gabriele D'Annunzio's Novel Le vergini delle rocce: "Una cosa naturale vista in un grande specchio."' *Stanford Italian Review* 3/2 (1983): 241–71.
– 'Gabriele D'Annunzio's Theater of Memory: Il Vittoriale degli Italiani/Il teatro della memoria di Gabriele D'Annunzio; Il Vittoriale degli Italiani.' *Journal of Decorative and Propaganda Arts* 3 (1987): 6–51.
– 'Per una lettura testualista del corpus dannunziano.' *Annali d'Italianistica* 5 (1987): 43–59.
– 'Valentine de Saint-Point, Ricciotto Canudo, F.T. Marinetti: Eroticism, Violence and Feminism from Prewar Paris to Colonial Cairo.' *Quaderni d'Italianistica: Official Journal of the Canadian Society for Italian Studies* 24/2 (2003): 37–69.
– 'D'Annunzio, Duse, Wilde, Bernhardt: Author and Actress between Decadence and Modernity.' In Luca Somigli and Mario Moroni, eds., *Italian Modernism: Italian Culture between Decadentism and Avant-Garde*, 86–129. Toronto: University of Toronto Press, 2004.
– 'Futurism, Seduction, and the Strange Sublimity of War.' *Italian Studies* 59 (2004): 83–111.
Redfern, W.D. 'Against Nature: Jean Giono and Le Grand Troupeau.' In Holger Klein, ed., *The First World War in Fiction: A Collection of Critical Essays*, 73–83. London: Macmillan, 1976.
Renga, Dana. 'Staging Memory and Trauma in French and Italian Holocaust Film.' *Romanic Review* 97/3–4 (2006): 461–82.
Ricciardi, Alessia. *The Ends of Mourning: Psychoanalysis, Literature, Film.* Stanford: Stanford University Press, 2003.
Ridolfi, Maurizio. *Le feste nazionali.* Bologna: Il Mulino, 2003.
Rifkin, Jeremy. *The Empathic Civilization: The Race to Global Consciousness in a World of Crisis.* New York: J.P. Tarcher/Penguin, 2009.
Riva, Massimo. 'Ortis, o dell'ombra amorosa: Le "Ultime lettere" e la genesi del simbolismo sepolcrale.' *Italian Quarterly* 29/111 (1988): 15–39.
– '1888–1988: Some Remarks on Nihilism and Secularization.' *History of European Ideas* 11 (1989): 979–88.
– *Malinconie del moderno: Critica dell'incivilimento e disagio della nazionalità nella letteratura italiana del XIX secolo.* Ravenna: Longo, 2001.
Rizzelli, Stefano. *Dalla sindone a Padre Pio: I segreti del sacro, Viaggio nell'Italia dei miracoli e dei misteri.* Rome: Newton Compton, 1999.

Rochat, Giorgio. *Gli arditi della Grande guerra: Origini, battaglie e miti.* Milan: Feltrinelli economica, 1981.

Rosch, Eleanor. 'How to Catch James' Mystic Germ: Religious Experience, Buddhist Meditation, and Psychology.' *Journal of Consciousness Studies* 9/9–10 (2002): 37–56.

Sabbatucci, Giovanni. *La Stampa del combattentismo.* Bologna: Cappelli, 1980.

Sacher-Masoch, Leopold von, and Gilles Deleuze. *Masochism: Coldness and Cruelty Followed by Venus in Furs.* New York: Zone Books, 1989.

Salaris, Claudia. *Alla festa della rivoluzione: Artisti e libertari con D'Annunzio a Fiume.* Bologna: Il Mulino, 2002.

Salotti, Guglielmo. 'I rapporti d'Annunzio-Giulietti dall' "idillio" fiumano al "pactum sine nomine."' In Renzo De Felice, ed., *D'Annunzio politico, 1918–1938*, 237–96. Rome and Bari: Laterza, 1978.

Sartre, Jean-Paul. *La nausée: Roman.* Paris: Gallimard, 1938. Translated by Robert Baldick as *Nausea* (Norfolk: New Directions, 1959).

Saunders, Nicholas J. *Matters of Conflict: Material, Memory, and the First World War.* London and New York: Routledge, 2004.

Scarry, Elaine. *The Body in Pain: The Making and Unmaking of the World.* New York: Oxford University Press, 1987.

Schnapp, Jeffrey T. 'Politics and Poetics in Marinetti's Zang Tumb Tuuum.' *Stanford Italian Review* 5/1 (1985): 75–92.

– 'Propeller Talk.' *Modernism/Modernity* 1/3 (1994): 153–78.

– *Staging Fascism: 18 BL and the Theater of Masses for Masses.* Stanford: Stanford University Press, 1996.

– 'Worte des Schweigens in D'Annunzios Fiume-Reden.' In Hans Ulrich Gumbrecht, Friedrich Kittler, and Bernhard Siegert, eds., *Der Dichter als Kommandant: D'Annunzio erobert Fiume*, 133–45. Munich: Fink, 1996.

– 'Fascism after Fascism.' In Richard J. Golsan, ed., *Fascism's Return: Scandal, Revision, and Ideology since 1980*, 63–85. Lincoln: University of Nebraska Press, 1998.

– *A Primer of Italian Fascism.* Lincoln: University of Nebraska Press, 2000.

– 'The Mass Panorama.' *Modernism/Modernity* 9/2 (2002): 243–81.

– *In cima – Giuseppe Terragni per Margherita Sarfatti (Architetture della memoria nel '900).* Milan: Marsilio, 2004.

Schnapp, Jeffrey T., and Matthew Tiews. *Crowds.* Stanford: Stanford University Press, 2006.

Semeria, Giovanni. *Venticinque anni del cristianesimo nascente.* Rome: F. Pustet, 1900.

Serres, Michel. 'L'homme est un loup pour l'homme.' In Michel Deguy and

Jean-Pierre Dupuy, eds., *René Girard et le problème du mal*, 301–9. Paris: Grasset, 1982.
– *Les cinq sens.* Paris: Grasset, 1985.
– *Hominescence: Essais.* Paris: Pommier, 2001.
– *La guerre mondiale.* Paris: Pommier, 2008.
Sheehan, Thomas. *The First Coming: How the Kingdom of God Became Christianity.* New York: Random House, 1986.
Sherman, Daniel. 'Bodies and Names: The Emergence of Commemoration in Interwar France.' *American Historical Review* 102/2 (1998): 443–6.
Siegel, Mona. 'To the Unknown Mother of the Unknown Soldier: Pacifism, Feminism, and the Politics of Sexual Difference among French Institutrices between the Wars.' *French Historical Studies* 22/3 (1999): 421–51.
– '"History Is the Opposite of Forgetting": The Limits of Memory and the Lessons of History in Interwar France.' *Journal of Modern History* 74/3 (2002): 770–800.
Somigli, Luca. *Legitimizing the Artist: Manifesto Writing and European Modernism, 1885–1915.* Toronto: University of Toronto Press, 2003.
– 'The Mirror of Modernity: Marinetti's Early Criticism between Decadence and "Renaissance latine."' *Romanic Review* 97/3–4 (2006): 331–52.
Sontag, Susan. 'Regarding the Torture of Others.' *New York Times*, 23 May 2004.
Soudagne, Jean-Pascal. *L'histoire incroyable du soldat inconnu.* Rennes: Ouest-France, 2008.
Spackman, Barbara. 'Il verbo (e)sangue: Gabriele D'Annunzio and the Ritualization of Violence.' *Quaderni d'Italianistica: Official Journal of the Canadian Society for Italian Studies* 4/2 (1983): 218–29.
– *Decadent Genealogies: The Rhetoric of Sickness from Baudelaire to d'Annunzio.* Ithaca: Cornell University Press, 1989.
– 'Mafarka and Son: Marinetti's Homophobic Economics.' *Modernism/Modernity* 1/3 (1994): 89–107.
– *Fascist Virilities: Rhetoric, Ideology, and Social Fantasy in Italy.* Minneapolis and London: University of Minnesota Press, 1996.
Springer, Carolyn. '"L'urne de' forti": Foscolo's Sepulchral Archaeology.' *Quaderni d'Italianistica: Official Journal of the Canadian Society for Italian Studies* 4/1 (1983): 26–46.
– *The Marble Wilderness: Ruins and Representation in Italian Romanticism, 1775–1850.* Cambridge: Cambridge University Press, 1987.
Staid, Andrea. *Gli Arditi del popolo: La prima lotta armata contro il fascismo, 1921–1922.* Ragusa: La Fiaccola, 2007.

Strenski, Ivan. *Contesting Sacrifice: Religion, Nationalism, and Social Thought in France.* Chicago: University of Chicago Press, 2002.

Sundholm, John. '"The Unknown Soldier": Film as a Founding Trauma and National Monument.' In John Sundholm, Conny Mithander, and Maria Holmgren Troy, eds., *Collective Traumas: Memories of War and Conflict in 20th-Century Europe,* 111–41. Brussels: Presses Interuniversitaires Européennes (P.I.E.)–Peter Lang, 2007.

Talamo, Giuseppe, and Marco Pizzo, eds. *Dalla vittoria al milite ignoto: 4 novembre 1918 – 4 novembre 1921.* Rome: Ministero per i beni e le attività culturali, 2002.

Tandonnet, Maxime. *1940, un autre 11 novembre: Étudiant de France, malgré l'ordre des autorités opprimantes, tu iras honorer le soldat inconnu.* Paris: Tallandier, 2009.

Tarkovskij, Arsenij. *Poesie Scelte.* Translated by Gario Zappi. Milan: Scheiwiller, 1989.

Tarkovsky, Arsenii. 'Poems from films, translated by Alex Nemser and Nariman Skakov.' Retrieved from http://people.ucalgary.ca/~tstronds/nostalghia.com/TheTopics/Tarkovsky_Poems.html.

'Tarpeian Rock – Wikipedia.' Retrieved from http://en.wikipedia.org/wiki/Tarpeian_Rock.

Tauriac, M. 'La bataille inconnue du soldat inconnu.' *Historia,* no. 474 (1986): 47–54.

Taylor, Mark C. *Disfiguring: Art, Architecture, Religion.* Chicago: University of Chicago Press, 1992.

Ténor, Augsute. *Il s'appelait ... le soldat inconnu.* Paris: Gallimard, 2004.

Testa, Gian Pietro. *Io sono il milite ignoto.* Bologna: Minerva, 2006.

– *Il terrorismo: La strategia che viene dall'alto.* Bologna: Thyrus, 1986.

'The Burial of the Unknown Warrior.' Retrieved from http://www.westernfrontassociation.com/great-war-people/memorials/400-burial-unkown.html.

'The Unknown Warrior – Wikipedia.' Retrieved from http://en.wikipedia.org/wiki/The_Unknown_Warrior.

Thomas, Susanne Sara. 'Petrarch, Tournier, Photography, and Fetishism: The Veil in the Rime Sparse.' *Romance Quarterly* 43 (1996): 131–42.

Thucydides. *History of the Peloponnesian War.* Translated by Richard Crawley. Forgotten Books, 2008. Retrieved from http://www.forgottenbooks.org.

Tick, Edward. *War and the Soul: Healing Our Nation's Veterans from Post-Taumatic Stress Disorder.* Wheaton, Ill.: Quest Books, 2005.

Tobia, Bruno. *L'altare della patria.* Bologna: Il Mulino, 1998.

Tognasso, Augusto. *Ignoto Militi.* Milan: Tipografia Magnani, 1922.

– *Ignoto Militi.* Milan: Tipografia A. Zanoli, 1962.

'Tomb of the Unknowns, Arlington National Cemetery.' Retrieved from http://www.arlingtoncemetery.net/tombofun.htm.

'Tomb of the Unknowns: "Here Rests In Honored Glory An American Soldier Known But To God."' Retrieved from http://www.military.com/veterans-day/tomb-of-the-unknowns.htm.

Tournier, Michel. *Vendredi ou les limbes du Pacifique.* Paris: Gallimard, 1967.

– 'Les suaires de Véronique.' In *Le Coq de bruyère,* 155–72. Paris: Gallimard, 1978.

Tricerri, Rosa. 'Il linguaggio della tradizione cristiana nelle scritture del tempo di guerra.' *Quaderni del Vittoriale: D'Annunzio e la religiosità,* no. 28 (1981): 97–109.

Tyrrell, George. *External Religion: Its Use and Abuse.* London: Sands, 1899.

Ulivi, Ferruccio. *La maschera senza il volto.* Naples: Istituto Suor Orsola Benincasa, 1989.

Underhill, Evelyn. *Mysticism.* [1911]. New York: Doubleday, 1990.

Ungaretti, Giuseppe. *Vita d'un uomo: Tutte le poesie.* Milan: Mondadori, 1969.

– *Vita d'un uomo: Saggi e interventi.* Milan: Mondadori, 1974.

– *A Major Selection of the Poetry of Giuseppe Ungaretti.* Translated by Diego Louis Bastianutti. Toronto: Exile Editions, 1997.

– *Giuseppe Ungaretti: Selected Poems.* Translated by Andrew Frisardi. New York: Farrar, Straus, and Giroux, 2002.

Unia, Marco. 'Il soldato torna in Patria: E adesso pover'uomo? [book review of Hobsbawm, Sabbatucci, Gentile, Isnenghi, Gibelli, Leed, and Winter].' Retrieved from http://www.storiain.net/arret/num74/artic3.asp.

Valesio, Paolo. 'Pax Italiae and the Literature of Politics.' *Yale Italian Studies* 2 (1978): 143–67.

– 'Pasolini come sintomo.' *Italian Quarterly* 21–2/82–3 (1980): 31–42.

– 'Declensions: D'Annunzio after the Sublime.' *New Literary History: A Journal of Theory and Interpretation* 16/2 (1985): 401–15.

– *Ascoltare il silenzio: La retorica come teoria.* Bologna: Il mulino, 1986.

– 'The Beautiful Lie: Heroic Individuality and Fascism.' In *Reconstructing Individualism: Autonomy, Individuality, and the Self in Western Thought,* 163–83. Stanford: Stanford University Press, 1986.

– 'Ungaretti and the Miles Patiens: Dannunzian Genealogies.' *Stanford Italian Review* 8/1–2 (1990): 103–37.

– *The Dark Flame.* New Haven: Yale University Press, 1992.

– 'Il metodo come strada accidentata.' *Lettore di Provincia* 30/104 (1999): 47–53.

– 'The Most Enduring and Honored Name.' In *Selected Poems and Related Prose [of F.T. Marinetti],* i–xix. New Haven: Yale University Press, 2002.

– 'Foreword: After *The Conquest of the Stars.*' In Luca Somigli and Mario Moroni, eds., *Italian Modernism: Italian Culture between Decadentism and Avant-Garde*, ix–xxiii. Toronto: University of Toronto Press, 2004.
– 'La "macchina morbida" di Marinetti.' *Annali d'Italianistica* 27 (2009): 243–62.
Van Gennep, Arnold. *Les rites de passage.* Paris: Nourry, 1909. Translated by Monika B. Vizedom and Gabrielle L. Caffee as *The Rites of Passage* (Chicago: University of Chicago Press, 1960).
Vander, Fabio. *L'estetizzazione della politica: Il fascismo come anti-Italia.* Bari: Dedalo, 2001.
Varela, Francisco J., Evan Thompson, and Eleanor Rosch. *The Embodied Mind: Cognitive Science and Human Experience.* Cambridge, Mass.: MIT Press, 1991.
Verdery, Katherine. *The Political Lives of Dead Bodies: Reburial and Postsocialist Change.* New York: Columbia University Press, 1999.
Vighi, Fabio. 'On the Real Limits of Self-Consciousness: Gazing Back at the Subversive Subject with Marco Bellocchio.' *Culture, Theory and Critique* 46/2 (2005): 147–61.
Vilain, Charles. *Le Soldat inconnu: Histoire et culte.* Paris: Éditions Maurice d'Hartoy, 1933.
Virilio, Paul. 'Le soldat inconnu.' *Temps Modernes* 31/360 (1976): 2334–53.
Vovelle, Michel. *La mort et l'Occident de 1300 à nos jours.* Paris: Gallimard, 1983.
Winter, Jay M. *Sites of Memory, Sites of Mourning: The Great War in European Cultural History.* Cambridge and New York: Cambridge University Press, 1995.
Wittman, Laura. '*Omnes velut aqua dilabimur*: Antonio Fogazzaro, the Saint, and Catholic Modernism.' In Luca Somigli and Mario Moroni, eds., *Italian Modernism: Italian Culture between Decadentism and Avant-Garde*, 130–66. Toronto: University of Toronto Press, 2004.
– 'Imagined Geographies.' In 'Imagined Geographies: Italy and France,' special issue, ed. by Laura Wittman. *Romanic Review* 97/3–4 (2006): 265–74.
– 'Mystical Insight and Psychoanalysis in the Fin-de-Siècle Novel: Huysmans, Bourget, d'Annunzio.' *Forum Italicum* 42/1 (2008): 30–51.
– 'Introduction to Part Three: Stars-in-Freedom and the Dark Night of Futurism.' In Lawrence Rainey, Christine Poggi, and Laura Wittman, eds., *Futurism: An Anthology*, 409–18. New Haven: Yale University Press, 2009.
– '"Cuore-proiettile": From Marinetti's Mafarka to World War One and Ungaretti.' *L'anello che non tiene* 24–25 (2011).
Wittman, Richard. *Architecture, Print Culture, and the Public Sphere in Eighteenth-Century France.* New York: Routledge, 2007.
Woodhouse, John Robert. *Gabriele D'Annunzio: Defiant Archangel.* Oxford and New York: Clarendon Press, 1998.

Wyschogrod, Edith. *The Phenomenon of Death: Faces of Mortality*. New York: Harper and Row, 1973.
– *Spirit in Ashes: Hegel, Heidegger, and Man-Made Mass Death*. New Haven: Yale University Press, 1985.
– *Saints and Postmodernism: Revisioning Moral Philosophy*. Chicago: University of Chicago Press, 1990.
– *An Ethics of Remembering: History, Heterology, and the Nameless Others*. Chicago: University of Chicago Press, 1998.
Yalom, Irvin D. *Staring at the Sun: Overcoming the Terror of Death*. San Francisco: Jossey-Bass, 2008.
Yalom, Marilyn. *The American Resting Place: Four Hundred Years of History through Our Cemeteries and Burial Grounds*. Boston: Houghton Mifflin, 2008.
Young, Allan. *The Harmony of Illusions: Inventing Post-Traumatic Stress Disorder*. Princeton: Princeton University Press, 1995.
Zago, S., G. Sartori, and G. Scarlat. 'Malingering and Retrograde Amnesia: The Historic Case of the Collegno Amnesic.' *Cortex* 40/3 (2004): 519–32.
Žižek, Slavoj. *The Sublime Object of Ideology*. London: Verso, 1989.
– 'Neighbors and Other Monsters: A Plea for Ethical Violence.' In Slavoj Žižek, Eric L. Santner, and Kenneth Reinhard, *The Neighbor: Three Inquiries in Political Theology*, 134–237. Chicago: University of Chicago Press, 2005.

Index

Page numbers in italics refer to illustrations.

FSC
www.fsc.org
MIX
Papier | Fördert gute Waldnutzung
FSC® C083411

Kolibri 360 GmbH
Ferdinand-Jühlke-Straße 7
99095 Erfurt, Deutschland
produktsicherheit@kolibri360.de